This item must be returned or renewed by the last date shown below. The loan period may be shortened if it is reserved by another reader. A fine will be due if it is not returned on time.

DATE OF RETURN

1 4 MAY 2009

Nuclear Waste Research: Siting, Technology and Treatment

NUCLEAR WASTE RESEARCH: SITING, TECHNOLOGY AND TREATMENT

ARNOLD P. LATTEFER

EDITOR

Nova Science Publishers, Inc.

New York

NOTICE TO THE READER

The Publisher has taken reasonable care in the preparation of this book, but makes no expressed or implied warranty of any kind and assumes no responsibility for any errors or omissions. No liability is assumed for incidental or consequential damages in connection with or arising out of information contained in this book. The Publisher shall not be liable for any special, consequential, or exemplary damages resulting, in whole or in part, from the readers' use of, or reliance upon, this material. Any parts of this book based on government reports are so indicated and copyright is claimed for those parts to the extent applicable to compilations of such works.

Independent verification should be sought for any data, advice or recommendations contained in this book. In addition, no responsibility is assumed by the publisher for any injury and/or damage to persons or property arising from any methods, products, instructions, ideas or otherwise contained in this publication.

This publication is designed to provide accurate and authoritative information with regard to the subject matter covered herein. It is sold with the clear understanding that the Publisher is not engaged in rendering legal or any other professional services. If legal or any other expert assistance is required, the services of a competent person should be sought. FROM A DECLARATION OF PARTICIPANTS JOINTLY ADOPTED BY A COMMITTEE OF THE AMERICAN BAR ASSOCIATION AND A COMMITTEE OF PUBLISHERS.

LIBRARY OF CONGRESS CATALOGING-IN-PUBLICATION DATA

Nuclear waste research / Arnold P. Lattefer, editor.
   p. cm.
 Includes bibliographical references and index.
 ISBN 978-1-60456-184-5 (hardcover)
 1. Radioactive waste disposal. I. Lattefer, Arnold P.
TD898.N838 2007
621.48'38--dc22

2007048149

Published by Nova Science Publishers, Inc. ✦ New York

CONTENTS

PREFACE

Radioactive wastes are waste types containing radioactive chemical elements that do not have a practical purpose. They are sometimes the products of a nuclear processes, such as nuclear fission. However, other industries not directly connected to the nuclear industry can produce large quantities of radioactive waste. For instance, over the past 20 years it is estimated that just the oil-producing endeavors of the US have accumulated 8 million tons of radioactive wastes. The majority of radioactive waste is "low-level waste", meaning it has low levels of radioactivity per mass or volume. This type of waste often consists of used protective clothing, which is only slightly contaminated but still dangerous in case of radioactive contamination of a human body through ingestion, inhalation, absorption, or injection.

The issue of disposal methods for nuclear waste was one of the most pressing current problems the international nuclear industry faced when trying to establish a long term energy production plan, yet there was hope it could be safely solved. In the U.S., the DOE acknowledged much progress in addressing the waste problems of the industry, and successful remediation of some contaminated sites, yet also major uncertainties and sometimes complications and setbacks in handling the issue properly, cost effectively, and in the projected time frame. In other countries with lower ability or will to maintain environmental integrity the issue would be more problematic. This new book presents the latest research in the field.

Short Communication A - The leaching behavior and structure of magnesium phosphate glasses containing 45-55 mol% MgO incorporated with simulated high level nuclear wastes (HLW) were studied. The leach rate of the waste glasses decrease with increasing the simulated HLW content. The gross leach rate of the glass waste form containing 50 mol% MgO and 45 mass% simulated HLW is of the order of 10^{-6} g/cm2·day at 90°C, which is small enough as compared with the corresponding release from a currently used borosilicate glass waste form. The isolated ions such as dimeric $(P_2O_7)^{4-}$ and monomeric $(PO_4)^{3-}$ ions increase upon as increasing the incorporating amount of the simulated HLW. The changes in properties can be attributed to the structure changes owing to the incorporation of the simulated HLW. For the comparison, the leaching behavior and structure of calcium phosphate glasses containing 45-55 mol% CaO incorporated with the simulated HLW were also studied.

Short Communication B – A method for solidification of water-soluble Cs^+ ion in a chemically and physically stable form of single-crystalline titanates is presented.

Electrochemical reduction of molten molybdates containing Cs^+ and Ti^{4+} ions has lead to the formation of single crystals of Cesium Titanate Hollandite, $Cs_{1.35}Ti_8O_{16}$, at ambient pressure. X-ray diffraction (XRD) has showed that the single crystals of $Cs_{1.35}Ti_8O_{16}$ have high crystallinity. The electrochemically prepared $Cs_{1.35}Ti_8O_{16}$ has advantage as a matrix material of radioactive ^{137}Cs in high-level radioactive wastes (radwastes) over the conventional solid-state matrices including borosilicate glass. Firstly, bulk $Cs_{1.35}Ti_8O_{16}$ shows much higher leaching resistance against hot water under high pressure than borosilicate glass. Secondly, single-crystalline $Cs_{1.35}Ti_8O_{16}$ has much lower specific surface area and higher crystallinity than the bulk $Cs_{1.35}Ti_8O_{16}$, which can minimize the leaching of ^{137}Cs to the environment.

Short Communication C - Alloy 22 (UNS 06022) has been selected as the Corrosion Resistant Material (CRM) for the Waste Package (WP) outer barrier for the proposed High Level Nuclear Waste (HLNW) repository at Yucca Mountain (YM), Nevada, USA. A heated electrode technique has been developed for studying the corrosion behavior of Alloy 22 under simulated YM service conditions. The following tests were conducted: (1) potentiodynamic cyclic polarization data were collected to investigate electrochemical properties of Alloy 22 immersed in simulated ground waters; (2) extended immersion tests were carried out to investigate corrosion behavior of Alloy 22 under the expected ground water seepage conditions in the repository drifts ; (3) a long-term time delayed dripping test was run for studying susceptibility to localized corrosion under accumulated multi-component salt deposits with cyclic wet-dry conditions. The effects of temperature, pH, relative humidity (RH), and the accumulation of salt deposits on corrosion resistance of Alloy 22 are discussed in detail. Of the three simulated ground waters tested, Simulated Concentrated Water (SCW) was found to be the most corrosive media. Heat treatment of Alloy 22 for 100 hours at 800°C, cyclic wet-dry conditions, and the uneven crevices formed between the salt deposits and the hot metal surface, all contributed to accelerated corrosion of Alloy 22.

Short Communication D - Radioactive waste is highly dangerous and long-lived. Any attempt by the current generation to safely dispose or store this waste is racked by unsolvable technical difficulties such that future generations deserve to be warned about the characteristics and whereabouts of the waste. Since the information about the waste may be more susceptible to degradation than the actual waste a number of issues emerge in this study that throw doubt upon the morality of producing the waste in the first place.

Chapter 1 - Previous studies have shown that experts and the public assess risks of nuclear waste at very different levels. In the present study, a broad range of risk perception and attitude dimensions was investigated by means of a mail survey directed to experts in the field of nuclear waste. Comparative data was available from a random sample of the Swedish population and from a sample of graduate engineers working in other fields than the nuclear one. The analysis of risk perceptions shows that experts rated nuclear risks as much smaller than members of the public did. They also rated risk aspects of nuclear waste as much less worrisome, with the exception of Voluntariness, Dread and Novel risk. Engineers rated risks similar to experts, but they were closer to the public. Both experts and engineers showed a rhetorical contrast in their ratings, since they expressed more worry about lifestyle risks than the public did. There was a small, but consistent, tendency for female experts to judge risks as larger than their male colleagues, but still much smaller than women in the public did. The structure of perceived risk was similar for the three groups, contrary to the notion that experts' risk perception is qualitatively different from that of the public. It was found that experts and the public held very different policy attitudes. The policy differences among groups could be accounted for by demographics to some extent, about 1/3 of the group differences. The remaining intergroup variance (2/3) was accounted for by attitude to nuclear power, trust in institutions, experts and Science, and to some extent by perceived risk of nuclear waste. These explanatory variables were also powerful factors in accounting for within-group variation of policy attitudes explaining 56 (public) to 40 (experts) percent of the variance. It is concluded that the most important factors behind the difference between experts and the public were attitudes likely to reflect processes of professional socialization and social validation of beliefs.

Chapter 2 - Nuclear waste vitrification melters currently lack sophisticated on-line monitoring capability to ensure efficient operation and to guarantee in real time the production of a high quality glass waste product that meets stringent regulatory and disposal requirements. This lack of on-line monitoring capability is of particular concern for the costly multi-decade national effort in the U. S. to clean up cold war legacy waste, where the waste stream is not well characterized and it continues to vary significantly over the long time period it has been produced and stored. Current operations at the Defense Waste Processing Facility (DWPF) at the Savannah River Site and future operations at the Waste Treatment Plant (WTP), now under construction at Hanford, rely on predictive models for vitrification because of the lack of on-line monitoring of most melter processing parameters. This necessitates conservative operation to take into account uncertainties in the modeling and the ability to control actual input parameters. It also makes the operations susceptible to unpredictable anomalies such as foaming, liquidus crystallization, noble metals build up, and salt layer formation. Predictive modeling will become increasingly more difficult in time as the waste glass chemistry evolves with changes in the waste feed compositions. Additionally, the predictive models are limited by experimental validation, thus building more conservatism into the models. Comprehensive on-line real time monitoring of the vitrification process would alleviate the uncertainties, improve processing efficiencies (waste loading and throughput), and safeguard the facilities from anomalies. Advanced high temperature thermal analytical tools would also address the needs identified at the Hanford site and at the Savannah River Site (SRS) for increased waste loading and accelerated cleanup.

Much of the desired on-line monitoring capabilities can be realized through the use of millimeter-wave (MMW) technologies. Electromagnetic radiation in the 10 – 0.3 mm (30 – 1000 GHz) range of the spectrum is ideally suited for remote measurements in harsh, optically unclean, industrial, and unstable processing environments. Millimeter waves are long enough to penetrate optical/infrared obscured viewing paths through dust, smoke, and debris, but short enough to provide spatially resolved point measurements for profile information. Another important advantage is the ability to fabricate efficient MMW melter viewing components from refractory materials because optical or infrared quality is not required in the millimeter-wave wavelength range. The same ceramics and alloys from which the melter is constructed can be used to fabricate MMW waveguide/mirror components that go into the melter for long life survivability.

MMW techniques and technologies therefore make possible new robust diagnostic tools for glass melt monitoring in a high temperature and radioactive melter environment that other technologies can not access. These new tools will also make possible new insights into the dynamics of molten glass science and into nuclear waste glass melt property modeling because of the new way the melt properties are measured. In addition, these new monitoring developments will contribute toward modernizing nuclear waste glass vitrification operations to industrial standards of feed-back control as employed by most industries in manufacturing.

Chapter 3 - The principle of work and feature of a design of a pulsed reactor are the reason for essential fluctuations of power pulse amplitudes generated by a reactor. These fluctuations reaching 40% are caused by high sensitivity of a pulsed reactor to changes of reactivity. The general problem of definition of optimal (in statistical sense) automatic control algorithms is considered for various operating modes of a pulsed reactor. A minimum of the expected mean-square deviation of a control parameter of the future power pulse from base value of this parameter is accepted as the optimality criterion. To define optimal algorithm elements of the theory of optimal systems are used. The optimal algorithm is received in the assumption, that in a regulator any information on the previous pulses can be used. A feature of the definition of algorithm is the use of an added concept of a degree of ageing of the information. This concept has a clear enough physical sense. The decision of a problem in such a general statement gives the basis to simplify the algorithm and to lead it to the kind more convenient for realization (when in formation of control action the parameters accessible to direct measurement are used only). The algorithms achieved as a result of physically proved simplifications essentially do not miss traditional algorithms. On the one hand, it specifies a correctness of the considered method of a choice of algorithm, and with another, once again confirms efficiency of the engineering intuitive approach demanded from a regulator certain inertia. In the model of the reactor modeling transients are realized at random and regular disturbances of reactivity for the mode of power stabilization and for the mode of reactor going up on the set period. The estimation of influence of the parameter of the regulator on transients is given. The expediency of a compromise choice of a degree of regulator inertance is confirmed.

Chapter 4 - Accelerator Driven Systems (ADS) are innovative nuclear concepts that are currently under study for its application to nuclear waste management. ADS are subcritical devices driven by a high power external neutron source, usually generated by spallation reactions in heavy materials induced by accelerated protons. The subcritical operation mode of the nuclear core opens the possibility to use fuel matrixes with high minor actinide contents in a safe way. In that sense, ADS are applied as nuclear waste burners, getting rid of long lived radioactive waste as Pu, Np isotopes, and some long lived fission fragments as ^{99}Tc, producing a certain amount of energy, contributing to the sustainability of nuclear energy by a better energetic profit of the nuclear materials. The development of ADS has also as an objective the reduction of the nuclear waste stockpile and, once integrated into the nuclear fuel cycle, reduce the requirements in monitoring time and volume of final disposal repositories.

The ADS development is encountering several technological challenges, as it is foreseen in the arising of an innovative machine that integrate different fields that has been traditionally separated, as the accelerator, the high energy physics and the reactor physics. The integration of those technologies implies the overcoming of some uncertainties with material damage, accelerator reliability or the fabrication and high-burn up integrity of new high actinide nuclear fuels.

Nevertheless, those technological uncertainties seem to be solved in the short-medium term. In the meantime, nuclear knowledge is reviewed to establish the best options for the practical application of ADS. One of these technologies is based on the experience of gas cooled reactors. Graphite as neutron moderator permits certain spectrum modulation in function of the moderator/fuel ratio to obtain, thermal, fast or epithermal spectrum, and therefore permits spectrum adaptation to the targeted nuclear species to increase transmutation efficiency. The high burn-up that can be obtained by coated particle ceramic fuels, the possibility of high temperature operation and the inherent safety of subcritical systems, are also positive inputs for the deployment of gas-cooled ADS. In this chapter, the current state-of-the-art of this technology will be given, including a discussion about its future development and limitations.

Chapter 5 - Deep argillaceous formations are potential host rocks for high-level radioactive waste repositories due to favourable properties, particularly their low permeabilities and high sorption capacities for radionuclides. Disposal concepts considered in several countries involve steel and concrete components, which could react with such clayey material and thus induce changes in the containment properties of clays.

This safety issue is addressed by the French Institute for Radiological Protection and Nuclear Safety (IRSN) through an experimental and modelling program aiming at assessing the intensity and expansion of such geochemical perturbations. For this purpose, IRSN Underground Research Laboratory (URL) in Tournemire (Aveyron, France) which includes several man-made facilities (century-old railway tunnel, 1996 and 2003 drifts, many boreholes) crossing the Toarcian argillite offers various opportunities. The present paper goes over the main outlines of the program developed in that context since several years on indurated argillite sampled in Tournemire URL.

The research program devoted to argillite/concrete interactions is twofold: laboratory experiments (batch and diffusion) with alkaline fluids and field investigations dealing with so-called "engineered analogues" of pre-existing argillite/concrete interfaces for times up to 125 years. This experimental platform offers two kinds of hydraulic conditions: (1) zones with water circulating from the upper aquifer ("wet" context) and (2) zones excluding any influence of this aquifer (so-called "dry" context though argillite is partly of fully water saturated).

Regarding the perturbations undergone by argillite in contact with steels, the program is focused on other "engineered analogues" sampled in the Tournemire tunnel. Three types of steels were introduced in boreholes drilled in 1998 in a zone located far from any mechanical disturbance in "dry" conditions as well as in two different areas of the Excavated Disturbed Zone (EDZ) around the tunnel. After two or six years of contact, both steels and clay were recovered by overcoring.

This paper illustrates the added value provided by the Tournemire URL, allowing to combine various experimental approaches as well as different interaction time scales, completed with modelling which help us understanding what processes are responsible for the propagation of disturbances induced in the argillite by concretes or steels. Though some features worth being further investigated, the data acquired in the framework of this program combined to others reported in the literature provide with valuable information regarding the extrapolation to the geochemical evolution of clay materials due to the presence of concrete or steel for periods lasting several tens of thousands of years.

Chapter 6 - In order to effectively eliminate or reduce the hazards of long-lived activity from nuclear waste, an approach of laser Compton scattering gamma-ray based nuclear transmutation is proposed. Comparing with the conventional bremsstrahlung gamma-ray, laser Compton scattering gamma-ray features a peak in energy spectrum that can merit the coupling of gamma-ray and nuclear giant resonance to induce the photonuclear reaction. The advantage of laser Compton scattering gamma-ray in the application of transmutation is discussed, and the reaction rate of transmutation is theoretically analyzed. According to the proposal, a laser Compton scattering gamma-ray facility has been developed on New SUBARU storage ring; fundamental experiments concerning nuclear transmutation is carried out; the measurement of reaction rate based on $^{23}Na^{127}I$ target is described and the experimental data is close to the simulation result. To improve the transmutation efficiency, the second target is considered by using the neutrons generated from the first target, therefore, a complex targets set is proposed. The related investigation is introduced.

Chapter 7 - A novel method of exploring the uppermost 100-200 km of the Earth is examined. A small, dense, heat-generating probe melts its way down through the crust and mantle while its position and progress are tracked by acoustic signals generated in the rocks. The data from the descending probe and the signals themselves will yield new insights into the physical properties of the rocks through which they pass. These, when combined with other geophysical methods, should provide unequivocal information on the nature and composition of the Earth's interior. The probe consists of an outer sphere of tungsten ~ 1m in diameter inside which is a ^{60}Co radioactive heat source. The authors calculate that such a probe will reach the oceanic Moho in less than 6 months and attain minimum depths of well over 100 km in a few decades beneath both oceanic and continental lithosphere.

Chapter 8 - The usual approach to nuclear waste fixation is to incorporate radioactive elements in borosilicate glasses. However, researches are in progress to incorporate long lived nuclear wastes like actinides in a glass matrix with a higher chemical durability. Good chemical durability is generally achieved by increasing the silica content in the glass composition. Associated to high structural stability and thermal shock resistance, silica glass will optimize the properties which characterize a desirable material for the actinide fixation. But, actinide containment in silica glass involves melting the glass at high temperature ($\approx$ 2000°C) giving rise to various problems (evaporation losses, interaction between crucible and molten materials...).

According to an easy sintering stage, the sol-gel process is a new way to synthesize silica glasses at low temperature ($\approx$1000°C). In the first part of the paper the authors explain what is the sol-gel process and why it is an interesting way to prepare glasses at a temperature two times lower than by the conventional melting process.

Generally nuclear wastes exist as aqueous salt solutions and in the second part of the paper, the authors propose to use the totally open pore structure of the gel to allow migration of that liquid throughout the silica gel. They investigate the physical properties (mechanical strength, pore volume, permeability) of different gels porous structure (xerogels, aerogels, and composite gels). With this approach they show that thanks to a higher permeability and good mechanical properties, the porous network of composite gels can be easily used as a host matrix for the actinides simulating salts (Nd and Ce nitrates dissolved in water). After soaking and drying, the loaded material is sintered in the temperature range 1100°C-1200°C. Nd_2O_3 and CeO_2 loading in the range 0-20% can be achieved with this process.

The last part of the paper shows that the final structure of the fully sintered materials is that of a glass ceramic (silica glass+lanthanide oxides) and the results prove clearly the improvement of the chemical durability glass ceramic compared to the conventional nuclear glass (borosilicate glass). Owing to its simple structure, the corrosion rate of the glass-ceramics is close to that of the pure silica, almost 2 orders of magnitude lower than that of the borosilicate nuclear glass. Beside the good chemical durability the glass ceramics present also interesting mechanical properties.

Chapter 9 - This commentary describes the development of a family of calixbiscrowns for nuclear waste treatment. The development has been world wide with the involvement of scientists and governments. This development previously devoted to application in nuclear wastes has sparked continuing fundamental works in chemistry. This is one striking example of an applied research that spanned and opened new fundamental research fields. In a more general point of view it is shown how supramolecular chemistry has reached nanochemistry.

In: Nuclear Waste Research: Siting, Technology and Treatment   ISBN 978-1-60456-184-5
Editor: Arnold P. Lattefer, pp. 1-15

Short Communication A

NEW MAGNESIUM PHOSPHATE GLASSES FOR HIGH LEVEL NUCLEAR WASTE IMMOBILIZATION

Toshinori Okura and Hideki Monma*
Department of Materials Science and Technology,
Faculty of Engineering, Kogakuin University,
2665-1, Nakano, Hachioji,
Tokyo 192-0015, Japan

ABSTRACT

The leaching behavior and structure of magnesium phosphate glasses containing 45-55 mol% MgO incorporated with simulated high level nuclear wastes (HLW) were studied. The leach rate of the waste glasses decrease with increasing the simulated HLW content. The gross leach rate of the glass waste form containing 50 mol% MgO and 45 mass% simulated HLW is of the order of $10{-}6$ g/cm2·day at 90°C, which is small enough as compared with the corresponding release from a currently used borosilicate glass waste form. The isolated ions such as dimeric $(P2O7)4{-}$ and monomeric $(PO4)3{-}$ ions increase upon as increasing the incorporating amount of the simulated HLW. The changes in properties can be attributed to the structure changes owing to the incorporation of the simulated HLW. For the comparison, the leaching behavior and structure of calcium phosphate glasses containing 45-55 mol% CaO incorporated with the simulated HLW were also studied.

Keywords: *Nuclear waste immobilization; Magnesium phosphate glass; Leach rates; Thermal properties; Vibrational spectra; Microstructure.*

---

* Corresponding author: Tel.: +81-42-628-4149; fax: +81-42-628-4149. *E-mail address:* okura@cc.kogakuin.ac.jp (T. Okura).

1. INTRODUCTION

The disposal of radioactive waste generated by the nuclear fuel cycle is among the most pressing and potentially costly environmental problems. The high level nuclear wastes (HLW) are immobilized in a stable solid state and completely isolated from the biosphere.

Nuclear waste glasses are typically borosilicate glasses, and these glass compositions can experience phase separation at elevated concentrations of P_2O_5. The maximum P_2O_5 concentrations must be limited to between 1 and 3 mass%. For some waste streams, this can require considerable dilution and a substantial increase in the volume of the waste glass produced. Hence, there has been a continuing interest in developing phosphate glasses as waste forms. Furthermore, typical borosilicate glasses are limited to no more than 5 mass% actinides (2 mass% for plutonium). In contrast, iron phosphate glass with up to 15 mass% P_2O_5 can accommodate up to 40 mass% of simulated HLW [1, 2].

Phosphate glasses have some advantages over borosilicate glasses, such as a lower melting temperature and higher solubility for problematic elements, such as sulfur, and were investigated as early as the 1960s. Later work on sodium-aluminum phosphate glass [3] and iron-aluminum phosphate glass [4] showed that some of these glasses have comparable or better chemical durability than the borosilicate glasses. Present efforts are focused on the development of lead-iron phosphate glasses [1, 5-10]. The main disadvantage of phosphate glass is that the melts are highly corrosive. Still, a number of the engineering problems were overcome and in the 1980s at Mayak in the Urals, considerable amounts of waste, approximately 1000 m^3, were immobilized in a phosphate glass [11]. Vitrification of wastes with Na-Al phosphate glass matrix continues today at the Mayak Production Association in Chelyabinsk where 300 million Curies of activity of HLW have been immobilized in glass. There have been studies to investigate the immobilization of Cs [12], CsCl and SrF_2 [13], mixed-waste sludge [14] and spent nuclear fuel [2] in iron phosphate glass compositions.

Magnesium phosphate glasses are classified as 'anomalous phosphate glasses', which exhibit anomalies in the relationship between physical properties, such as density and refractive index, and MgO/P_2O_5 (M/P) molar ratio around the metaphosphate composition (M/P=1). The structures of M-P glasses have been studied [15]. Most of the phosphate glasses form high polyphosphate consisting of chains of phosphate ions, while the structures of M-P glasses are of two types, one includes four membered rings of PO_4 tetrahedra at M/P<1 (type T) and the other contains dimers of PO_4 tetrahedra at M/P>1 (type P).

In this study, M-P glasses are chosen as the base glass [16]. Calcium phosphate glasses classified as 'normal phosphate glasses' are also chosen for the comparison. Mixed metal oxide, which acts as the simulated HLW (radioactive isotopes were not employed) [17] was incorporated into the base glass to study its effects on the properties of the glasses. The present article reports on the leach rates to water, some of thermal properties and microstructure. The variations of the glass structure due to the incorporation of the simulated HLW are also examined by Fourier-transformed infrared (FT-IR) and Raman spectra.

2. EXPERIMENTAL

2.1. Sample Preparation

The M-P or $CaO-P_2O_5$ (C-P) glass frit that is used to produce glass waste form can be prepared by combining appropriate amounts of magnesium or calcium oxide and phosphoric acid and by heating at 1250 °C for 1 h. The powder mixtures of the glasses containing 0, 25 and 45 mass% of simulated HLW were melted at 1250 °C for 2 h. The melt waste glass was poured into a stainless plate. The composition of the simulated HLW is shown in Table 1.

Table 1. Composition of simulated nuclear wastes

Waste element	Raw material	mass%
Na	$NaNO_3$	64.8
Sr	SrO	2.9
La	La_2O_3	16.1
Mo	MoO_3	7.5
Mn	MnO_2	1.2
Fe	Fe_2O_3	6.6
Te	TeO_2	0.9

2.2. Leach Test

According to the technique of MCC-2 [18], the leach test for the glass waste forms was conducted in distilled water. About 1 g of each sample crushed to 10-20 mesh was dipped into 50 ml of water in a Teflon mini-autoclave beaker within an oven kept at 90 °C for 20 days. The total surface area of the grains was estimated by the following:

$$S = W S_0 / \rho \tag{1}$$

where W and ρ are the mass in g and the density in g/cm^3 of sample, and S_0 the specific surface of crushed specimen, respectively. The leach rates of gross and each constituent element were determined from the total weight loss of the specimen and the leachate analysed by inductively coupled argon plasma spectroscopy (ICP).

2.3. Density, XRD, DTA and SEM

The density of the waste forms was measured at room temperature using the Archimedes method with kerosene as the immersion fluid. Powder X-ray diffraction (XRD) analysis of the as-quenched melt was used to verify the amorphous state of the samples. The differential thermal analyses (DTA) were performed in flowing air at a heating rate of 20 °C/min. The microstructure of the waste forms was investigated with the scanning electron microscope (SEM).

2.4. FT-IR and Raman Spectra

The FT-IR spectra were measured using the KBr pellet technique in the frequency range 400-4000 cm^{-1} at room temperature. The Raman spectra were measured using a double grating spectrometer with an argon ion laser, scattered radiation being collected at 90 ° to the incident beam. The spectra were recorded over the 400-1400 cm^{-1} range at room temperature.

3. RESULTS AND DISCUSSION

3.1. Vitrification of Wastes with Glass Matrix

The composition and density of the glass waste forms prepared in this study were listed in Table 2, where the structure of 45M55P0W (M/P<1) glass is referred to as the type T, and that of 50M50P0W (M/P=1) and 55M45P0W (M/P>1) glass is referred to as the type P. The density increases with increase in simulated HLW content. The appearance of the base glass is colorless and transparent. The colors of the glasses, which contain the simulated HLW, are dark brown and turn darker with increasing the simulated HLW content.

Table 2. Composition and density of glass waste forms prepared in this study

Composition (mol%) (A=Mg,Ca)	Simulated Waste Content (mass%)	MgO-P_2O_5		CaO-P_2O_5	
		Glass waste form	Density (g/cm³)	Glass waste form	Density (g/cm³)
AO:P_2O_5=45:55	0	45M55P 0W	2.45	45C55P 0W	2.67
	25	45M55P25W	2.67	45C55P25W	2.85
	45	45M55P45W	2.91	45C55P45W	2.99
AO:P_2O_5=50:50	0	50M50P 0W	2.25	50C50P 0W	2.75
	25	50M50P25W	2.73	50C50P25W	2.93
	45	50M50P45W	2.89	50C50P45W	3.04
AO:P_2O_5=55:45	0	55M45P 0W	2.51	55C45P 0W	2.91
	25	55M45P25W	2.76	55C45P25W	2.96
	45	55M45P45W	3.02	55C45P45W	3.14

The glass states were confirmed by the absence of XRD peaks. The results of XRD measurements are shown in Table 3. Only the 55M45P25W glass waste form partially crystallized during cooling. To understand the effect of temperature on the crystallization of the glass waste forms, the samples were annealed isothermally at 500 °C for 2h. The glass forming tendency of the compositions with high simulated HLW contents decreased and these samples partially crystallized during cooling, except the sample 40M60P45W.

Table 3. The results of XRD measurements of glass waste forms

Glass waste forms	Vitreous state (V) or Crystalline state (C)	
	Before heat-treatment	After heat-treatment at 500 °C for 2h
45M55P 0W	V	V
45M55P25W	V	V
45M55P45W	V	V
50M50P 0W	V	V
50M50P25W	V	C
50M50P45W	V	C
55M45P 0W	V	V
55M45P25W	C	C
55M45P45W	V	C
45C55P 0W	V	C
45C55P25W	V	C
45C55P45W	C	C
50C50P 0W	V	C
50C50P25W	C	C
50C50P45W	C	C
55C45P 0W	V	C
55C45P25W	C	C
55C45P45W	C	C

The borosilicate glasses are limited to no more than 5 mass% actinides. In contrast, M-P glass with up to 55 mol% P_2O_5 can accommodate up to 45 mass% of simulated HLW.

3.2. Leach Rates of Samples in Water

The gross leach rates and the leach rates of each constituent element of the sample in water at 90 °C were determined from the total weight loss of the specimen and chemical analysis of leachate solution. The results are summarized in Figures 1 and 2 and Table 4. The chemical durability of the glasses was greatly improved as the addition of simulated HLW. Figure 1 shows that 50M50P45W has the gross leach rate of the order of 10^{-6} $g/cm^2 \cdot day$, which is fairly low as compared with that of the borosilicate waste glass. Of the elements in most phosphate glasses, Na shows higher leach rate than others, probably because of rather higher solubility of its polyphosphate consisting of chains of phosphate ions. No effect of Na in the M-P glass waste form on its leachability was found. These results can be attributed to the glass structure.

Table 4. Leach rate of each constituent element of 55M45P45W glass waste form

Element	55M45P45W $(g/cm^2 \cdot day)$
Mg	4.79×10^{-7}
P	1.08×10^{-6}
Na	7.35×10^{-7}
Sr	1.00×10^{-9}
La	n.d.
Mo	1.80×10^{-7}
Mn	2.00×10^{-9}
Fe	1.60×10^{-8}
Te	1.00×10^{-9}

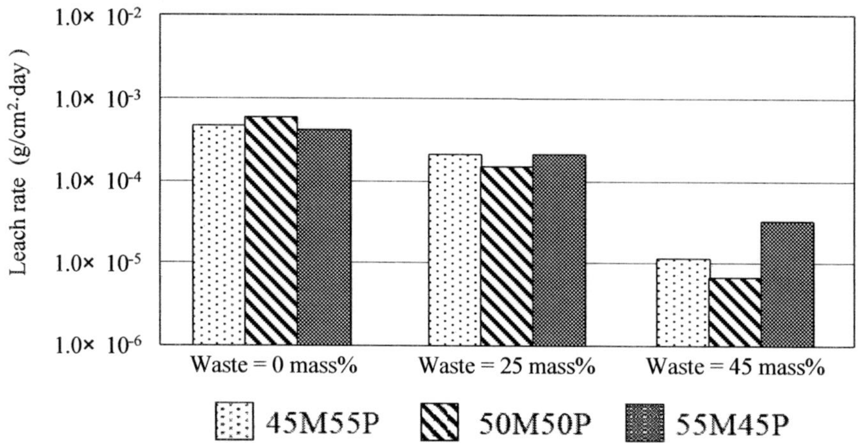

Figure 1. Gross leach rate of M-P glass waste forms.

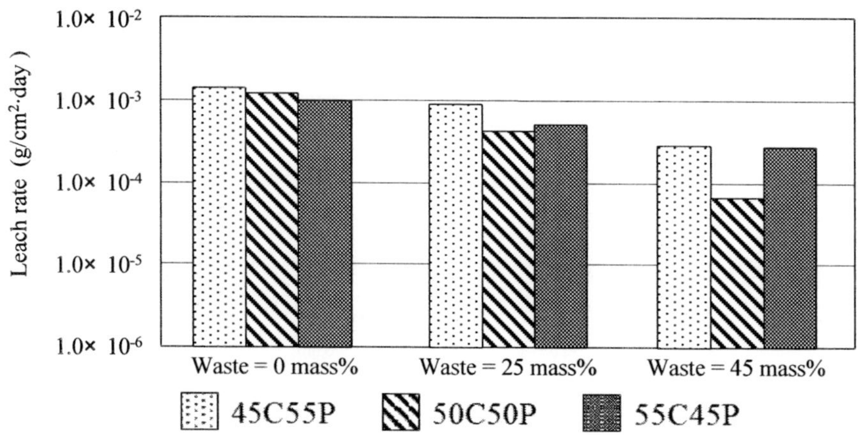

Figure 2. Gross leach rate of C-P glass waste forms.

3.3. Thermal Properties of Glass Waste Form

It is important to obtain information about the thermal stability of the glass waste form, since the crystallization of glass waste can mostly increase the undesirable aqueous corrosion rate of the form, probably due to the formation of somewhat more soluble crystals or the increase in surface area.

Figure 3 shows the DTA curves of the samples. The starting temperature of the crystallization peaks (Tx) and glass transition temperature (Tg) determined from the DTA curves and the stability of the waste forms (Tx-Tg) are listed in Table 5 and Figures 4 and 5. It indicates that both Tx and Tg decrease with increasing simulated HLW content. The stability of the waste forms with the type P glass matrix does not change with increasing the simulated HLW content.

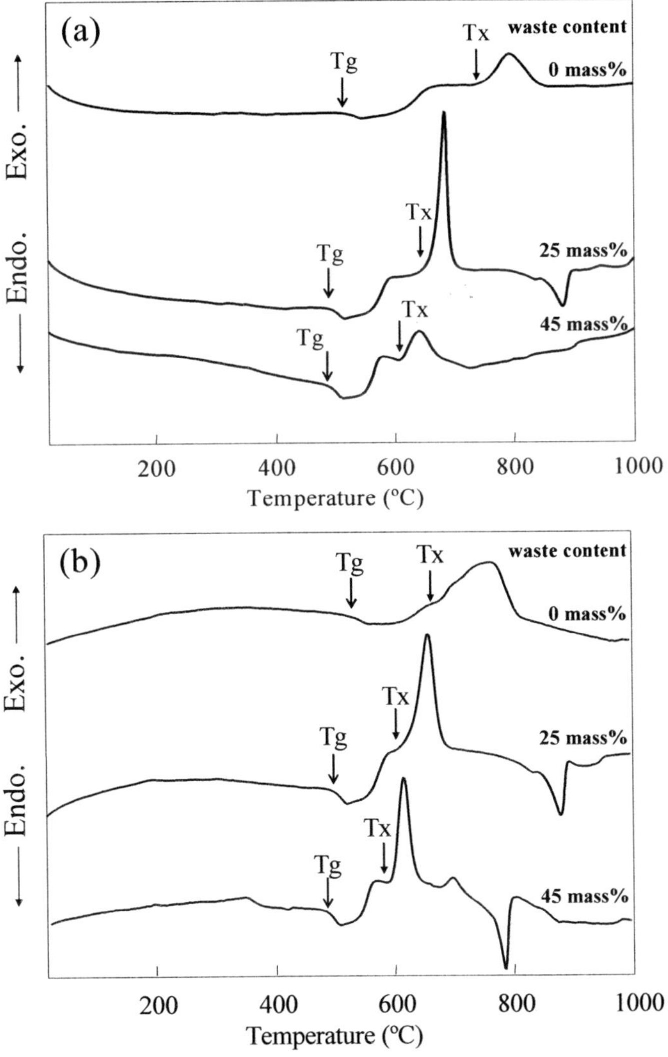

Figure 3. (Continued).

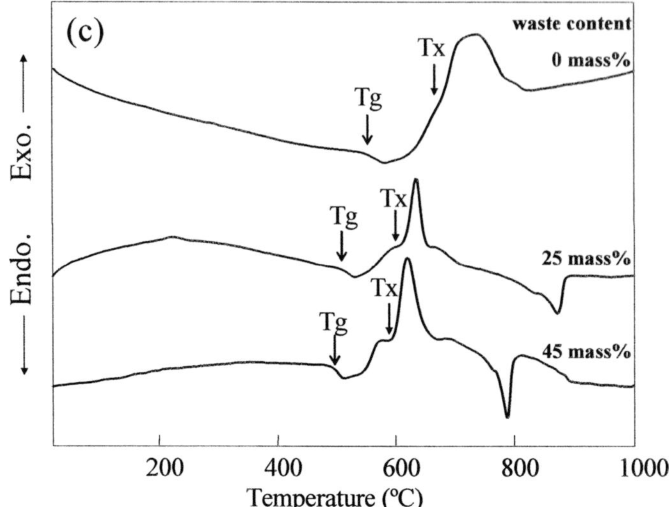

(a) 45M55P, (b) 50M50P, (c) 55M45P.

Figure 3. DTA curves for samples with simulated waste content (0, 25 and 45 mass%).

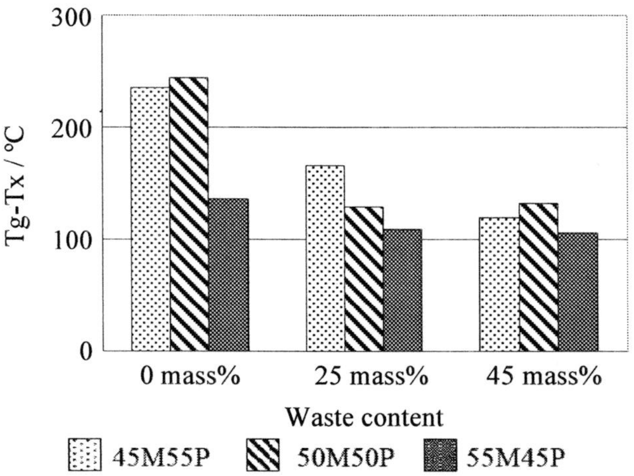

Figure 4. The stability of M-P glass waste forms.

Table 5. The results of DTA measurements of glass waste forms

Glass waste forms	45M55P			50M50P			55M45P		
Waste	0W	25W	45W	0W	25W	45W	0W	25W	45W
Tg (°C)	519	492	490	528	506	495	547	508	490
Tc (°C)	754	658	609	772	635	617	683	617	596
Glass waste forms	45C55P			50C50P			55C45P		
Waste	0W	25W	45W	0W	25W	45W	0W	25W	45W
Tg (°C)	496	453	400	526	474	386	541	389	377
Tc (°C)	666	623	491	609	580	484	618	529	472

Tg ; glass transition temperature   Tc ; starting temperature of crystallization.

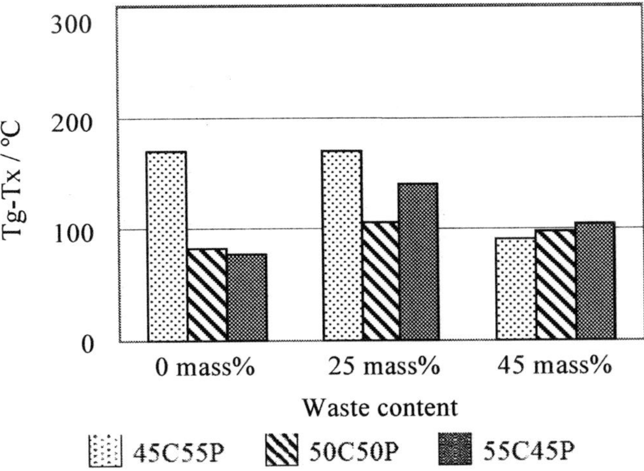

Figure 5. The stability of C-P glass waste forms.

3.4. Microstructure

Figure 6 shows SEM photographs for M-P samples with simulated waste content of 25 and 45 mass%. Crystalline grains were observed in the 55M45P25W form. The 45M55P25W, 50M50P25W, 45M55P45W, 50M50P45W and 55M45P45W forms were vitreous states. These facts correspond to the results of XRD measurements well. In the 45M55P25W form, phase separation was observed.

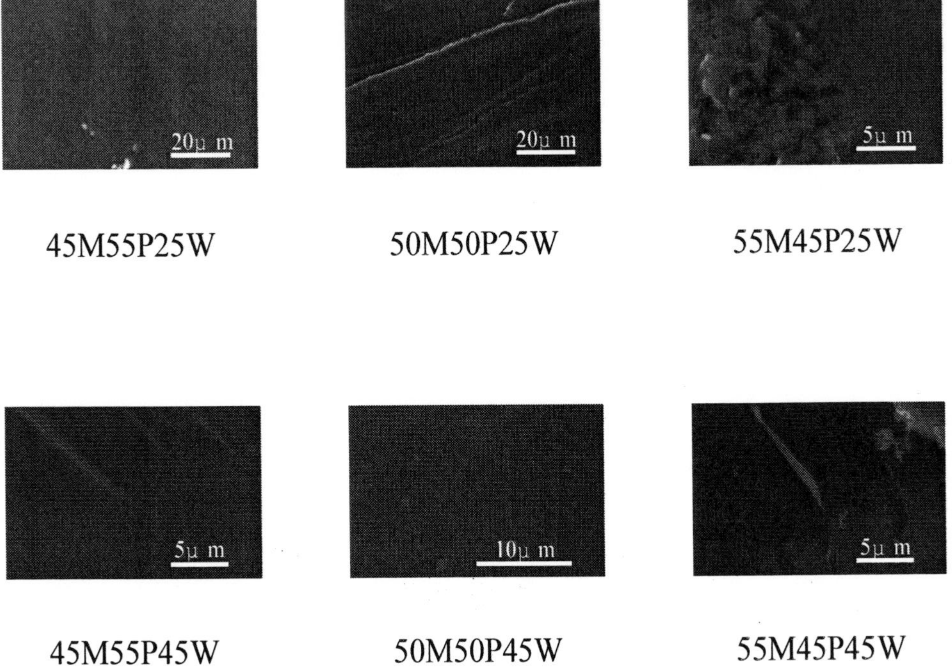

Figure 6. SEM photographs of M-P glass waste forms.

3.5. FT-IR Spectra of Glass Waste Form

The results of FT-IR measurements of the samples listed in Table 2 are shown in Figure 7 and Table 6. The strong band near 1300 cm^{-1} is attributed to (P=O) stretching mode. The band obviously becomes smaller indicating a decrease in the double bond character and in the effective force constant of the (P-O) bond as found in ultraphosphates with increasing the simulated HLW content.

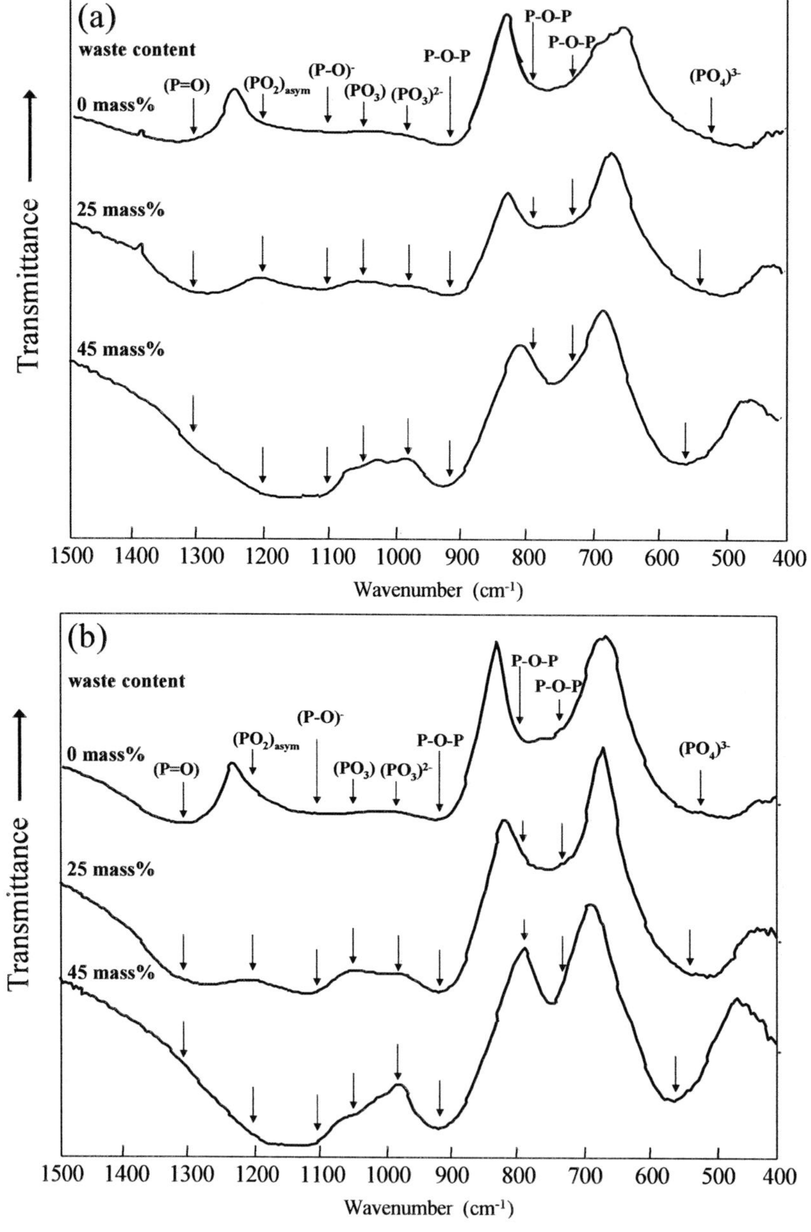

Figure 7. (Continues)

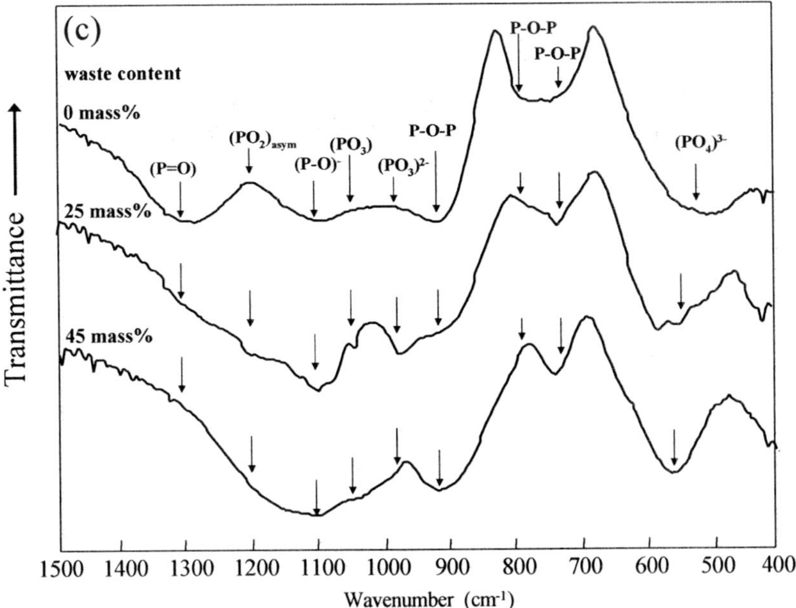

(a) 45M55P, (b) 50M50P, (c) 55M45P.

Figure 7. FT-IR spectra of glass waste forms.

The band near 1200 cm^{-1} is assigned to asymmetric stretching modes of the two non-bridging oxygen atoms bonded to phosphorus atoms, $(PO_2)_{asym}$ or Q^2 units, in the phosphate tetrahedra. Their amplitudes obviously increase with increasing waste content. This result indicates that the phosphate linkages are shortened as the simulated HLW incorporate into the glass structure and leading to decrease the relative content of the Q^2 units.

The absorption bands near 1100 cm^{-1} have been assigned to $(P-O)^-$ groups, and its amplitudes increase with increasing waste content. It is suggested that the absorption bands of (R-O-P) (R = waste element) also locate at near 1100 cm^{-1}, and the relative content of these bonds increase with increasing waste content leading to increased intensity of 1100 cm^{-1} band.

Table 6. The results of FT-IR measurements of glass waste forms

Wave number (cm^{-1})	1300	1200	1100	1050	980	920	720-790	480-570
	(P=O)	$(PO_2)_{asym}$	$(P-O)^-$	$(PO_3)^-$	$(PO_4)^{3-}$	$(P-O-P)_{asym}$	$(P-O-P)_{sym}$	$(PO_4)^{3-}$
M-P glass	decrease	increase	increase	decrease	decrease	shift*	decrease	shift*
C-P glass	decrease	increase	unchanged	increase	increase	unchanged	unchanged	shift*

*;shift to higher frequency.

The intensity of the band near 1050 cm^{-1}, which is assigned to (PO_3) end groups (Q^1), tends to decrease with increasing waste content. The absorption bands near 980 cm^{-1} and 480-570 cm^{-1} are assigned to the stretching and deformation modes of $(PO_4)^{3-}$ groups (Q^0), respectively. It is shown that the absorption bands of the deformation modes of $(PO_4)^{3-}$ group shift to higher frequencies, and the amplitudes of $(PO_4)^{3-}$ group's absorption bands near 980 cm^{-1} decrease with increasing waste content. The Q^1 and Q^0 groups decrease with increasing waste content. These results indicate that the relative content of non-bridging oxygen (P-O$^-$), which may be replaced by the formation of (R-O-P) bonds, decreases as the incorporation of the simulated HLW.

The absorption bands near 920 and 790-790 cm^{-1} are assigned to the asymmetric and symmetric stretching modes of the (P-O-P) linkages, respectively. The asymmetric stretching band of (P-O-P) near 920 cm^{-1} initially shifts to higher frequencies as the amount of the waste increases. The larger wavenumber of the (P-O-P) band is a result of the smaller (P-O-P) bond angle, which results from shorter phosphate linkages or smaller metal cation size. The phosphate linkages of the glasses with higher waste content are shorter due to the depolymerization of the glass structure.

The formation of asymmetric bridging oxygen (R-O-P) would increase the cross-link density of the glass network, improving the chemical durability of the glasses.

3.6. Raman Spectra of Glass Waste Form

The results of Laser Raman measurements of the samples listed in Table 2 are shown in Figure 8 and Table 7. The band near 1300 cm^{-1} is assigned to the symmetric stretching mode of terminal oxygen (P=O). The bands near 1200 and 700 cm^{-1} are due to the symmetric stretching mode of non-bridging oxygen (PO$_2$)$_{sym}$ on each tetrahedron and to the symmetric stretching mode of bridging oxygen between two tetrahedral (P-O-P)$_{sym}$, respectively. The bands near 1050 and 760 cm^{-1}, which are characteristic of pyrophosphate groups, are attributed to (P-O) symmetric stretching mode of non-bridging oxygen and (P-O-P) symmetric stretching mode of bridging oxygen, respectively. There are also some traces of orthophosphate groups (PO$_4$)$^{3-}$ as indicated by the band near 950 cm^{-1}. With increase waste content, the (P=O) stretching mode bands becomes smaller indicating a decrease in the double bond character and in the effective force constant of the (P-O) bond as found in ultraphosphates. Also, with increasing waste content, the bands near 1200 and 700 cm^{-1} have disappeared from the spectrum and the bands near 1050, 760 and 950 cm^{-1} are predominate, that is, the bands characteristic of the pyrophosphate dimer (P$_2$O$_7$)$^{4-}$ and orthophosphate monomer (PO$_4$)$^{3-}$ species increased.

Table 7. The results of Laser Raman measurements of glass waste forms

Wave number (cm^{-1})	1300	1200	1050	950	760	700
	(P=O)	(PO$_2$)$_{asym}$	(PO)$_{sym}$	(PO$_4$)$^{3-}$	(P-O-P)$_{sym}$	(P-O-P)$_{sym}$
M-P glass	decrease	decrease	increase	increase	increase	decrease
C-P glass	decrease	decrease	increase	increase	increase	decrease

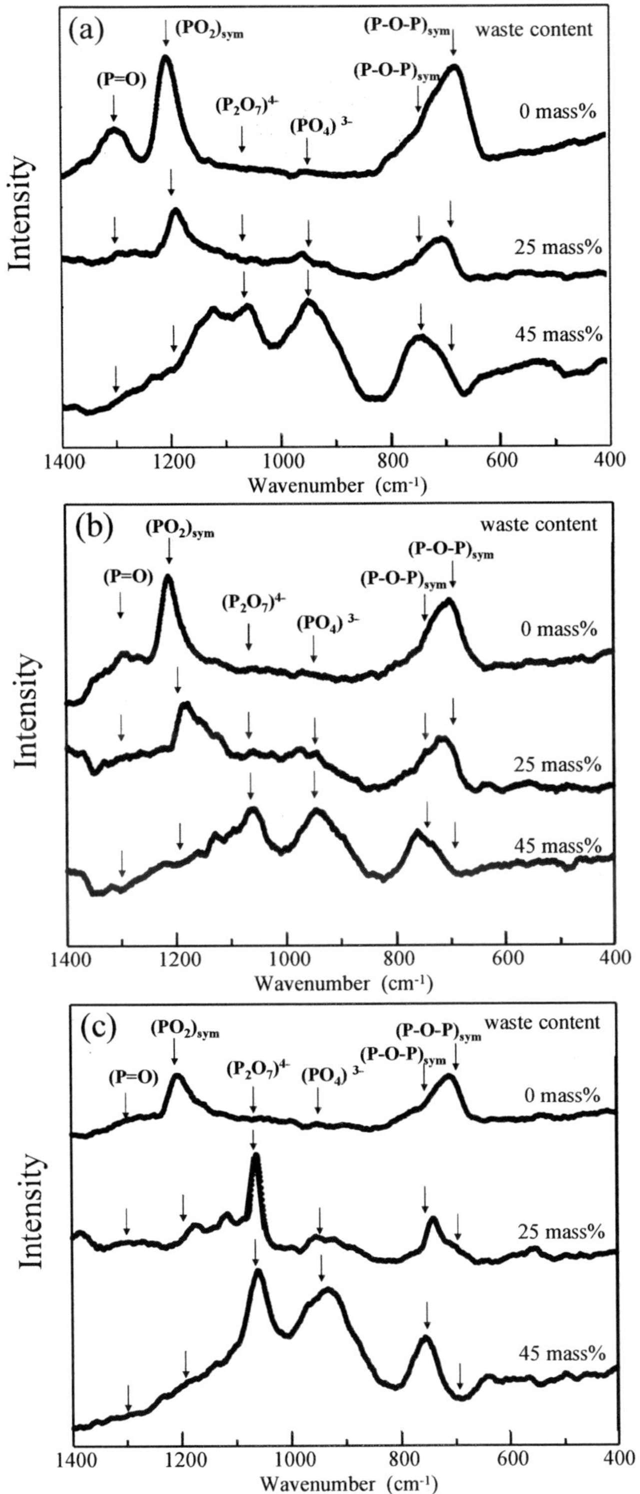

(a) 45M55P, (b) 50M50P, (c) 55M45P.

Figure 8. Laser Raman spectra of glass waste forms.

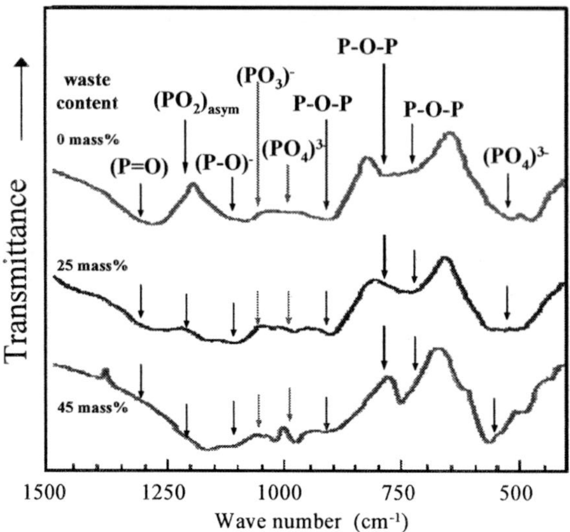

Figure 9. FT-IR spectra of 50C50P glass waste form.

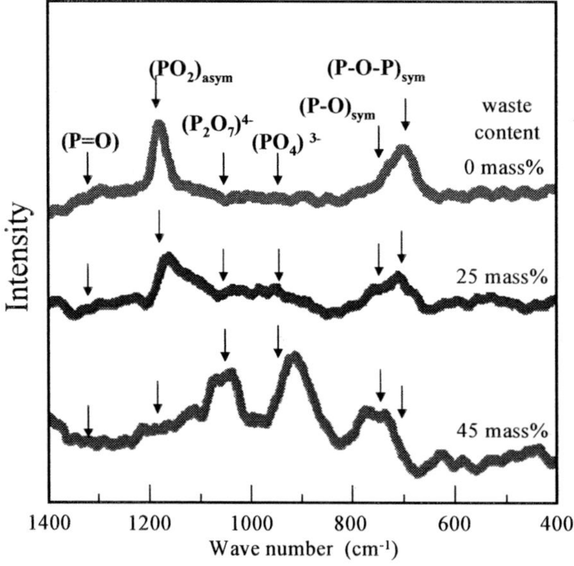

Figure 10. Laser Raman spectra of 50C50P glass waste form.

4. CONCLUSION

M-P glasses were proposed as the potential nuclear waste glasses. The leach rates and structure of M-P glasses loaded with the simulated HLW were examined. The main features of this work are as follows:

(1) Up to 45 mass% loadings of the simulated HLW can be incorporated into the base M-P glasses.

(2) The leach rate of the waste glasses decreases with increasing the simulated HLW content. The gross leach rate of the 50M50P45W waste form with the type P glass matrix is of the order of 10^{-6} g/cm^2·day at 90°C, which is fairly low as compared to that of the borosilicate waste glass.

(3) The stability of the waste forms with the type P glass matrix does not change with increasing the simulated HLW content.

(4) With increasing waste content, the bands characteristic of the pyrophosphate dimer $(P_2O_7)^{4-}$ and orthophosphate monomer $(PO_4)^{3-}$ species increased.

(5) The formation of asymmetric bridging oxygen (R-O-P) increases the cross-link density of the glass network, improving the chemical durability of the glasses.

(6) The type P 50M50P glass with low density is most suitable for vitrification of wastes.

REFERENCES

[1]   D. E. Day, Z. Wu, C. S. Ray and P. Hrma. *J. Non-Cryst. Solids*, 241, 1-12 (1998).

[2]   M. G. Mesko and D. E. Day, *J. Nucl. Mater.*, 273, 27-36 (1999).

[3]   J. Van Geel, H. Eschrich, W. Heimerl and P. Grziwa, *IAEA, Vienna*, 22-26 (1976).

[4]   B. Grambow and W. Lutze, *"Scientific Basis for Nuclear Waste Management*, Vol. 2. Northrup", ed. by CJM, Jr, Plenum Press, New York (1979) pp. 109-116.

[5]   B. C. Sales and L. A. Boatner, *Science*, 226, 45-48 (1984).

[6]   B. C. Sales and L. A. Boatner, *"Radioactive Waste Forms for the Future"*, ed. by W. Lutze and R. C. Ewing, North-Holland, Amsterdam (1988) pp. 193-231.

[7]   T. Yanagi, M. Yoshizoe and N. Nakatsuka, *J. Nucl. Sci. Technol.*, 25, 661-666 (1988).

[8]   T. Yanagi, M. Yoshizoe and K. Kuramoto, *J. Nucl. Sci. Technol.*, 26, 948-954 (1989).

[9]   S. T. Reis, M. Karabulut and D. E. Day, *J. Nucl. Mater.*, 304, 87-95 (2002).

[10] P. Y. Shin, *Mater. Chem. Phys.*, 80, 299-304 (2003).

[11] W. Lutze, *"Radioactive Waste Forms for the Future"*, ed. by W. Lutze and R. C. Ewing, North-Holland, Amsterdam (1988) pp. 1-159.

[12] S. T. Reis and J. R. Martinelli, *J. Non-Cryst. Solids*, 247, 241-247 (1998).

[13] M. G. Mesko, D. E. Day and B. C. Bunker, *Waste Management*, 20, 271-278 (2000).

[14] R. D. Spence, T. M. Gilliam, C. H. Mattus and A. J. Mattus, *Waste Management*, 19, 453-465 (1999).

[15] T. Okura, K. Yamashita and T. Kanazawa, *Phys. Chem. Glasses*, 29, 13-17 (1988).

[16] T. Okura, T. Miyachi and H. Monma, *Trans. Mater. Res. Soc. Jpn.*, 29[5], 2175-2178 (2004).

[17] M. Ishida, T. Yanagi and R. Terai, *J. Nucl. Sci. Technol.*, 24, 404-408 (1987).

[18] D. M. Strachan, "Scientific basis for nuclear waste management", ed. By J. D. Moor., Plenum Press, New York (1980) pp. 347-348.

In: Nuclear Waste Research: Siting, Technology and Treatment   ISBN 978-1-60456-184-5
Editor: Arnold P. Lattefer, pp. 17-23   © 2008 Nova Science Publishers, Inc.

Short Communication B

ELECTROCHEMICAL SOLIDIFICATION OF RADWASTES IN SINGLE-CRYSTALLINE TITANATES

Hideki Abe, Akira Satoh, Kenji Nishida and Hideaki Kitazawa*

National Institute for Materials Science (NIMS),
Sengen 1-2-1, Tsukuba, Ibaraki 305-0047, Japan

ABSTRACT

A method for solidification of water-soluble Cs^+ ion in a chemically and physically stable form of single-crystalline titanates is presented. Electrochemical reduction of molten molybdates containing Cs^+ and Ti^{4+} ions has lead to the formation of single crystals of Cesium Titanate Hollandite, $Cs_{1.35}Ti_8O_{16}$, at ambient pressure. X-ray diffraction (XRD) has showed that the single crystals of $Cs_{1.35}Ti_8O_{16}$ have high crystallinity. The electrochemically prepared $Cs_{1.35}Ti_8O_{16}$ has advantage as a matrix material of radioactive ^{137}Cs in high-level radioactive wastes (radwastes) over the conventional solid-state matrices including borosilicate glass. Firstly, bulk $Cs_{1.35}Ti_8O_{16}$ shows much higher leaching resistance against hot water under high pressure than borosilicate glass. Secondly, single-crystalline $Cs_{1.35}Ti_8O_{16}$ has much lower specific surface area and higher crystallinity than the bulk $Cs_{1.35}Ti_8O_{16}$, which can minimize the leaching of ^{137}Cs to the environment.

INTRODUCTION

The human kind of the 21st century is facing a number of challenges caused by the huge consumption of fossil fuels through the past two centuries. One of the greatest challenges is the global warming that is giving rise to the global-scale climate change. The global warming is surmised to be caused mainly by the rapid increase of the CO_2 content in air as the result of

---

* Corresponding author: Hideki Abe National Institute for Materials Science (NIMS), Sengen 1-2-1, Tsukuba, Ibaraki 305-0047, Japan. Tel: +81-29-859-2732; Fax: +81-29-859-2801; E-mail Address: ABE.Hideki@nims.go.jp

combustion of fossil fuels. Besides the global warming, the supply of the fossil fuels itself is becoming unstable mainly due to the political situation of the petroleum-producing countries. Development of an alternative energy source for the fossil fuels is one of the most urgent issues to establish a sustainable society.

The nuclear fuel is the most highly established alternative for the fossil fuels. The nuclear energy is "clean" in terms of generation of electricity without emission of waste gases including CO_2. The nuclear energy has much higher figure of merit than the other "green" alternative energies such as the wind- or solar powers in terms of controllability and space-use efficiency.

The keystone of the usage of the nuclear energy is the development of safe and effective disposal methods of the radioactive wastes (radwastes) that are unavoidably generated through the consumption of nuclear fuels. Radwastes contain various kinds of radiotoxic elements such as ^{137}Cs, ^{90}Sr, ^{137}Ba or ^{90}Y, which have very different chemical and/or physical properties. ^{137}Cs is the most radiotoxic element among the radioactive elements contained in radwastes. ^{137}Cs has a relatively short lifetime that is close to the life spans of higher living organisms [1]. ^{137}Cs disperses widely into the environment due to its high water solubility and high volatility.

One of the most widely used methods for the disposal of nuclear wastes is the glass vitrification method that vitrifies the radwastes in chemically and physically stable Borosilicate glass matrices [2]. Borosilicate glass has high capacity to incorporate various kinds of radioactive elements coexisting in radwastes. The comparably low melting point of Borosilicate glass also is attractive from the viewpoint of processing costs for vitrification. Borosilicate glass, however, has a great demerit that the vitrified radwastes can be easily corroded in hot basic water at high pressures, since the Borosilicate shows high solubility toward basic water at high temperatures and pressures [2].

A group of Australian researchers has been developing another type of solid matrix to solidify radwastes in more chemically and physically stable form than the Borosilicate glass [2-5] Ringwood et al focused on the fact that the alkaline or alkaline earth elements contained in radwastes are often mined in the form of Titanates. This suggests that the radioactive elements in radwastes might be stably immobilized for a geographic span in the Titanates-based mimic of natural minerals, SYNROC (SYNthetic ROCks) [2-4]. Unlike the Borosilicate glass, SYNROC is a generic term to describe of a number of crystals that incorporate the radioactive elements in the thermodynamically stable form. Cesium Titanate Hollandite, $Cs_xTi_8O_{16}$, is a relevant SYNROC that immobilizes ^{137}Cs.[2,5-7] It has proven that $^{137}Cs_xTi_8O_{16}$ indeed shows much higher leach resistance against hot basic water at high pressures than ^{137}Cs-containing Borosilicate glass [2, 5-7].

Figure 1 shows the crystal strucrue of Cesium Titanate Hollandite ($Cs_xTi_8O_{16}$, $x \leq 1.35$; $I4/m$, $a = 0.3$ nm, $c = 1.0$ nm) [8]. The crystal structure of $Cs_xTi_8O_{16}$ consists of the rutile-type one-dimensional arrays of TiO_2 (Figure 1 left). The TiO_2 arrays form one-dimensional oxygen cavities by sharing the oxygen atoms on the apexes of the TiO_6 octahedra. The Cs^+ ions are incorporated in the oxygen cavities along the c-axis (Figure 1 right). The high leach resistance of $^{137}Cs_xTi_8O_{16}$ is accounted for on the basis of the high thermal and chemical stability of the rutile-type TiO_2 frame that encapsulate the $^{137}Cs^+$ ions.

The conventional synthetic method of $Cs_xTi_8O_{16}$ as well as the other SYNROC has been based on high-pressure sintering of TiO_2 and the precursors containing the targeted cations, which gives rise to polycrystals [2].

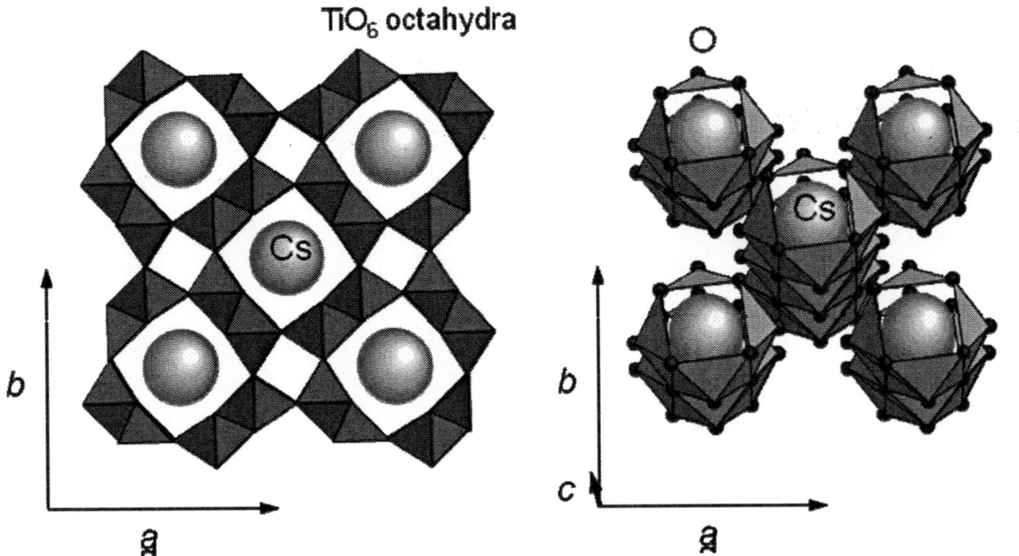

Figure 1. Crystal structure of Cesium Titanate Hollandite (CsxTi8O16, x < 1.35).

In this article, we report a new method to synthesize $Cs_xTi_8O_{16}$ in the form of single crystals. Electrolysis of a molten mixture of Cs_2CO_3, MoO_3, and TiO_2 enables the formation of single-crystalline $Cs_{1.35}Ti_8O_{16}$ at ambient pressure [9]. Electrochemically synthesized $Cs_{1.35}Ti_8O_{16}$ has advantage as a matrix material of ^{137}Cs over the conventional Borosilicate glass or SYNROC, since its small specific surface area can minimize the leaching of ^{137}Cs to the environment.

Experimental Section

Synthetic Methods

Cs_2CO_3 (99.99 % purity, Fruuchi Chemicals Co., Ltd.) was used for the Cs source. TiO_2 (99.98 % purity, Soekawa Chemicals Co., Ltd.) and MoO_3 (99.99 % purity, Furuuchi Chemicals Co., Ltd.) were used as purchased. Cs_2CO_3 and MoO_3 were mixed in a molar ratio of 1:1 and calcinated in air at 600 °C to obtain Cs_2MoO_4. An alumina boat with a dimension of $100 \times 10 \times 10$ mm^3 is filled with 5.0 grams of Cs_2MoO_4. Pt wires with diameters of 1.0 mm were placed at the both ends of the alumina boat as the cathode and anode. An aliquot of 0.1 grams of TiO_2 was put on Cs_2MoO_4 near the anode. Both the anode and cathode were connected to an electric power supply with Pt wires with diameters of 0.3 mm.

The alumina boat containing Cs_2MoO_4 and TiO_2 was heated in air up to 950 °C. Cs_2MoO_4 turned into a transparent melt. TiO_2 was sparingly soluble in the Cs_2MoO_4 melt. Electrolysis of the melt was performed at 950 °C in air at a constant electrolysis current of 10 mA. The potential between the anode and cathode was around 2.5 V during the electrolysis. After a duration of 1.5 hours, the cathode was removed from the melt and cooled to room temperature. Blue-black needle crystals precipitated on the cathode. The obtained crystals were washed thoroughly with a water solution of sodium ethylenediaminetetraacetate (EDTA) and $NaHCO_3$ (1 weight % each) to remove the residual electrolyte.

Characterization

Powder X-ray diffraction (pXRD; RIGAKU RINT2000; Cu $K\alpha$-radiation (λ = 0.1541 nm)) was performed for chemical/structural identification of the obtained crystals. The crystals were ground with an agate mortar for pXRD. Four-axis XRD (Bruker SMART APEX CCD area-detector diffractometer with graphite-monochromatized Mo $K\alpha$-radiation (λ = 0.071073 nm)) was performed at room temperature to determine the chemical composition as well as the structure of the crystals. Electron probe microanalysis (EPMA: JEOL SM-09010) was carried out to determine the chemical composition of the crystals.

Experimental Results

Figure 2 shows an optical microscope image of the crystals that precipitated on the cathode. The crystals were thin, blue-black needles with the average dimension of $2 \times 0.02 \times 0.02$ mm^3 in length, width and depth, respectively.

Figure 3 shows the pXRD profile of the crystals corresponding to Figure 3. All the visible peaks are indexed on the basis of the crystallographic parameters of the Cesium Titanate Hollandite, $Cs_xTi_8O_{16}$, ($I4/m$, a = 0.297 nm, c = 1.03 nm; $x \leq 1.35$). The chemical composition evaluated with EPMA, Cs:Ti:O = 2.0:8.0:17, is roughly consistent with the value expected by pXRD.

Figure 2. Optical microscope image of the crystals that precipitated on the Pt cathode.

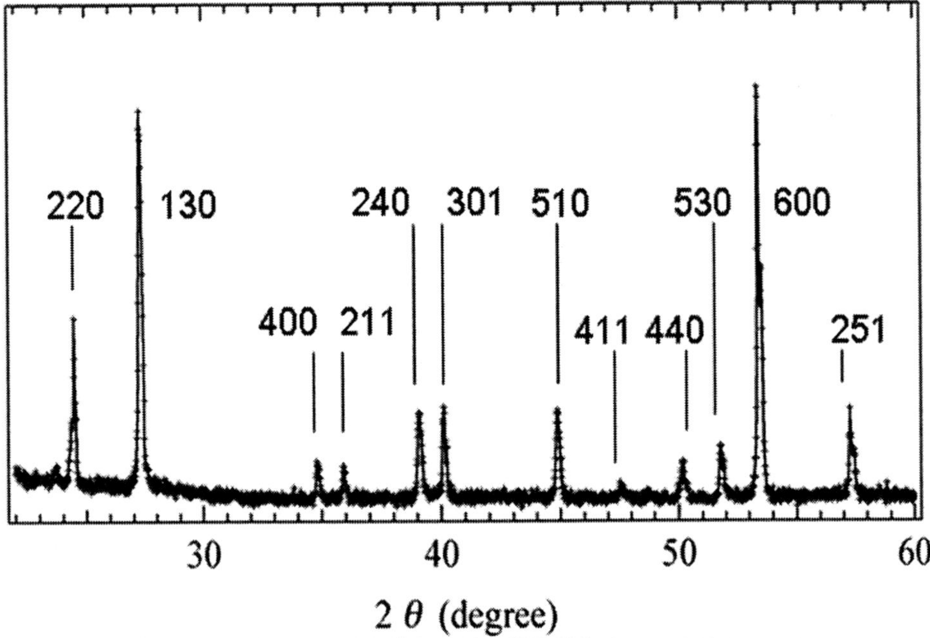

Figure 3. pXRD profile of the crystals corresponding to Figure 2. The reflection peaks are indexed with the crystallographic parameter of CsxTi8O16 (I4/m, a = nm, c = nm).

Four-axis XRD was performed on a single crystal of $Cs_xTi_8O_{16}$ to quantitatively determine the content of Cs, x. The structure was refined using the SHELX97 program on the basis of the XRD data assuming the atomic positions reported in the reference [8]. The final refinement converged to $wR(F_0^2) = 0.0961$ on all reflections. The low R-factor indicates the high crystallinity of the obtained crystals. The crystallographic parameters are listed in Table. The Ti and O sites (O1 and O2) are fully occupied by Ti and O, respectively.

Table. Positional parameters and equivalent isotropic displacement parameters for the single-crystalline $Cs_{1.35}Ti_8O_{16}$. a: Reference [8]

Atom	x	y	z	$U_{eq}(\text{Å}^2)$	SOF
Cs	0.00000	0.00000	0.6238(3) (0.627(3)) ^{a}	0.0180(2)	0.0841(4)
Ti	0.16487(3) (0.1652(3)) ^{a}	0.34837(3) (0.3485(2)) ^{a}	0.00000	0.0076(1)	0.50000
O1	0.2088(1) (0.2084(1)) ^{a}	0.1580(1) (0.1580(2)) ^{a}	0.00000	0.0066(2)	0.50000
O2	0.1671(1) (0.1664(2)) ^{a}	0.5379(1) (0.5377(1)) ^{a}	0.00000	0.0077(2)	0.50000

In contrast, only 2/3 of the Cs sites are occupied by Cs. The content of Cs, x, is calculated to be $x = 1.35$ on the basis of the occupancy. The atomic positions of Cs, Ti and O are well consistent with the reported data for $Cs_{1.35}Ti_8O_{16}$ (see the values indicated in parentheses).

DISCUSSION

The electrochemically synthesized $Cs_{1.35}Ti_8O_{16}$ (called electrochemical Hollandite hereafter) have two advantages over the polycrystalline SYNROC in terms of immobilization of Cs^{137}. Firstly, the electrochemical Hollandite has the much smaller specific surface area (SA/V) than the polycrystalline SYNROC. Indeed, SA/V of the single crystals of the electrochemical Hollandite is 30 times smaller than that of the polycrystalline SYNROC that has average grain size of 1 μm.[2,7] The electrochemical Hollandite that actually incorporates ^{137}Cs would show much higher leach resistance than the polycrystalline SYNROC, since the leach rate of soluble ions from matrix materials is generally proportional to SA/V of the matrix materials.

Secondly, ^{137}Cs might potentially be solidified in the electrochemical Hollandite directly from the radwastes that contain various kinds of radioactive elements other than [137]Cs. Figure 4 shows a schematic diagram of the selective electrochemical solidification of ^{137}Cs from actual radwastes. Radwastes are mixed with the flux material, MoO_3, together with the matrix material, TiO_2 (1). The materials are molten at high temperatures at ambient pressure into a molten electrolyte (2). The molten electrolyte is electrolyzed at the precipitation potential of the electrochemical Hollandite (3). The electrochemical Hollandite is precipitates on the electrode as a pure phase.

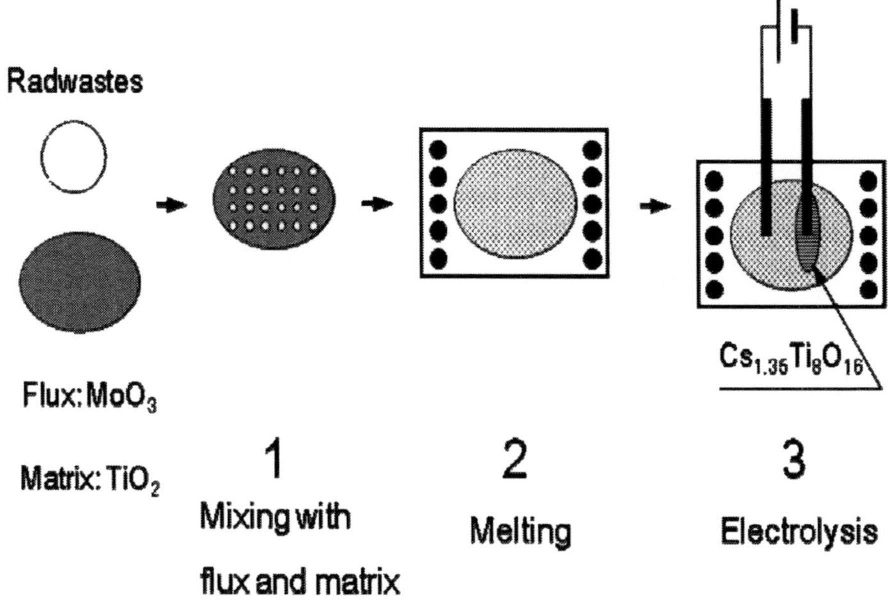

Figure 4. Schematic diagram of electrochemical solidification applied to actual radwastes.

The selective electrochemical solidification would enable the individual control of the radioactive elements that are immobilized in the single-crystalline, pure-phase titatnates. This can be a great advantage over the conventional indiscriminate solidification methods including glass vitrification or SYNROC in terms of storage costs. The development of the electrochemical solidification of radwastes is just beginning. It would contribute to the sustainable society in the future based on the nuclear energy.

ACKNOWLEDGMENTS

This work was supported by the Ministry of Education, Culture, Sports, Science and Technology (MEXT) and Japan Society for the Promotion of Science (JSPS) through Grant-in-Aid 19560845. This work was carried out at the Analysis Laboratory of the National Institute for Materials Science (NIMS).

REFERENCES

[1]   Nikula, E. J.; Muggenburg, B. A.; Griffith, W. Z.; Carlton, T. E.; Fritz, T. E.; Boecker, B. B. *Radiat. Res. 1996*, 146, 536-547.

[2]   Lutze, W.; Ewing, R. C. (eds.), Radioactive Waste Forms for the Future, *Elsevier,* Amsterdam, 1988.

[3]   Ringwood, A. E.; Kesson, S. E.; Ware, N. G.; Hibberson, W.; Major, A. *Nature* 1979, 309, 776-778.

[4]   Deivns, D. M.; Smart, R. S. C. *Nature* 1984, 309, 776-778.

[5]   Kesson, S. E. Radioactive Waste Manage. *Nucl. Fuel Cycle* 1983, 4, 53-72.

[6]   Kesson, S. E.; White, T. *J. Proc. R. Soc.* London A 1986, 405, 73-101.

[7]   Smith, K. L.; Lumpkin, G. R.; Blackford, M. G.; Day, R. A.; Hart, K. P. *Journal of Nuclear Materials* 1992, 190, 287-294.

[8]   Cheary, R. W. *Acta Crystallogr.* B 1991, 47, 325-333.

[9]   Abe, H.; Satoh, A.; Nishida, K.; Abe, E.; Naka, T.; Imai, M,; Kitazawa, H. *J. Solid State Chem.* 2006, 179, 1521-1524.

In: Nuclear Waste Research: Siting, Technology and Treatment   ISBN 978-1-60456-184-5
Editor: Arnold P. Lattefer, pp. 25-36   © 2008 Nova Science Publishers, Inc.

Short Communication C

A HEATED ELECTRODE TEST SYSTEM FOR STUDYING CORROSION BEHAVIOR OF ALLOY 22

Yugo Ashida, L. Glen McMillion and Manoranjan Misra
University of Nevada, Reno
Department of Chemical and Metallurgical Engineering
1664 North Virginia Street, Reno, NV 89557

ABSTRACT

Alloy 22 (UNS 06022) has been selected as the Corrosion Resistant Material (CRM) for the Waste Package (WP) outer barrier for the proposed High Level Nuclear Waste (HLNW) repository at Yucca Mountain (YM), Nevada, USA. A heated electrode technique has been developed for studying the corrosion behavior of Alloy 22 under simulated YM service conditions. The following tests were conducted: (1) potentiodynamic cyclic polarization data were collected to investigate electrochemical properties of Alloy 22 immersed in simulated ground waters; (2) extended immersion tests were carried out to investigate corrosion behavior of Alloy 22 under the expected ground water seepage conditions in the repository drifts ; (3) a long-term time delayed dripping test was run for studying susceptibility to localized corrosion under accumulated multi-component salt deposits with cyclic wet-dry conditions. The effects of temperature, pH, relative humidity (RH), and the accumulation of salt deposits on corrosion resistance of Alloy 22 are discussed in detail. Of the three simulated ground waters tested, Simulated Concentrated Water (SCW) was found to be the most corrosive media. Heat treatment of Alloy 22 for 100 hours at 800°C, cyclic wet-dry conditions, and the uneven crevices formed between the salt deposits and the hot metal surface, all contributed to accelerated corrosion of Alloy 22.

INTRODUCTION

Alloy 22 has been specified as the Corrosion Resistant Material (CRM) for the outer shell of the Waste Package (WP) for the proposed High Level Nuclear Waste (HLNW) repository at Yucca Mountain, Nevada, USA. The corrosion behavior of this highly corrosion-resistant alloy has been studied by research groups from different points of view [1-11]. Dunn et al. [5-9] addressed the localized corrosion susceptibility of Alloy 22 by comparing the crevice corrosion repassivation potentials with corrosion potentials in chloride solutions. They found that the corrosion potential of Alloy 22 is strongly dependent on pH, and the corrosion potentials in acidic solution are significantly greater than the repassivation potential in anodic direction. Evans et al. [10] evaluated the long- term corrosion potential and corrosion rate of creviced Alloy 22 samples in chloride and nitrate solutions. The data show that no reduced resistance of Alloy 22 to localized corrosion was found even after 250 days immersion. Miyagusuku and Devine [11] investigated the passive films formed on Alloy 22 in acidic solution at room temperature and at 90°C. The results imply that the passive film is approximately less than 1 nm thick and shows n-type semiconductor behavior. However, insufficient experiments were conducted in the solutions with ground water compositions at predicted container surface temperatures; the evolution of repository environments including temperature, relative humidity, and solution pH was not adequately considered in experimental design.

The waste package surface temperature is predicted to decrease from 160°C to 55°C in the high-temperature operating mode and from 85°C to 50°C in the low-temperature mode during the regulatory lifetime of the repository [2]. Changes in humidity or the dripping of ground water with high ionic strength onto the waste package surface may induce repeatable wet-dry conditions. A heated electrode corrosion test method was developed [12, 13] to experimentally study the corrosion behavior of Alloy 22 under conditions that simulate expected service conditions in closed Yucca Mountain repository drifts. The effect of temperature fluctuation on corrosion reaction rates under periodic wet-dry conditions may be accurately assessed using the heated electrode technique at temperatures up to the boiling point of the solution [13]. In this chapter, the heated electrode test system was used for typical electrochemical cyclic polarization and extended immersion tests in an integrated electrochemical corrosion cell. In addition, the heated electrode method also worked for understanding corrosion susceptibility under the condition of water evaporation induced accumulation of hygroscopic salt deposits on a hot surface.

EXPERIMENTAL TECHNIQUE

Most electrochemical corrosion experiments are conducted by immersing samples in heated electrolyte. The method presented here heats the electrode instead of the solution. A schematic of the apparatus is shown in Figure 1. An electrochemical cell was built on a support vessel. The heater and working electrode were insulated from the support vessel, and a PTFE holder with an O-ring exposed a limited surface area of the Alloy 22 working electrode to the test solution. The potentiostat was integrated with testing and analytical software on Windows XP (Gamry PC4/750). A programmable, proportional-integral-derivative (PID) temperature controller (Omega CN9300) was connected to the RS-232C port of the potentiostat computer and the temperature was remotely controlled. A J-type thermocouple was located 2 mm under the test surface, where the temperature is considered to be the same as that on reaction surface. The measured temperature was used as feedback for the PID controller and was continuously recorded during the tests. The test materials were mill annealed and 800°C, 100 hours heat treated Alloy 22. The nominal composition (wt%) of Alloy 22 is 56Ni, max. 2.5 Co, 22Cr, 13Mo, 3W, 3Fe, max. 0.08 Si, max. 0.5 Mn, max. 0.01 C, and max. 0.35 V. The surface was wet-polished up to 0.3 μm using Al_2O_3 particles, and was thoroughly cleaned with ethanol and distilled water before use.

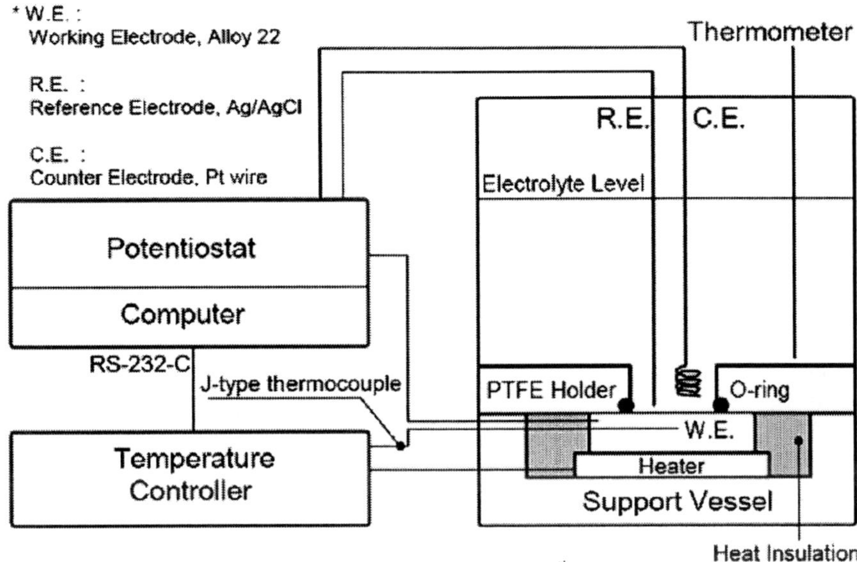

Figure 1. Schematic of heated electrode test method [13].

The electrolytes, which simulate Yucca Mountain ground waters [2], were prepared using analytical-grade chemicals: KCl, NaCl, NaF, $NaHCO_3$, $Na_2SiO_3 \cdot 9H_2O$, Na_2SO_4, Na_2NO_3, $MgCl_2 \cdot 6H_2O$, $CaCl_2 \cdot 2H_2O$, and 18.2 MΩ/cm deionized water. The molarities of anion and cation were 0.3558 Na^+, 0.0002 Si (aq.), 0.005 Ca^{2+}, 0.0174 K^+, 0.0082 Mg^{2+}, 0.1368 Cl^-, 0.0742 NO_3^-, 0.0804 SO_4^{2-} for Simulated Acidified Water (SAW) solution, 0.3558 Na^+, 0.0002 Si (aq.), 4.5×10^{-6} Ca^{2+}, 0.0174 K^+, 7.4×10^{-6} Mg^{2+}, 0.0147 F^-, 0.0378 Cl^-, 0.0206 NO_3^-, 0.2294 HCO_3^-, 0.0348 SO_4^{2-}, 0.229 HCO_3^- for SCW solution, and 2.011 Na^+, 0.0669 Si (aq.), 0.4170 K^+, 0.0170 F^-, 0.9545 Cl^-, 0.5716 NO_3^-, 0.0352 SO_4^{2-}, 0.3513 HCO_3^- for Basic Saturated Water (BSW) solution. In some tests, the acidity or alkalinity of solution was adjusted with HCl or NaOH.

The electrochemical cell consisted of an Alloy 22 sample as the working electrode, a coiled platinum wire as the counter electrode, and a miniature silver/silver chloride reference electrode (Ag/AgCl, 3.5M KCl, 205 mV vs. SHE at 25°C) which was placed in a 2 mm diameter PTFE tube with a fine porous fritted end. The distance between the porous tip and specimen surface was kept at about 2 mm; no salt bridge was used.

All of the electrochemical tests were performed under ambient air conditions, and were started after one hour of immersion in order to attain a stable temperature. Cyclic polarization was conducted potentiodynamically with a potential scan rate of 10 mV/min, as recommended in ASTM-G 61-86 (Reapproved 1998). The potential scan started at -100 mV below the corrosion potential, and the reverse scan was started when the current density reached 1.6 mA/cm^2. The test was terminated when the anodic current fell below 0.2 μA/cm^2. In the prolonged immersion test, corrosion potential was continuously monitored. After the system was dried with only salt deposit remaining on the hot surface, deionized water was added to observe the corrosion potential change. In the dripping test, one drop was 0.1 mL in volume, and the same volume of water was repeatedly dropped on the hot surface after 3, 5, 10, and 20 minutes. After each test, the surface of the test specimen was examined under an optical microscope. The dripping test sample was also examined by Scanning Electron Microscopy (SEM).

This work was performed under control of the Nevada System of Higher Education (NSHE) Quality Assurance (QA) Program.

RESULTS AND DISCUSSION

1. Cyclic Polarization Curves in Simulated YM Ground Waters

Figure 2 shows the polarization curves obtained by heated electrode method in SAW, BSW, and SCW solution. The value of pH was 2.8 for SAW, 10.3 for SCW, and 12.2 for BSW, respectively. The passive current density measured in SCW solution was approximately 10 times higher than that in acidic solution SAW, and nearly 5 times higher than that in strong alkaline solution BSW. In addition, a current peak due to some oxidative process was observed at 0.2 V in SCW. This result suggests that Alloy 22 could be more sensitive to corrosion in SCW. On the other hand, repassivation potential E_R is considered to be a critical potential for localized corrosion. Pitting and crevice corrosion cannot occur at a potential lower than E_R. Dunn et al. [6, 7] defined E_R of Alloy 22 as the potential where current density remained below 2 $\mu A/cm^2$.

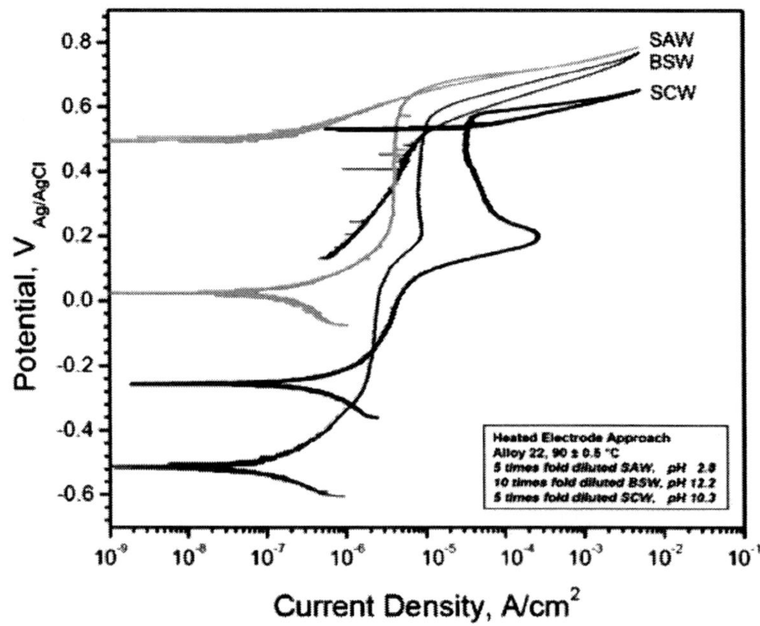

Figure 2. Cyclic polarization curves in simulated YM ground waters SAW, BSW, and SCW.

It seems that repassivation potentials in alkaline solutions BSW and SCW are lower than that in SAW solution. The alkalinity of ground water on Alloy 22 is variable during long term exposure. The corrosion susceptibility in SCW was expected to be interesting in extended immersion tests and dripping tests on heated Alloy 22 surfaces.

2. Corrosion Potential Monitoring during the Long-term Immersion Test

Figure 3 shows the evolution of corrosion potential for Alloy 22 heated to 90°C during exposure to SAW and SCW. At the beginning, immersion tests were conducted on a heated Alloy 22 surface with the test cell open to ambient air. Evaporation concentration effects resulted in increased ionic strength and changes in corrosion potential and pH. In SAW solution, corrosion potential constantly increased during the immersion.

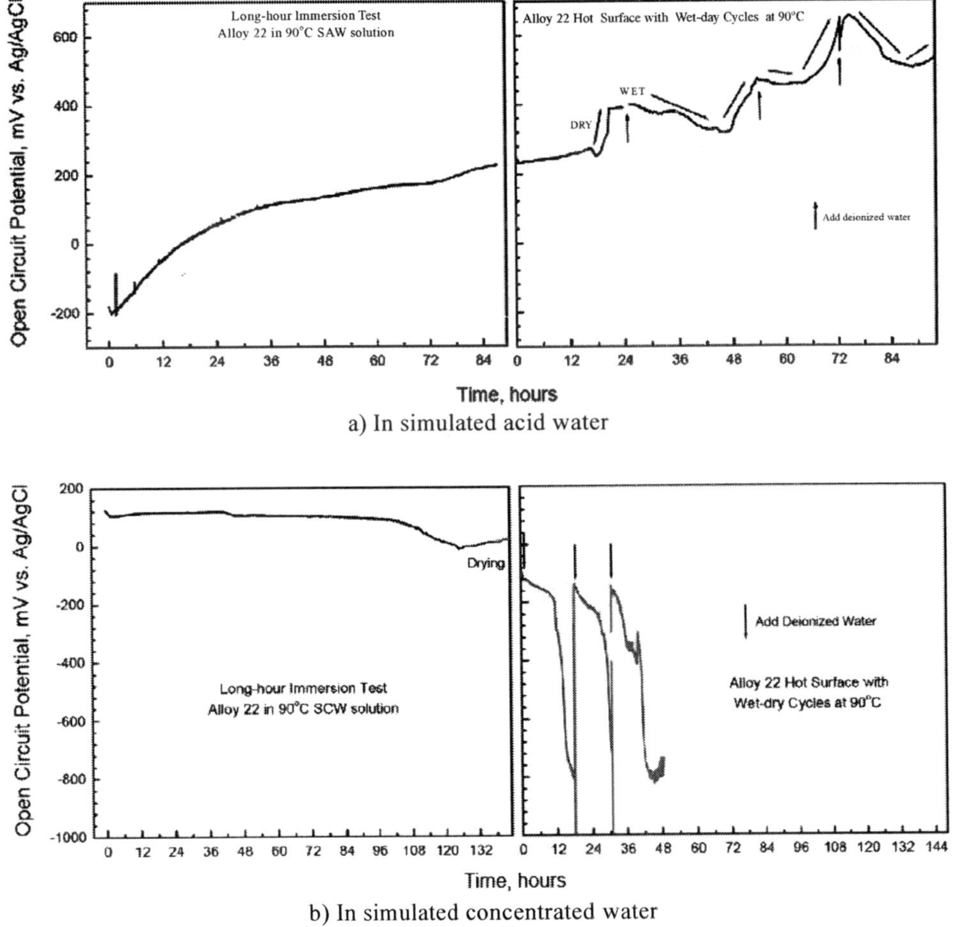

Figure 3. Corrosion potential of Alloy 22 hot surface during long-term immersion and subsequent wet-dry cycles.

However, the decreasing potential was observed in SCW solution, which implies a sensitive corrosion system formed during the exposure. Subsequently, water was cyclically dripped onto the hot surface to produce a cyclic wet-dry condition after creating a salt deposit by complete evaporation of the original simulated water volume. Corrosion potential was still measurable during the entire wet-dry cycle. In SAW solution, potential displayed an increasing trend with superimposed periods of more rapidly increasing potential as the salt deposit approached complete dryness, and decreasing potential as the salt deposit was re-wetted. In the case of SCW, a range of more than 600 mV was observed in a decreasing trend during the drying process. This may be explained by the work of Nishikata et al. [14-16]. They conducted AC impedance corrosion monitoring and found that the generation and growth of localized corrosion occur preferentially just before the metal surface completely dries, and repassivation occurs preferentially during complete drying and subsequent immersion.

3. Crevice Corrosion under the Accumulated Salt Deposit

In a dripping test using heat treated Alloy 22 (800°C for 100 hours), SCW was dripped onto a 90°C Alloy 22 surface with a time-delay dripping pattern for 40 days. Dripping intervals of 3, 5, 10, and 20 minutes were applied in order for simulating the possible dripping pattern in the YM repository drift. Figure 4 shows the experimental set up and the morphology of the salt deposit at the beginning of the dripping. Figure 5 shows the resulting salt deposit on the Alloy 22 hot surface at the end of the dripping test. The tip of the dripping tube was located about 1 inch above the hot surface. A cavity was formed under the tube at the beginning of the test. The accumulated salts around the cavity were dispersed later because of the dripping force and evaporation. Finally, a dome-like block of salt deposits formed on the metal surface.

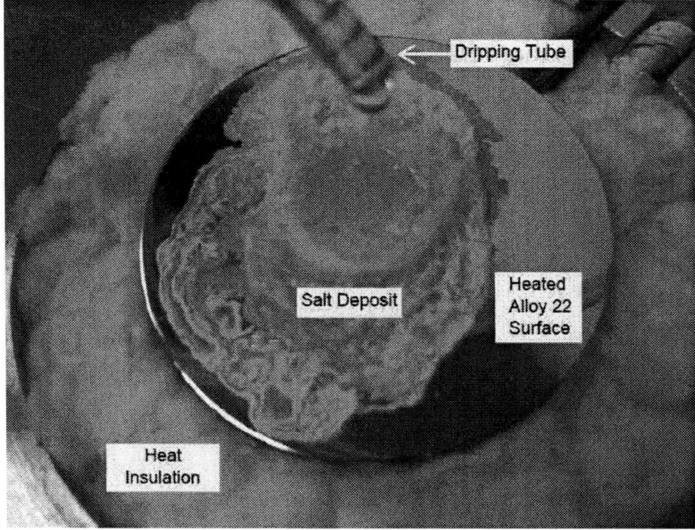

Figure 4. Dripping test setup and salt deposits on the Alloy 22 hot surface at the beginning.

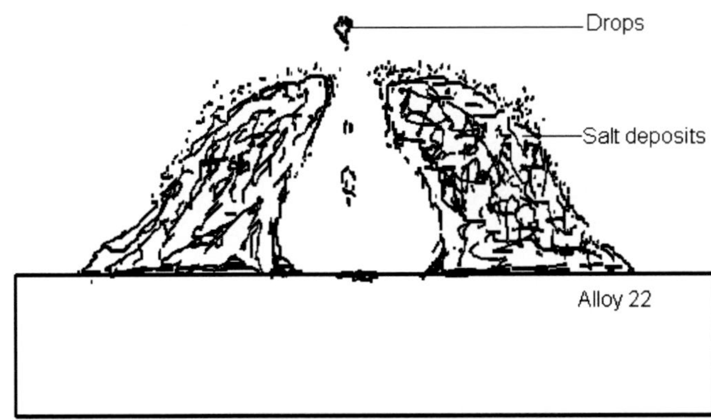

(a) Dome-like salt deposit block.

(b) Different view of the accumulated salt deposit block

Figure 5. Salt deposits on the hot surface of Alloy 22 at the end of the dripping test.

Figure 6 shows the profile of the sample surface after removing the salt deposit. The surface appearance is mixed with dark colors (brown or brown on blue) at the dripping site and under the accumulated salt deposits. There is a bright area between the dripping site and salt deposits, where the surface was open to air during the dripping process. As shown in Figure 7, the dark areas are etched revealing the microstructure including grain boundary precipitates, which are the result of the heat treatment of the specimen (Figure 7 (b), (c)). In addition, a spheroidal appearance and changes in area fraction were observed. Only a small amount of color change was found on the bright area (Figure 7 (a)).

This evidence suggests that corrosive solution was concentrated, heated and evaporated in these areas. Accordingly, a hot etching effect appeared on the sample surface. It is believed that tight crevices formed between the salt deposits and the hot surface, which are suitable sites for corrosion initiation. The bright area could be the evidence of the path for solution and air passing through the space between the salt deposit and specimen surface. Typical localized differential aeration cells could be formed in many locations. As shown in Figure 8, intergranular attack, micro pits, and micro-scale localized corrosion were observed by SEM. Differential temperature cells or differential concentration cells might be formed due to the localized differentials in temperature, ion concentrations, and potential. These results suggest that the accumulation of salt deposit on the Alloy 22 hot surface increased the susceptibility to localized corrosion.

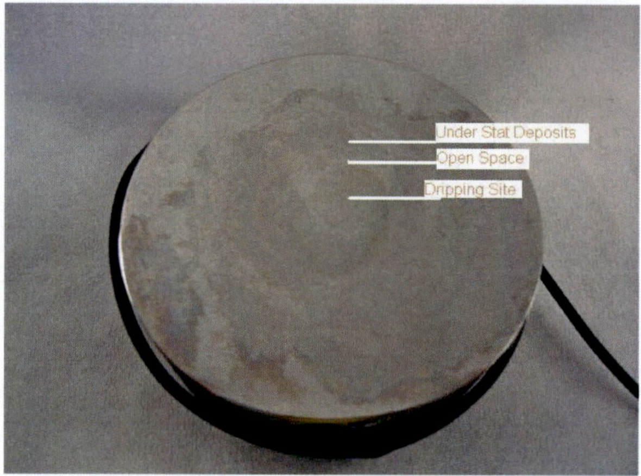

Figure 6. Color changing, bright and dark areas on the surface of Alloy 22.

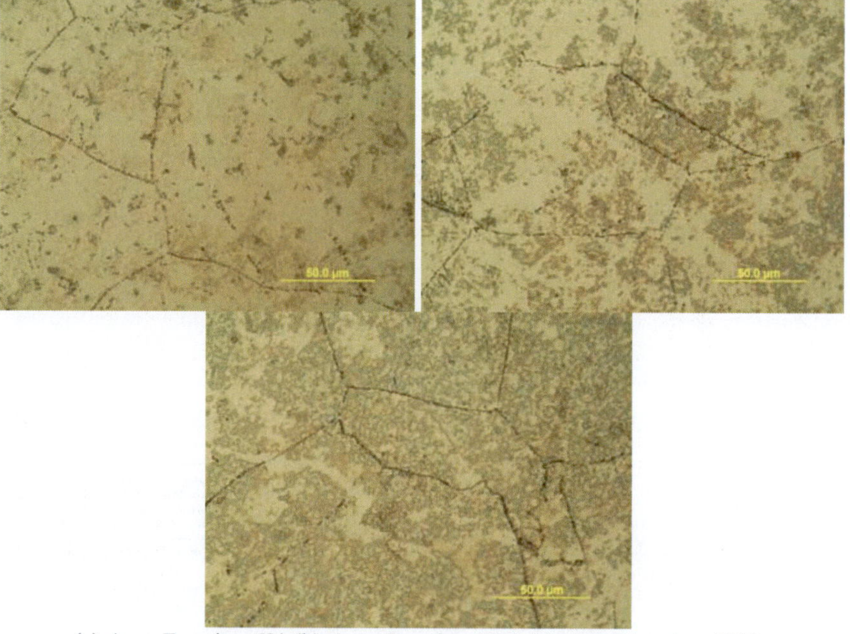

(a) Area Fraction 6% (b) Area Fraction 38% (c) Area Fraction 57%.

Figure 7. Bead-like appearance and the area fraction on the surface of Alloy 22.

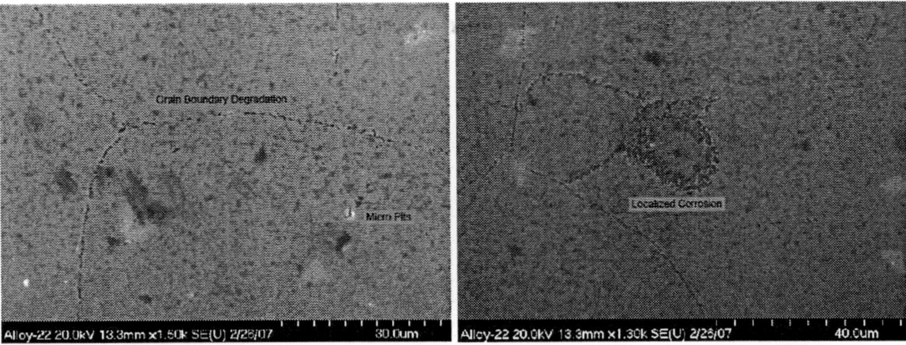

(a) Grain Boundary Degradation and Micro Pits (b) Micro-Scale Localized Corrosion

Figure 8. Micro-scale localized corrosion on heat treated Alloy 22.

4. Temperature Fluctuation Pattern Related to Water Dripping and Oscillating Polarization Curve

As shown in Figure 9, temperature variation with the wet-dry cycles has been observed during the dripping test. The dripping interval was 3 minutes for Figure 9 (a) and was 10 minutes for Figure 9 (b). The dripping volume was about 0.1 mL each time and surface temperature was 90°C. In this case, a complete drying needed about 10 minutes after each drop and the temperature increased on average 3°C during the process. In order to understand if the temperature fluctuation has any effect on corrosion properties of Alloy 22, a temperature-oscillating method was utilized for the investigation of polarization behavior. An example is shown in Figure 10. The temperature was varied between 78°C and 94°C. The upper and lower temperature points are marked on the figure. The test was able to distinguish effects on current due to temperature variations within temperature ranges of approximately 10°C or greater.

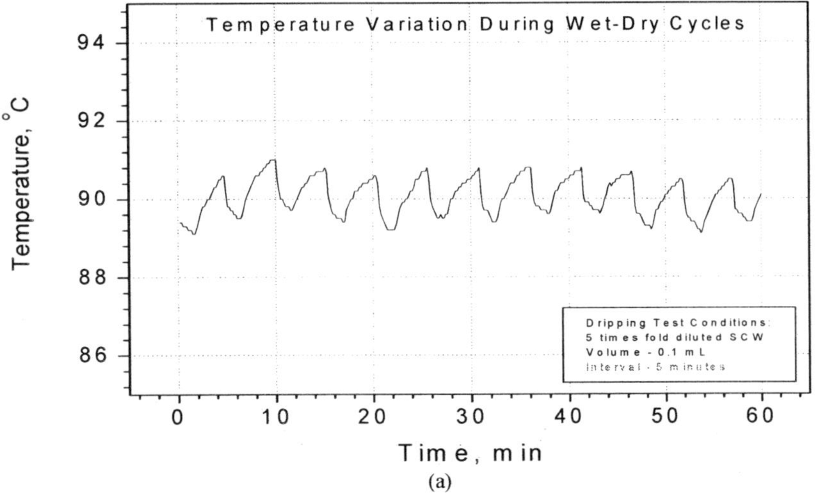

Figure 9. (Continued).

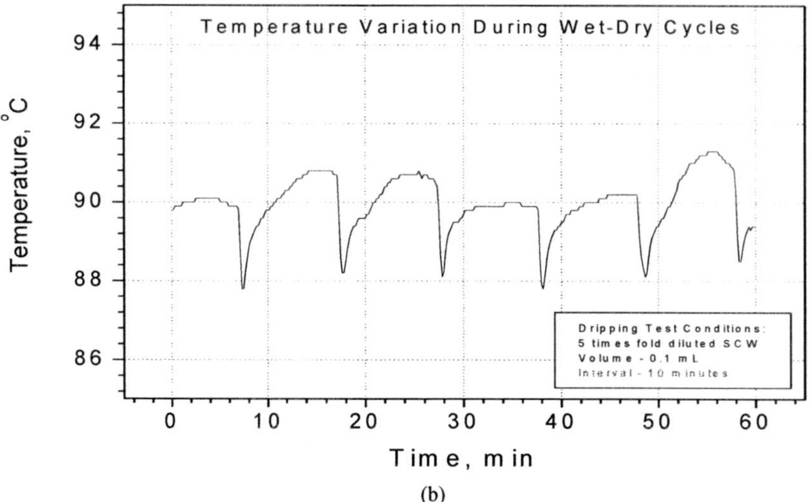

Figure 9. Temperature variations during SCW dripping test with different dripping intervals.

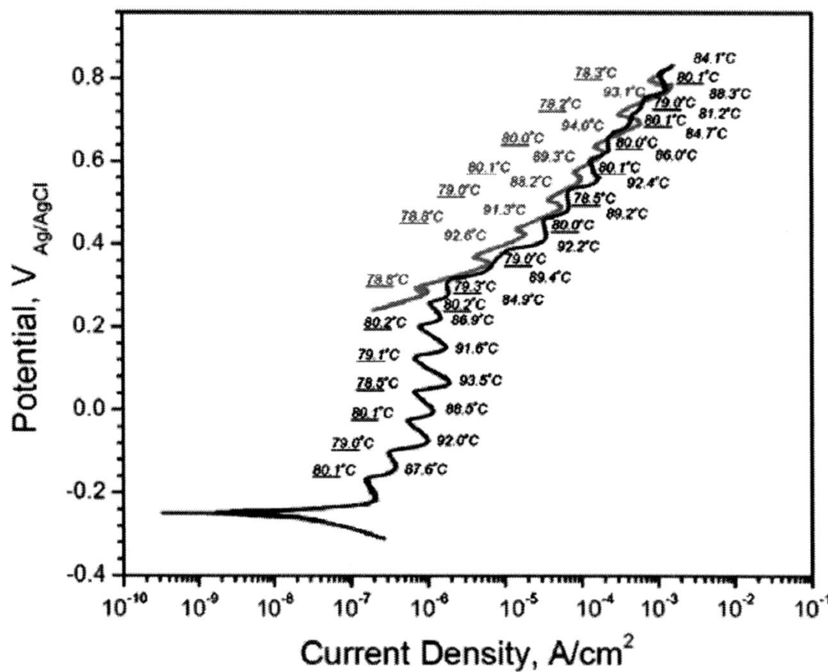

Figure 10. Cyclic polarization curve with the effect of temperature oscillation [13].

The results show that passive current density at approximately 90°C was 1–2 $\mu A/cm^2$, which is about twice that which was observed at approximately 80°C. It is clear that temperature oscillation affected polarization behavior. In the dripping test, temperature variation around the dripping site could be one of the localized environment conditions accelerating corrosion initiation. At the same time, humidity at the location close to dripping site varies cyclically, which should also be taken into account when considering the accelerated corrosion.

5. The Effect of pH on the Corrosion of Alloy 22

Cyclic polarization curves of heat treated Alloy 22 (100 hours at 800°C) in SCW solution (pH 10.3) and in two other pH-adjusted SCW solutions (pH 8.5 and pH 9.5) are shown in Figure 11. Passive current density increased with decreasing pH in the range of pH 10.3 and pH 8.5. This suggests that the corrosion rate of the heat treated Alloy 22 was higher in the less alkaline solution. Also, the potential of the dissolution peak on the polarization curve shifted to the anodic direction with decreasing pH, and repassivation potential was lower at the high pH value. As described in immersion test and dripping test discussion, the solution pH could be changed due to water evaporation and/or the evolution of crevice solution between salt deposit and hot surface. The changing of pH from strong alkalinity to weak alkalinity in SCW is probably a reason of increasing corrosion susceptibility.

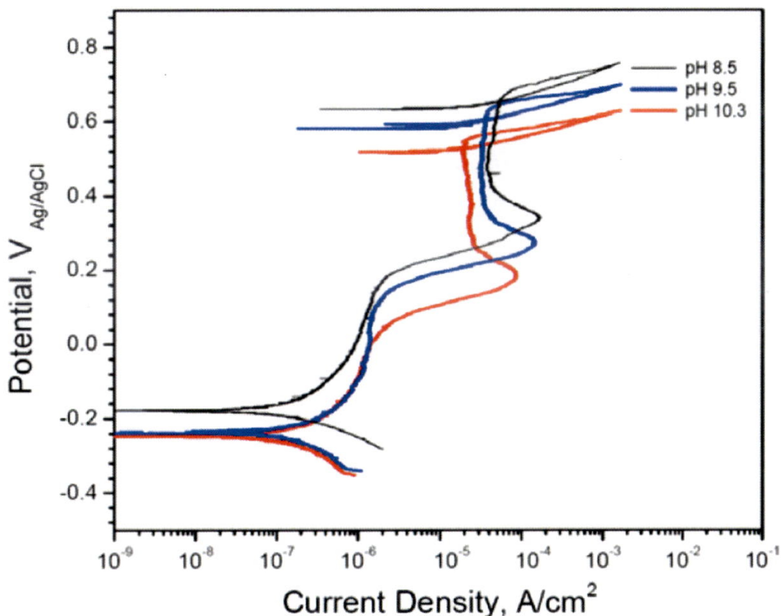

Figure 11. Polarization curves of heat treated Alloy 22 in SCW solution with different pH.

CONCLUSION

In this chapter, the corrosion behavior of Alloy 22 in simulated ground waters was experimentally investigated by a newly developed heated electrode technique. Cyclic polarization tests, long term immersion and dripping tests were conducted. Through comparing polarization curves in SCW with those in SAW and BSW, SCW is considered as a relatively corrosive solution. In SCW solution, corrosion potential decreased under immersion condition and a 600 mV decrease in potential was observed during the drying process. After dripping SCW for 40 days, intergranular attack, micro pits, and micro-scale localized corrosion are observed on the heat treated Alloy 22 sample. These results acquired by applying the heated electrode technique imply that periodic changes in localized environment, such as the evolution of temperature and pH, accelerated corrosion initiation. Although the long term corrosion resistance needs to be examined for more detail, the possibility of corrosion initiation on heat treated Alloy 22 especially under accumulated multi-component salt deposits cannot be completely dismissed.

ACKNOWLEDGMENTS

The authors gratefully acknowledge the support of this work by the US Department of Energy (DOE) Office of Civilian Radioactive Waste Management (OCRWM) through DE-FC28-04RW12232, Task ORD-UNR-014 at the University of Nevada, Reno.

REFERENCES

[1] Shoesmith, D. W. *Corrosion* 2006, 62, 703-722.

[2] Gordon, G. M. *Corrosion* 2002, 58, 811-825.

[3] Rebak, R.; Efien, R. A.; Gordon, S. R.; llevbare, G. O. *Corrosion* 2006, 62, 967-980.

[4] Windisch Jr, C. F.; Baer, D. R.; Jones, R. H.; Engelhard, M. H. *Corro. Sci.* 2007, 49, 2497-2511.

[5] He, X.; Dunn, D. S. *Corrosion* 2007, 63, 145-158.

[6] Dunn, D. S.; Pan, Y.-M.; Yang, L.; Cragnolino, G. A. *Corrosion* 2006, 62, 3-12.

[7] Dunn, D. S.; Pan, Y.-M.; Yang, L.; Cragnolino, G. A. *Corrosion* 2005, 61, 1078-1085.

[8] Dunn, D. S.; Pan, Y.-M.; Chiang, K. T.; Yang, L.; Cragnolino, G. A.; He, X. *JOM* 2005, 57, 49-55.

[9] Evans, K. J.; Yilmaz, A.; Day, S. D.; Wong, L. L.; Estill, J. C.; Rebak, R. B. *JOM* 2005, 57, 56-61.

[10] Evans, K. J.; Stuart, M. L.; Etien, R. A.; Hust, G. A.; Estill, J. C.; Rebak, R.B. In *CORROSION*/06; NACE: Houston, TX, 2006; Paper06623.

[11] Miyagusuku, M.; Devine, T. M. In CORROSION/07; NACE: Houston, TX, 2007; Paper07586.

[12] Badwe, S.; Raja, K. S.; Misra, M. Electrochimica Acta 2006, 51, 5836-5844.

[13] Ashida, Y; McMillion, L. G.; Taylor, M. L. Electrochem. Commun. 2007, 9, 1102-1106.

[14] Cruz, R. P. V.; Nishikata, A.; Tsuru, T. Corro. Sci. 1998, 40, 125-139.

[15] Cruz, R. P. V.; Nishikata, A.; and Tsuru, T. Corro. Sci. 1996, 38, 1397-1406.

[16] Nishikata, A.; Yamashita, Y.; Katayama, H.; Tsuru, T.; Usami, A.; Tanabe, K.; Mabuchi, H. *Corro. Sci.* 1995, 37, 2059-2069.

In: Nuclear Waste Research: Siting, Technology and Treatment   ISBN 978-1-60456-184-5
Editor: Arnold P. Lattefer, pp. 37-45   © 2008 Nova Science Publishers, Inc.

Short Communication D

LEAVING MESSAGES ABOUT OUR RADIOACTIVE WASTE FOR FUTURE GENERATIONS

Alan Marshall

Sustaining Gondwana Initiative,
Curtin University of Technology
GPO Box U1987, Perth,
Western Australia, 6845
Australia

ABSTRACT

Radioactive waste is highly dangerous and long-lived. Any attempt by the current generation to safely dispose or store this waste is racked by unsolvable technical difficulties such that future generations deserve to be warned about the characteristics and whereabouts of the waste. Since the information about the waste may be more susceptible to degradation than the actual waste a number of issues emerge in this study that throw doubt upon the morality of producing the waste in the first place.

Keywords: *radioactive, waste, warning, intergenerational, communication, ethics, future.*

Of all the myriad types of human-made waste that has been introduced into the world since the dawn of the industrial revolution, perhaps the longest lived is radioactive waste. This waste is so long lived that it will outlast this generation by between 100 and a million years. This means that the electricity produced by modern nuclear power stations will deliver upon the Earth waste that will linger on for some 200,000 more generations. The long-lived nature of nuclear waste is equalled by it pernicious character. Plutonium, for example, is one of the most toxic substances humans have ever produced [1]. The not-so-dangerous nuclear waste is still pretty dangerous. Strontium 90, for example, is dangerous for only 600 years but if it manages to leak out of a rusty waste container and then leach into the water table it could eventually find its way into drinking water and accumulate in the bones and teeth of children, vastly increasing their chance of contracting leukaemia and other cancers.

The favourite candidate solution for handling long-lived high-level waste is to put it into canisters and then emplace these canisters in repositories built deep within the earth (some 100 to 1000 metres). Whilst some [2] believe this to be a permanent solution to the problem others [3] believe that all sorts of things could go wrong:

- The containers may corrode and leak (this is pretty much a certainty given the long lifetime of the wastes).
- An earthquake could trigger a collapse of the repository, or cause an underground flood, which then brings radioactive materials to the surface.
- The heat of high level waste could start an underground fire during the decades-long emplacement phase, triggering airborne release of radioactivity.
- Plutonium pellets may migrate together within the repository to produce a critical mass; resulting in an underground nuclear reaction.
- Humans may intrude upon the repositories by accident or on purpose and possibly seek to use the waste for sinister/military/terrorist purposes.

If future generations are to be able to care for or avoid the radioactive waste that this generation produces, then some way of communicating the dangers of radioactive waste to these future generations has to be realised. However, any attempt to do this must be cognisant of the changing regimes of information storage. Mainstream manners of conveying information are obviously subject to change over long time periods. Many of the oral traditions and symbolic representations that were standard thousands of years ago are largely lost to or lost on the current generation. Similarly, the documents we produce now relating to the siting of dangerous waste are less likely to survive than the waste itself. The digital revolution may exacerbate this problem according to Ulrike Fink [4]:

> "data losses may take place even faster due to the rapid progress and subsequent incompatibility of computer systems: date files stored 20 years ago on magnetic tapes can't be read by modern computers anymore"

She goes on to give an example of information loss that she believes is somewhat analogous to what may happen with regards to radioactive waste:

> "Everyone knows that Germans are especially tidy and painstaking – but nevertheless, now and then it happens that old pits of abandoned coal mines are just drilled by chance! That means, either the knowledge has got lost during 100 years or the people didn't study the available, existing data – the people were not conscious of the problem."

This case, suggests Ulrike Fink, implies that despite our best record-keeping efforts, it is likely that the information we produce about our radioactive waste activities is probably going to be inaccessible to future generations.

To help present-day managers of radioactive waste to formulate a solution to such problems, Jensen and his colleagues [5] investigated the various problems of long-term upkeep and transferral of information under the guise of 6 key issues:

(1) the kind of responsibility that current society should accept toward warning future societies,

(2) the time period we judge to be necessary for protection of future societies,

(3) the resource expenditure that should be made to assure successful communication,

(4) the selection of the information that should be conveyed,

(5) the means of presenting the information, and

(6) determining the likelihood that the information will be understood.

With regards to number 1, it is accepted by numerous commentators that this generation of waste producers has some obligation to attempt to warn future generations of what we have produced and where it is located.

With regards to number two, Jensen himself feels that it is impractical to look too far into the future and he suggests that most predictive mechanisms decrease in accuracy with time.

Jensen also offers what he believes are important philosophical reasons why we should concentrate mainly on communicating to nearer generations rather than those in the far future:

"We are ignorant of both capabilities and needs of a society 1000 years (and more) into the future. For that reason alone, we are forced to design the information system to serve the interests of the societies in the nearer future. Generations are dependent upon each other. It is therefore in the interest of societies in the distant future that earlier societies receive the benefit of adequate information. A solution designed to serve the interest of societies in the first millennium will therefore in some sense also be beneficial for later societies. Societies in the future are in a better position to assess changing needs of their successors and adjust the information system according to those need [6].

In light of Jensen's comments we have to work out if we should:

a) set up a system for a rolling present, whereby successive generations relay information to the next, or

b) set up a final system of communication that is supposed to inform people once and forever more of the dangers of the waste.

To choose b) while excluding a) may be to rely too much upon the possibility of creating a near-universal language presented in a medium which is durable for the lifetime of the waste. The managers of the WIPP (Waste Isolation Pilot Plant) facility in New Mexico, USA, charged a panel of outside experts with the task of designing a marking system which was both durable and understandable over a 10,000 year period. They came up with a variety of designs of granite blocks, spike-sculptured fields, menacing earthworks as well as pictograms using trefoils and various images of dread such as 'skull and cross bones' and reproductions of Edvard Munch's painting 'The Scream' [7].

This form of information upkeep doesn't rely on the continual management of information throughout the ages and it avoids the problem of information extinction based upon the decaying operation of necessary institutional infrastructure. However, this approach doesn't convey comprehensive information and it does little to give our immediately succeeding generations all the information they might need in order to address the managerial and technical issues in their own chosen way.

Those advocating pictograms for WIPP and other nuclear waste sites declare that such symbols and the messages they accompany will be understood by future archaeologists simply because humans over time get smarter and smarter [8]. Given the varied history of humanity up until now, this is a debatable point.

Now if we choose 'a', from above, while excluding 'b' it would probably be because we are convinced that there is absolutely no way in which we can communicate unambiguous information to a particular generation at a desired point in the future. Choosing 'a' would mean to concentrate on preserving a system of transferral that guarantees that a regime is in place to transfer information to one or two generations ahead. These generations can then erect their own system of information transfer to convey information to the succeeding generation and so on and so on.

The advantage of this approach is that it could provide an impetus to construct a workable system of information transfer to the generation that is to follow ours. It also allows for the transfer of very detailed information; repository designs, physical characteristics of the waste, expected problems etc. However, it does put great faith in the transferability of intergenerational systems of such information, something which Fink [9] doubts can be achieved for very long based upon analogous interruptions of past information transfer.

What all this might suggest is that there may be the need for a multi-scale approach to information upkeep, an approach that would utilize both a) and b) from above. Givens [10] has formulated such a multi-scale approach and Jensen [11] relates these as follows:

"A Level I message is easy to comprehend but rudimentary. For example, a large earthwork of obviously unnatural origins may signal that something manmade is here. The Level I message may not inform the receiver of the message of the purpose of the marker nor even of any danger. It is an 'attention getter' designed to initiate the search for additional information."

"Level II messages, in Givens hierarchy, convey a warning through iconic, symbolic and/or linguistic signs. Examples of Level II messages include a sequence of drawings showing a human drinking from contaminated water and dying, an international emblem such as the radiation trefoil, or a written statement such as 'Danger radioactive waste. Do not drill here!'"

"Level III messages contain basic information explaining what (that radioactive waste is present), why (the waste is dangerous and should not be disturbed), when (when were the materials implanted and when will they be safe), where (a map and description of the repository), who (who fashioned the site), and how (instructions for maintaining and updating information)".

"Detailed information including inventories, technical specification, repair procedures, etc. constitute the Level IV information."

Jensen encourages us to believe that the advantage of the lower levels is their more robust longevity and universality while the advantages of the higher levels is their transfer of greater amounts of usable information.

With regards to key issue number 4 from Jensen's list (introduced earlier in this subsection), if the present generation is concerned to safeguard future peoples from its waste, it seems reasonable to utilize all of Givens' levels of communication when trying to convey warnings and information. However, the cost of doing so has to be assumed by someone. The higher the level of communication, the higher is the cost to future generations, since they will be the ones charged with relaying the information on. Level I and II messages can be constructed at any point in time and, theoretically, will not need any maintenance in the future, thus if we construct them now we might relieve future peoples of any burdens associated with their upkeep.

If we are to be fully cognizant of intergenerational equity, though, perhaps we should forthrightly claim that those generations that produce the waste should be the ones to bear most of the cost of transferring the information, including at levels III and IV. This might mean the establishment of some sort of trust fund so that we can compensate future generations for the cost of maintaining records on the wastes that we produce.

With regards to Jensen's key issues 5 and 6 listed earlier in this sub-section, there are a number of issues that will have to be kept in mind when working out how and who to communicate with in the future and what shall be conveyed in those communications. Whether we choose a multi-level approach or a single level approach to our information upkeep we shall probably come across a myriad of problems. These are summarized below.

THE INFIDELITY OF COMMUNICATION

In regard to Given's multi-scale approach, the infidelity of messages is most likely to be a problem when little context is available for interpreting the information and when the information is passing in a unidirectional way. Beyond the first generation, these problems become more and more apparent as the contextualising elements are lost or forgotten.

At levels I and II, where messages are transferred primarily for far generations, there seems to be a near instant loss of context as messages are boiled down to 'essential' information. Whereas for level III and IV there is the possibility that a rolling scheme of changing information regimes can maximize the retention, re-evaluation and evolution of contextualizing information.

The value-ladeness of communication. Garfield [12], reporting about the results of a meeting organized in Scandinavia [13] summarized one key thought that emerged:

"the focus should be on the danger of the waste, and not of the use".

This seems like sound advice but it may be impossible to achieve. Each message communicated by an individual or group to another contains a plethora of context-dependent values which the communicants may or may not be aware. So, if we undertake level III and IV approach to information transfer of radioactive waste information, we must ask ourselves how much of that information is laden with values that act to do the following:

- imply that the waste was derived from a uniformly-accepted and uncontentious technology
- imply that the waste has been and can be used for weapons.
- imply that it is reasonable for one section of humanity to burden another section of humanity with its problems
- imply that the waste area (and those who live near it) should be fetishized or stigmatized in some way.

We might believe that the messages contained within levels I and II are largely immune to values, or at least that they can be confidently used to communicate the dangerous nature of the waste but when scientists were charged with communicating information about humans on a plaque attached to a spaceprobe destined for intersteller space (in the hopes that a passing alien would understand it) there was criticism from some feminists in the way that woman were represented compared to men.

This example may show that any group who intends to communicate over long generations is also involved in the process of reflecting and representing their own time period. If we are to relinquish the social prejudices of our time, and if we seek to undermine rather than reinforce the prejudices, then any Level I or II message must pass through a rigorous process of social discussion. In light of this, the public participation processes planned for the management of radioactive waste might have to be repeated regarding the information to be transferred to the future about that waste.

THE DISTRIBUTION OF INFORMATION

If we pursue levels III and IV as processes of information upkeep, then we will soon have to decide where this information will be distributed. An open book policy may serve to encourage a decentralized archiving of the records in multiple locations so that even if official records are lost, some form of information regarding the radioactive waste may survive. Such an open book policy may also encourage a democratization of the future management of the waste since multiple sources will allow wider knowledge of the waste, more interpretations of the consequences of the waste, and more discussions of how to manage it.

There is the risk, of course, that this open book policy may increase the chance of future people to realize the military or terror potential of the radioactive waste, thus increasing the danger that stored or disposed waste will be interfered with and/or dispersed. To get around this threat, there is always the possibility to keep information about the waste both secret and centralized. However, this may mean that in the future, official authorities will be offered an opportunity to excavate and utilize the waste for their own military or terrorist purposes under welcome secrecy.

THE CULTURAL EMBEDEDNESS OF INFORMATION

One thing which a multi-scale approach allows for is the development of a network between the levels. This process will promote multiple pathways for the information to stay intact (to a greater or lesser extent) since levels will allow synergistic or redundant channels of communication to stay functional. For instance, if a large monument is placed next to a waste repository it may encourage future archaeologists to investigate relevant archives. It may also prompt popular storytelling about the waste so that even if the archives are lost, some form cultural communication is maintained. If a symbol, say the trefoil, is within the physical body of all the relevant documents and messages, then this may further encourage synergism between the various levels of information transfer by showing a connection between them.

According to some, like Garfield [14] documents and information systems are unlikely to last as long as various informal or cultural forms of information. If this is a correct assumption, then the dangers of the waste are more likely to be retained within the stories and myths of local peoples than in the archives of the institutions. Under this assumption, Garfield encourages us to keep alive an awareness of the dangers of radioactive waste by embedding them within our religious ideas and practices, including in our religious icons and symbols.

While Garfield's idea encourages us to think about radioactive waste in cultural terms, something which needs to be done if we are to be aware of all its ramifications, she gives the impression that followers of formal religions are open to instant acceptance of contemporary moral causes introduced like bills to a parliament. If this were so, religion might be too amenable to changing ideas to be as long lasting as Garfield assumes. However, perhaps the type of religion most likely to adopt and disseminate long-term knowledge about radioactive waste would be one with values somewhat akin to the New Age movement since New Agers are enveloped within a spiritual awareness of the sacred nature of the (Mother) Earth and all that beings that reside within her [15]. (Although the James Lovelock the inventor of the Gaia theory—which inspires many New Agers to value the Earth—is himself a proponent of nuclear energy [16]).

What Garfield also misses is the idea that documents and archives are *part* of a generation's culture. Thus if there is some reason for consecutive generations to maintain archives (if it contributes to the power of the government, for example, or if it fits in with a cherished social belief, say 'story-telling', 'fact-keeping', or 'ancestry worship') then it is likely that archives of one sort or another will be used to catalogue various social and technical problems, including the ongoing presence of radioactive waste. Equally likely, though, is that these archives will change in form and content and may be lost altogether before being built once again when the waste is accidentally discovered or brought back into the biosphere by natural processes.

SUMMARY AND CONCLUSION

If 21st century engineers could guarantee the physical containment of long-lived high-level waste then we would not need to worry about how future generations may suffer from such waste. However, it is quite possible that radioactive waste buried this very day could end up in the environment of some future community via various natural and/or non-natural processes. To enable future generations to avoid or manage the waste, it seems responsible for those that produced it to warn them of the dangers of the waste in some way. However, two problems arise that will make the warnings dangerous in themselves.

Firstly, the warnings are likely to be lost, abandoned or misconstrued over the hundred thousand year lifetime of the waste. Thus, time and tide will render our efforts of communication ineffectual at some point.

Secondly, our warnings may result in the waste being sought after and utilised for sinister purposes by certain members of future generations. Thus, the information we provide to future generations could be used by some people against others in ways not foreseen or condoned by us.

If we chose not to warn future generations of the nature and whereabouts of our radioactive waste then we present them with a strong chance of suffering from radioactive pollution but if we chose to warn them then we present them with the risk of (re-emergent) nuclear and radioactive arms.

Each generation has baggage that it bestows upon the next but the long-lived nature of our radioactive waste will leave a legacy of nuclear danger and pollution to future generations (which they neither expect nor deserve). For these reasons, we should only carry on producing radioactive waste if we do not need to leave messages for future generations (since such messages that will probably be lost or misused). Given that we have no way of conferring complete technical safety upon our stockpiles of radioactive waste to the point where messages needn't be sent, it seems a responsible course of action (as far as future generations are concerned) to stop producing the waste in the first place.

REFERENCES

[1] Arjun Makhijani (1997) *Health Effects of Plutonium*, IEER Energy and Security Report No 3.

[2] N.A Chapman (2006) Geological Disposal of Radioactive Waste—Concept, Status and trends, *Journal of Iberian Geology*, Vol 32 (1) 2006, pp7-14.

[3] K.S. Shrader-Frechette, (1993) *Burying Uncertainty: Risk and the Case Against Geological Disposal of Waste,* University of California Press, Berkeley.

[4] U. Fink (1992) "The Nuclear Guardianship: Concept for a Radioactive Future", *Day of the Deserts*, p136

[5] M. Jensen (1993), *Conservation and Retrieval of Information: Elements of a Strategy to Inform Future Societie*s, Elements of a Strategy to Inform Future Societies, Final Report of the NKS project KAN-1.3.

[6] ibid.,p16.

[7] F.d. Drake, *et al* (1992) The Development of Markers to Deter Inadvertent Human Intrusion into the Waste Isolation Pilot Plant. Presented to Sandia National Laboratories in Albuquerque, January 1992. (Team B Report).

[8] R. Colville (2006) *Quest for a Nuclear Waste Sign that Goes Beyond Words*, Telegraph online, 30th May 2006.

[9] U. Fink (1992) "The Nuclear Guardianship: Concept for a Radioactive Future", *Day of the Deserts*, p136.

[10] Givens, D.B. (1982) "From Here to Eternity: Communicating with the Distant Future", *Et Cetera*, Summer Volume., 159-179.

[11] M. Jensen (1993), *Conservation and Retrieval of Information: Elements of a Strategy to Inform Future Societie*s, Elements of a Strategy to Inform Future SocietiesFinal Report of the NKS project KAN-1.3,p12.

[12] A. Garfield, *Religion's Carry Nuclear Waste Danger Warnings into the Future*, NGF Website.

[13] SSI and Scandpower (1991) "Transmittal of Information Over Extremely Long Periods of Time", Oslo, Norway.

[14] A. Garfield, *Religion's Carry Nuclear Waste Danger Warnings into the Future*, NGF Website.

[15] E. MacGaa (1995) *Mother Earth Spirituality*, Harper: San Francisco.

[16] A.Marshall (2002) *The Unity of Nature*, Imperial College Press: London.

In: Nuclear Waste Research: Siting, Technology and Treatment   ISBN 978-1-60456-184-5
Editor: Arnold P. Lattefer, pp. 47-74

Chapter 1

ATTITUDES TOWARDS NUCLEAR WASTE AND SITING POLICY: EXPERTS AND THE PUBLIC

Lennart Sjöberg[1,2] and Britt-Marie Drottz-Sjöberg[2]

[1]Center for Risk Research, Stockholm School of Economics,
Stockholm, Sweden
[2]Center for Risk Psychology, Environment, and Safety,
Department of Psychology,
Norwegian University of Science and Technology,
Trondheim, Norway

ABSTRACT

Previous studies have shown that experts and the public assess risks of nuclear waste at very different levels. In the present study, a broad range of risk perception and attitude dimensions was investigated by means of a mail survey directed to experts in the field of nuclear waste. Comparative data was available from a random sample of the Swedish population and from a sample of graduate engineers working in other fields than the nuclear one. The analysis of risk perceptions shows that experts rated nuclear risks as much smaller than members of the public did. They also rated risk aspects of nuclear waste as much less worrisome, with the exception of Voluntariness, Dread and Novel risk. Engineers rated risks similar to experts, but they were closer to the public. Both experts and engineers showed a rhetorical contrast in their ratings, since they expressed more worry about lifestyle risks than the public did. There was a small, but consistent, tendency for female experts to judge risks as larger than their male colleagues, but still much smaller than women in the public did. The structure of perceived risk was similar for the three groups, contrary to the notion that experts' risk perception is qualitatively different from that of the public. It was found that experts and the public held very different policy attitudes. The policy differences among groups could be accounted for by demographics to some extent, about 1/3 of the group differences. The remaining intergroup variance (2/3) was accounted for by attitude to nuclear power, trust in institutions, experts and Science, and to some extent by perceived risk of nuclear waste. These explanatory variables were also powerful factors in accounting for within-group variation of policy attitudes explaining 56 (public) to 40 (experts) percent of the variance. It is concluded that the most important factors behind the difference between experts and the public were attitudes likely to reflect processes of professional socialization and social validation of beliefs.

Keywords: *risk perception, nuclear waste, experts, public.*

INTRODUCTION

Experts and the public have been found, in several studies, to have strongly different views with regard to the risk of ionizing radiation, including nuclear industry risks. Public risk perception has been found to be affected by many factors beyond "objective" risk (Slovic, 1987). People judge radiation very differently in different contexts (Sjöberg, 2000; Slovic, 1996). Experts' risk assessment is also affected by many factors, just as the public's risk perception (Becker, 2001; Shrader-Frechette, 1992), but not necessarily the same factors.

Conflicts between various groups over risk management and policies can lead to entrenched positions and the attribution of incompetence and vested interests in the opposite party (Sjöberg, 1980). The intense controversies about nuclear waste disposal (Jacob, 1990) are related to social trust, or lack of it; in the U.S. case trust in DOE (Bella, Mosher, and Calvo, 1988; Frey, 1993; Lemons and Malone, 1991).

In Sweden, the nuclear waste disposal issue was very important in the eyes of the public already in the 1970's when the basic decisions regarding the size of the country's nuclear program were taken (Anshelm, 2006; Sjöberg, 2004c; Sundqvist, 2002). The picture of intense controversy is similar all over the world (Carle, Charron, Milochevitch, and Hardeman, 2004; Openshaw, Carver, and Fernie, 1989). It is often argued that the root of the problem resides in the public's emotional reactions, or "radiophobia" (Becker, 2004). This is an unfortunate rhetorical standpoint (Drottz-Sjöberg and Persson, 1993), since the public is denigrated by it and the complex problem of expert risk perception grossly simplified. There is no good alternative, in a democracy, to taking the public's views seriously and treating them with respect (Sjöberg, 2001b, 2001c).

EMPIRICAL WORK ON EXPERTS' RISK PERCEPTION

Slovic et al. found that the "Dread factor" was most closely related to demand for risk reduction, in the public. For experts there were no close relations between risk dimensions and perceived risk. Experts tended to see risk as synonymous with expected annual mortality (Slovic, Fischhoff, and Lichtenstein, 1979). This work has recently been discussed critically (Rowe and Wright, 2001). It has a very weak empirical basis. It was based on a handful of risk assessment experts who were not substantial experts with regard to all of the hazards investigated (Sjöberg, 2002).

Rothman and Lichter (Rothman and Lichter, 1987) reported one of the few studies, which investigated a wide array of very different groups. They claimed that the negative attitude to nuclear power, and concern about its risks, could be construed as a surrogate for an attack on the core values of capitalism. Such an attack could not be carried out directly according to them because these values are so deeply rooted in American thinking, according to them.

Barke and Jenkins-Smith (Barke and Jenkins-Smith, 1993) reported a major attempt to investigate the differences between scientific experts and the public with regard to the perception of nuclear waste risk. They studied responses from scientists in several different fields of science (physicists and engineers as well as life scientists) and found the following:

- Sierra Club members and the public rated the risks as higher than all groups of scientist did. These differences were not checked for effects of background variables such as gender and age, however
- Physicists and engineers rated the risk as smallest, while life scientists gave a higher rating
- Scientists employed by universities gave the highest risk ratings, those employed by business consultants or federal research laboratories (mostly involved in the national weapons program) the lowest
- The differences were associated, across fields and institutions, with several attitude measures. These measures were based on survey questions measuring (a) the perception of how severe the environmental problems are in general, (b) the perceived voluntariness of technological risk (is risk necessary, should it be compensated, is it acceptable to impose a slight risk on people without their prior consent), and (c) values related to environmental managerial strategies (how can and should society address environmental problems)
- Investigating several other risks (e.g. genetic engineering, downhill skiing) they found that field of research was associated with size of the perceived risk. Those who were primarily involved in an activity associated with the risk rated it as lowest. For example, physicists rated the risk of genetic engineering higher than did those involved in biomedical research while the opposite was true for the nuclear waste risk.

These results agree well with the work by Lynn (Lynn, 1986) who found that both risk perception and political ideology of scientists varied as a function of where they were employed. A comparison of nuclear waste risk judgments of the public and 40 participants in a meeting of the American Nuclear Society (ANS) gave the following results (Flynn, Slovic, and Mertz, 1993):

- Experts rated most risks of nuclear waste handling and disposal as very much smaller than the public did
- However, experts rated, as did the public, the risk of a transport (rail or highway) accident as large
- Experts also had much more trust in the Department of Energy than the public had
- Experts and the public were equally likely to judge that a repository would create new jobs in the local community
- The public was much more likely than the experts were to expect that a repository would create a stigma for the local community.

Although it was concerned with a non-nuclear field (toxicology), the Kraus, Malmfors and Slovic (1993) study is of interest here. They found very large differences between the public and experts. They also found differences, which covaried with experts' employer, and, generally, that experts held quite different views on key risk management issues, such as the validity of animal data for inferring human risk.

A problem in research on experts' risk perception is that it is not obvious who should be considered an expert. A recent study on biotechnology illustrates the problem (Savadori et al., 2004). In this study, graduate students of biology constituted the group of "experts". It is questionable to argue that graduate students are already experts. Furthermore, they probably had no professional experience. Note also that they were not necessarily oriented towards biotechnology.

EXPLAINING THE DIFFERENCES BETWEEN EXPERTS AND THE PUBLIC

The differences between experts and the public have so far not been well explained. Various possibilities exist:

- *Demographics.* Gender may account for the apparent effect of expertise, since most experts are men in the various fields of technology. It is well known that gender is related to perceived risks of technology (Davidson and Freudenburg, 1996). Previous work has found that gender differences still exist within a group of experts (Barke, Jenkins-Smith, and Slovic, 1997). This finding would be of interest to replicate on nuclear waste experts in the present context. Do female experts perceive risks in a manner different from female members of the public? Do male and female experts have different risk perceptions? Age could also have an effect, since experts tend to be older than people are in general. Socio-economic status has also been found to be related to perceived risk (Sjöberg, 2003a). Level of education could also be important, since experts have a higher level of education in general, not only with regard to their field of specialization.
- *Realism.* The public may in fact be misinformed and the experts may be making realistic risk assessments. This is a common notion, which partly is based on the early results reported by Slovic et al. (1979). Lay risk perception is said to be complex and multidimensional, in contrast to the straightforward risk assessments made by experts
- *Different risk definitions.* Experts pay more attention to probability, the public to consequences (Drottz-Sjöberg, 1991)
- *Self-selection.* The differences may exist *before* scientists receive their professional training at college and graduate school. Drottz-Sjöberg and Sjöberg (1991) found that high school students (about 18 years old) in Sweden differed strongly with regard to perceived nuclear risks and these differences were clearly associated with chosen line of specialization in high school. The question remains why these adolescents had so different risk perceptions. Family influence is one likely candidate, peer influence another. Both may exist, of course, and have synergistic effects

- *Socialization of values and risk perception in professional training and work, and social validation.* The social environment is a very important source of beliefs and defines the "truth" on important issues (Cialdini and Trost, 1998; Fulk, 1993; Vishwanath, 2006). Conformity pressures and vested economic and career interests may also play a role. Some beliefs are internalized, others merely reflect compliance. Furthermore, one of the most important sources of information in validating one's beliefs is the social one. This is true for everyone, including experts
- *Perceived control and familiarity.* Experts directly involved in an area probably perceive that they have control over its risks, and long experience may have habituated them to these risks
- *Professional role.* Some experts have the role of protecting the public (e.g. physicians or fire fighters) while others are concerned also with the promotion of a certain technology (Fromm, 2006; Sjöberg, 1991). In the latter case, they enter into policy debates, which typically focus on risks, and they naturally argue that risks are smaller than some members of the public believe them to be
- *General political ideology.* This is a powerful factor in risk perception in general (Sjöberg, 2004b; Sjöberg and Wester-Herber, in press)
- *General tendency to dismiss risks.* It is possible that experts tend to rate all or most technology and environment risks as small. In most studies of the public's risk perception, it is possible to identify a group acting in such a way (Sjöberg, 2006b). However, experts in one specific area need not judge risks in *other* areas as small. Barny et al. (1990) found that experts on sanitation and hygiene made risks judgments quite similar to those made by members of the public, across a wide range of risks.

Some comments on these explanations are in order. Realism cannot be the whole story, since experts vary, also within the same field. They cannot all be right. In a comparative study involving experts in toxicology and the public, it was found that experts tended to disagree; the authors concluded that this was probably an important source of the public's lack of confidence (Kraus, Malmfors, and Slovic, 1992). Wright, Pearman and Yardley (2000) found in their study of off-shore risks that experts were as heterogeneous in their risk views as were members of the public. In addition, risk assessment is not only a question of factual judgment; values enter necessarily. At the same time, the large differences between experts and the public speak against a purely ideological explanation.

It is unlikely that experts are strongly atypical in their political attitudes, although this dimension may explain some of the differences among different groups[1]. Self-selection, on the other hand, cannot be ruled out, and neither can conformity pressures and perceived control and familiarity. Professional role is also a viable alternative in some situations. Social validation of beliefs is a factor, which is likely to play a major role, both with regard to the level of assessed risks and the homogeneity of a group of experts with regard to perceived risk.

---

[1] In the Rothman-Lichter study, elite groups varied even *more* than the public and included, e.g., both high ranking military leaders and Hollywood directors.

PRESENT STUDY

Do levels of perceived risk of nuclear waste experts differ from risk perceptions of the public? The first question of the present paper calls for analyzing experts and the public with regard to risk perception of a range of different hazards, including those associated with nuclear industry. Hazards are assessed both in terms of general and personal risks (Drottz-Sjöberg, 1993). In addition, detailed mapping of nuclear waste risks was called for, and will be performed with an extended version of the Psychometric Model risk dimensions (Fischhoff, Slovic, Lichtenstein, Read, and Combs, 1978).

Even if levels of perceived risk differ, an expected finding, the structure of risk perceptions may be the same, in the sense that similar models can be fitted to risk ratings made by experts and the public. Data obtained with the extended Psychometric Model will be used for analyzing structure. The analysis is important since there is a wide-spread belief that the gap between experts and the public is due to the risk perceptions being qualitatively different in experts and the public (Marshall, Picou, and Nicholls, 2006). One focal issue with regard to nuclear waste is the siting of a repository for high-level waste. Policy judgments, e.g. the intended vote in a local referendum on siting, were therefore used to form an index measuring policy attitude, which was singled out as a dependent variable. The purpose of the present study was to compare such policy judgments made by experts in the field of nuclear waste to those made by the public, published elsewhere (Sjöberg and Drottz-Sjöberg, in press), and to delineate factors responsible for the difference.

Previous work has documented the importance of several factors in accounting for risk perception and policy attitude. The factors investigated here were:

- Attitude to nuclear power (Sjöberg, 1992a)
- Social trust, i.e. trust in experts and institutions (Slovic, Flynn, and Layman, 1991)
- Epistemic trust, i.e. trust in science behind repository technology (Sjöberg, 2001a)
- Perceived risk of nuclear waste (Sjöberg, 2004a)
- Demographics (gender, age and level of education)
- General risk sensitivity, i. e. concern with risks in general, excluding nuclear risks (Sjöberg, 2004a).
- The traditional basic factors of the Psychometric Model (Fischhoff et al., 1978), Dread and Novelty.

These factors were all investigated in recent work on siting and policy attitude (Sjöberg, 2006a). However, that study only included members of the public and did not specifically sample experts or highly trained graduate engineers. The samples presented below involve all three groups.

METHOD

Data Collection

Each questionnaire had a fully visible label carrying the name and address of the respondent. It was stated in the instructions that this was merely for checking who replied and that all analysis and reporting of the study would fully respect the respondents' anonymity. It was also stated that the respondent could delete the label before mailing the questionnaire to us. About 10 percent did so[2].

The data were collected by means of a mailed questionnaire, in 1992. A preliminary technical report has been made available (Sjöberg and Drottz-Sjöberg, 1994). The present paper contains several new analyses and does not report all details of data. Changes in attitudes during the time since 1992 are briefly treated in the present paper in a separate results section, and commented upon in the conclusion section.

Respondents

Data from the public have been described and analyzed elsewhere (Sjöberg and Drottz-Sjöberg, in press). The response rate, in a random sample from the population was 64.6 %. The number of returned questionnaires was 1099. Demographics of the respondents from the public were as follows:

- 50.7 percent men, 49.3 percent women
- average age 41.3 years
- primary school only 29.3 percent, some kind of secondary school 46.1 percent, college or graduate school 20.5 percent
- blue-collar workers 35.1 percent, white-collar workers 33.0 percent, farmers 1.7 percent, entrepreneurs and self-employed persons 6.8 percent, unemployed 7.3 percent, students 13.3 percent
- married or co-habitants 74.2 percent, single 25.8 percent
- own children 70.8 percent
- grandchildren 20.0 percent
- rented apartment 27.4 percent, owned apartment 12.7 percent, one-family house 47.9 percent, farm 6.6 percent, others 3.5 percent
- political preferences were dominated by those who said that they preferred none of the existing parties (31.6 percent). The only two parties, which attracted a sizable number of respondents, were the social democrats (30.3 percent) and the conservatives (15.1 percent). Note that these political preferences were elicited by a question which only asked about preference, not voting intention, and which allowed explicitly for the response "none"
- most respondents lived in small or medium-sized towns (49.3 percent) between 5000 and 175 000 inhabitants

---

[2] This is a large number. In samples from the public it is much more uncommon for people to delete identification information from submitted questionnaires.

The population of nuclear experts was defined as all employees with at least a college degree employed by the regulatory authorities in the fields of radiation protection and nuclear power. Lists of their names were provided by the employers. In addition, we approached industry employees with the same qualifications at one nuclear power plant, at SKB (the Swedish nuclear waste management corporation) and at Vattenfall (a major energy producer, both nuclear and hydro) and members of a professional university network with a special interest in nuclear waste questions. In all, 237 persons received the questionnaire. One hundred and thirty-seven had replied after two reminders, yielding a response rate of 59 percent. All responses were anonymous. This group will be referred to in the following as "the experts".

We also obtained a random sample of 200 members of the Swedish union of graduate engineers who did not work in the nuclear field (in the following this group is referred to as "the engineers"). They responded under the same conditions as the nuclear experts. One hundred and thirty-eight responses were obtained after two reminders, a response rate of 69 percent.

The percentage of male respondents was 90.6 in the nuclear experts group, 91.2 among the engineers, as compared to 50.7 in the data from the public. Their average ages were 46.3 and 45.2 years, respectively (41.3 in the public). In the group of nuclear experts, 59.4 percent had a college diploma and 29.0 percent a Ph.D. or equivalent. The corresponding numbers for the engineers were 77.4 and 18.9 percent. (In the public sample, 20.5 percent had a college degree).

Hence, the two expert groups were predominantly male, somewhat older than the average respondent in the public was and had a more advanced education. The nuclear experts had the most advanced education, on the average, although this group also contained a few with only a high school degree (8.7 percent), in spite of the precautions taken to include only university educated persons. Since they had been defined by their employers as experts, they were retained in the sample. We assumed that they had obtained expert knowledge from experience and in-house education.

Of the experts, 47.1 percent worked with nuclear power, 31.9 with nuclear waste, and 21.0 with both. Safety was the main concern of 73.5 percent, while 8.1 percent indicated economy and production results as their main responsibility. They were asked if they considered themselves specialists in some area of importance for nuclear power and 81.8 percent stated that they were so to a rather or very large extent. As to expertise, 83.5 percent stated that they were experts. They were asked if they considered themselves well oriented about the current scientific and technological development in their most important area of specialization; 94.1 percent stated they were so to a rather or very large extent.

Almost all, 98.5 percent stated that they had opportunity to follow the scientific and technological development in their most important area of specialization. They were asked how much of their total work tasks required specialist competence. The median rating was 60 percent. They were also asked to what extent their specialist area required continuously broadened and/or deeper knowledge; 92.4 percent stated that it did.

In what way had they developed their specialist knowledge? More than one response category could be checked. The proportions choosing each of the categories specified were as follows:

- Education: 0.57
- Practical experience: 0.78
- Research: 0.38
- Literature in the area: 0.55

- Conference participation: 0.65
- Other: 0.08

The over-all picture was thus that the experts were confident in their specialist knowledge, that they considered it demanding in the sense that they had to develop their knowledge continuously, that they felt they had opportunities to do so and that they mostly did it on the basis of practical experience and conference participation. Research was checked by almost 40 percent as a means of developing their competence.

Questionnaire

The questionnaire was rather extensive, 21 pages. It was green and printed in A5 format. It included (page 2) an explanation of the present policy regarding the responsibility for regulating nuclear waste issues - at the time of the study under the authority of the Swedish Nuclear Power Inspectorate (SKI).

The questionnaire to the public called for a total of 237 responses or judgments (for the experts and engineers 166 responses). The contents of the questionnaire were as follows:

- Questions about nuclear power, its utility and risks
- Nuclear waste risks: managed and physical (Sjöberg and Drottz-Sjöberg, 1997)[3]
- Comparisons of nuclear waste risks and other waste risks
- Political behavior with regard to nuclear waste risks
- Confidence in experts and authorities in nuclear waste risk matters
- Judgments of 20 personal[4] risks
- Judgments of 20 general risks
- Judgments of 21 risk dimensions of nuclear waste risk
- 10 attitude questions
- Nuclear waste risks and technology in general
- Specific questions about expertise
- Evaluation of the questionnaire
- Time to respond
- Comments

The questionnaire sent to the public also contained a number of questions deleted for the experts and the engineers[5]. Otherwise, the questionnaire sent to the expert and engineer

---

[3] The nuclear waste risk as usually studied is the net risk resulting both from a physical danger *per se*, and the process of managing that risk. In this study, we included two measures of physical risk: the length of time during which nuclear waste in itself would be dangerous to human beings, and the depth in bedrock desired for a rock deposit of nuclear waste - a somewhat indirect measure of physical risk. These items were included also in our previous study of nuclear waste risk and attitudes (Sjöberg & Drottz-Sjöberg, 1993) where they were found not to correlate with managed nuclear waste risks.

[4] Personal risks refer to the respondent him- or herself as a target. General risks refer to people in general (Sjöberg, 2003a).

[5] They were lifestyle and anxiety questions and some background questions (civil status, children and grandchildren, housing and region), deleted in order to make it less likely that a respondent could be identified.

groups contained largely the same questions as the questionnaire sent to the public, except that 12 specific questions about expertise and job tasks were added.

It is our general policy to avoid including an explicit don't know (DK) response alternative, cp. Krosnick's lucid discussion of this issue (Krosnick et al., 2002). Respondents are told that they should try to answer all questions, even if they are uncertain, and that they should skip any question they could not answer. Few questions are skipped, as a rule. The exception to our policy is when we want to compare with a question used by a polling firm, which may have been posed in a face-to-face interview (or, of course, in a questionnaire with explicit DK alternatives).

The differences between the experts, the engineers, and the public groups with regard to questionnaire quality were, overall, quite small. There were a few large, and natural, differences, though. The experts were apparently not stimulated by the questionnaire to get more information, for example. It should be added that these evaluations, showing that both groups with advanced education evaluated the quality at about the same level as the public, strengthen the conclusions of the study.

The median response time for both expert groups was 30 minutes, which was much shorter than for the public (45 minutes). The reason may be that the experts responded to a shorter questionnaire.

Indices

A number of indices (see Table 1 for a summary and Cronbach alpha values[6]) were constructed:

- *Attitude to a local repository for spent nuclear fuel.* Seven items were included, viz. a question comparing the Swedish solution to that of other countries, acceptance of a Government decision of siting a repository to one' own municipality, acceptance given that the best site is in one's own municipality, intended choice in a local referendum about siting, whether a repository would cause negative health effects, whether a repository would cause positive or negative economic effects, and whether or not it would bring a bad reputation for the area. These items differed as to the number of response categories and they were standardized before being combined to an index.
- *General risk sensitivity.* The questionnaire included 20 hazards, to be rated for both personal risk and risk to others, i.e. general risk. The scale had 7 categories, from "no risk at all" to "extremely large risk". Fifteen of the hazards were not concerned with nuclear power or nuclear waste. The 30 ratings (15 personal and 15 general) were combined to form an index of general risk sensitivity. Risk ratings referring to hazards of handling and transport nuclear waste were not included in this index.
- *Personal risk of nuclear waste.* The average of 3 risk ratings referring to hazards of handling and transport nuclear waste.
- *General risk of nuclear waste.* The average of 3 risk ratings referring to hazards of handling and transport nuclear waste.

---

[6] Cronbach's alpha is a common measure of the reliability of an index, see e.g. Allen and Yen (2003).

- *Novelty of risk.* This is one of the basic dimensions of the Psychometric Model. It was measured by two items in a set of 20 aspects of the risk of nuclear waste. They were rated on a 7-category scale, from "not at all" to "to an extremely large extent". The two ratings were averaged to form an index.
- *Dread.* This is the second of the basic dimensions of the Psychometric Model. It was measured on the same rating scale as Novelty of risk. One item was included, measuring emotional reactions[7].
- *Social trust.* The respondents rated their trust in competence of 7 groups of experts and institutions and of 6 experts/institutions with regard to the public risk assessment of nuclear waste management. The ratings were made on 5-step category scales, from "none at all" to "very much".
- *Epistemic trust.* This index measured trust in the science behind the technical solution for nuclear waste management and storing. One question referred to whether there was a satisfactory solution to the question of a final repository, and one item was an aspect from the list of risk aspects ("The risk is not well known even by scientists"). Other items asked about how complete the scientific knowledge was with regard the nuclear waste issues, and whether leading experts in the area agreed or disagreed. These items differed as to the number of response categories and they were standardized before being combined to an index.
- *Attitude to nuclear power.* Five items were included: general evaluative attitude to nuclear power, the utility (both personal and societal) of nuclear power, the economic consequences of phasing out nuclear power, and the balance of utility vs. risk with regard to nuclear power. These items differed as to the number of response categories and they were standardized before being combined to an index.

A few additional indices were constructed for the analysis of perceived risk. They are described in the section dealing with that topic.

Table 1. Constructed indices, Cronbach's alpha and number of items

	Alpha	Number of items
Attitude to repository	0.88	7
General risk sensitivity	0.92	30
Novelty of risk	0.72	2
Dread	-	1
Social trust	0.91	13
Nuclear waste risk, personal	0.93	3
Nuclear waste risk, general	0.94	3
Epistemic trust	0.74	4
Attitude to nuclear power	0.88	5

---

[7] Dread is often measured with many more and non-emotional items, basically referring to severity of consequences. In the present study, we chose to restrict its content to emotional reactions.

RESULTS

Risk Perceptions: Level

The mean risk ratings varied across groups and genders, in many cases strongly so. Table 2 shows the results for general risk, rated by male participants. (Results for personal risk, and for females, showed the same trends).

Table 2. Mean general risk ratings, male participants, and results of one-way ANOVAS

	Group			Statistics		
	Experts	Engineers	Public	F	Proba-bility	Eta squared
Smoking	3.83	3.49	3.44	4.301	0.014	0.015
Alcohol consumption	3.09	3.06	2.80	3.635	0.027	0.013
Vehicle exhausts	2.85	2.84	2.83	0.009	0.991	0.000
Industrial pollution	2.64	2.76	3.05	6.055	0.003	0.021
Climate change	1.60	2.04	2.51	21.871	0.000	0.073
Radon - lung cancer	2.52	2.20	2.34	2.141	0.119	0.008
Unsound diet	2.88	2.66	2.47	4.968	0.007	0.017
Serious traffic accident	3.39	3.07	2.87	8.846	0.000	0.031
Hit by lightning	0.73	0.72	0.85	1.826	.162	.006
Ozon layer	1.63	2.08	2.42	15.679	0.000	0.053
Swedish nuclear power	0.84	1.32	2.00	50.201	0.000	0.152
Foreign nuclear power	1.59	2.41	3.33	84.040	0.000	0.230
Background radiation	0.98	1.04	1.57	17.901	0.000	0.060
High-level nuclear waste	0.62	1.14	1.95	69.479	0.000	0.198
Nuclear waste in transport	0.70	1.42	1.94	55.850	0.000	0.167
Waste disposal facility	0.54	1.14	1.88	69.430	0.000	0.198
Unemployment	3.53	3.49	3.81	4.849	0.008	0.017
Inadequate housing	2.99	3.02	3.04	0.072	0.930	0.000
Inadequate food	2.46	2.41	2.40	0.106	0.899	0.000
Inadequate medical care	2.98	2.63	2.48	5.786	0.003	0.020

As shown by the table, one-way ANOVA tests gave significant group differences for 15 of the 20 hazards. The table shows that all nuclear hazards were rated as much lower by experts than by the public, with engineers in between. Note also that the male member of the public gave considerably higher risk ratings on environmental items, i.e. industrial pollution, climate change and the ozone layer, compared to the two other groups. Differences between groups are organized differently in Figure 1. This figure shows a tendency for experts to rate lifestyle risks (smoking, alcohol, etc) as larger than the public, as well as striking differences especially with regard to nuclear technology, where the public gave much higher risk ratings than experts and engineers. Figure 1 also shows that engineers gave risk ratings similar to those of the experts, but regressed towards the public, i.e. their ratings were less extreme than the experts were. It is also interesting that the three groups gave almost equal risk ratings for some hazards, such as unemployment and vehicle exhausts. This finding suggests that experts do not rate all risks different than the public does, and hence that the differences we observe are not due to variations in scale use or general risk sensitivity (Sjöberg, 2004a).

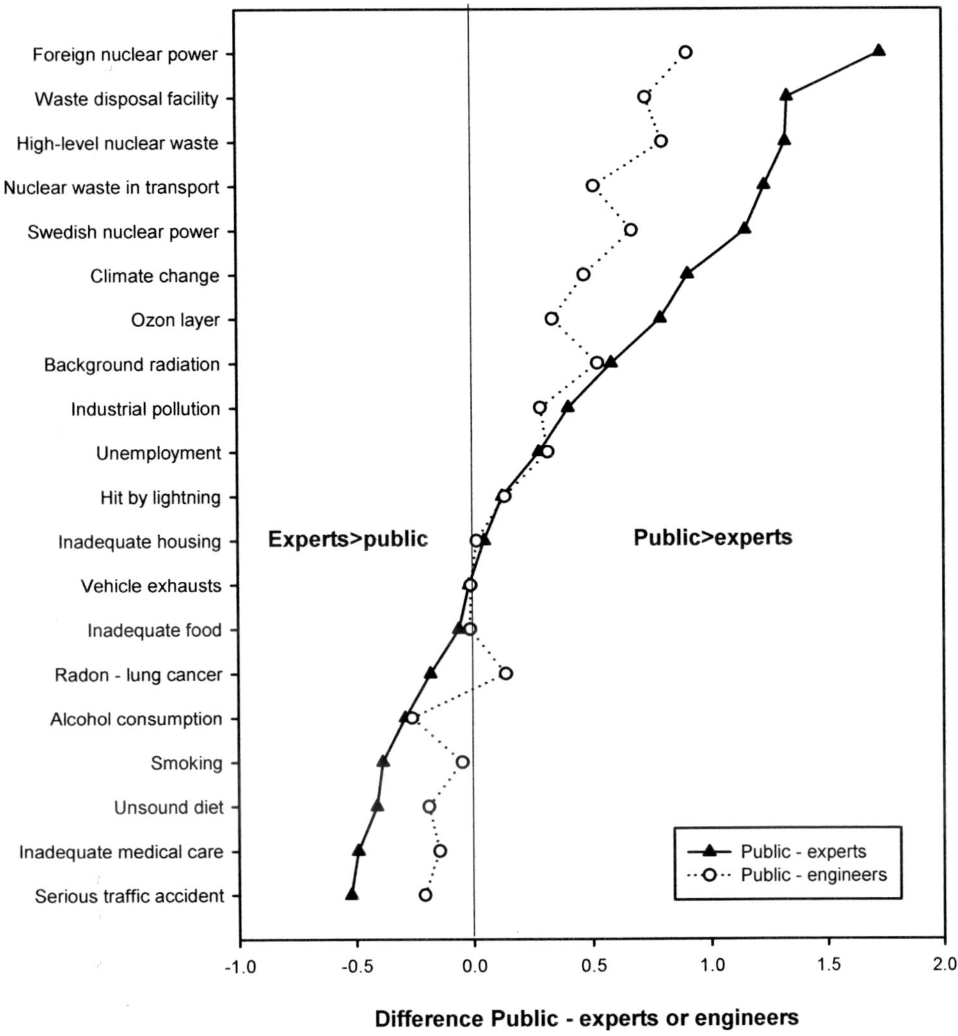

Figure 1. Differences between risk ratings made by the public and experts and engineers, 20 hazards.

The 21 risk aspects of nuclear waste were analyzed for the three groups. Table 3 gives the means and results of one-way ANOVA tests for male participants. There were large and statistically significant differences in most cases.

In Figure 2, we have plotted the differences between the mean values for the public and those from the experts and the engineers, respectively. Consistent and large positive differences were present, except for some of the traditionally emphasized dimensions in the Psychometric Model, such as Novel risk (e.g. "hard to understand") and Dread. Voluntariness, another traditional explanatory factor, also differentiated little among the groups. Experts and engineers tended to diverge from the public in similar ways, but experts did so much more strongly.

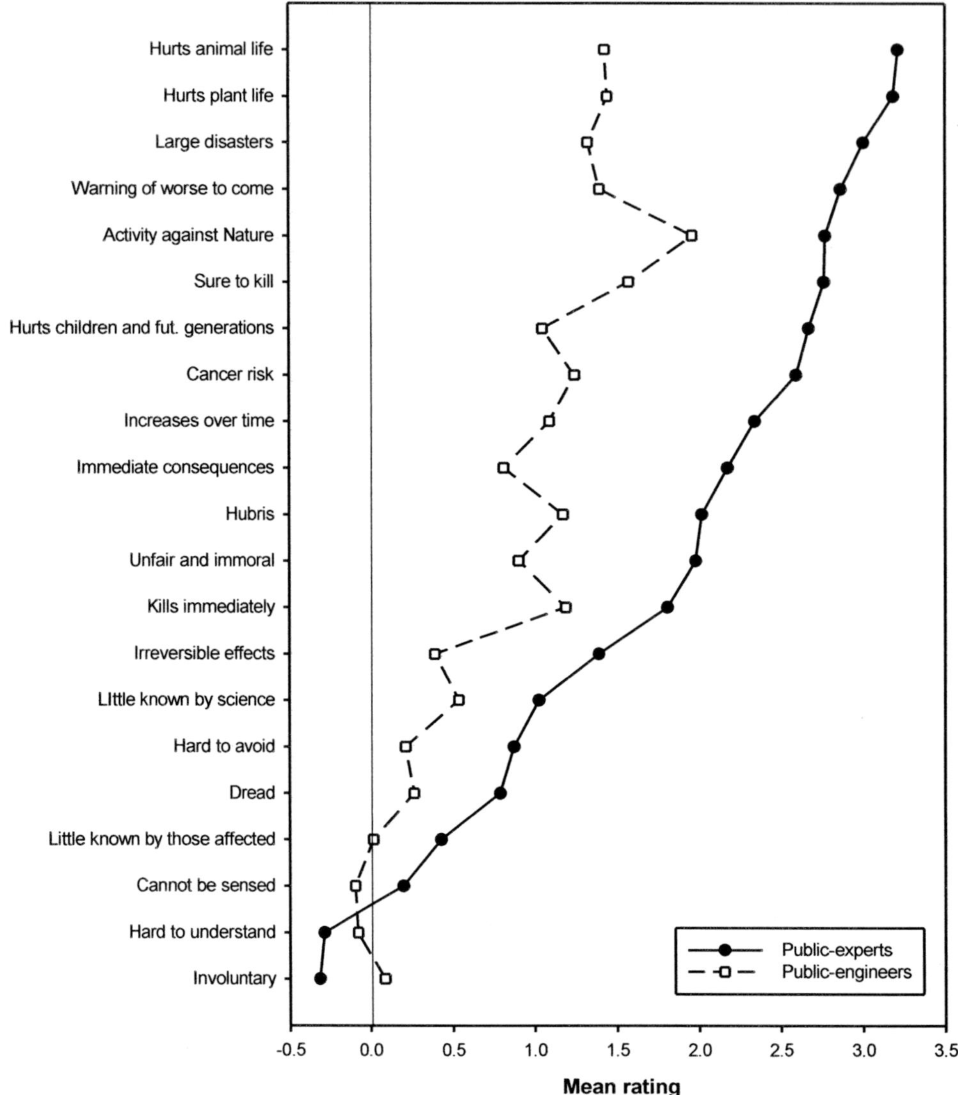

Figure 2. Differences between risk ratings made by the public and experts and engineers, 21 aspects of nuclear waste risks.

Differences between experts and the public were large for both male and female experts. See Figure 3. Figure 4 shows that there was a tendency towards a gender difference in the expert group, 16 of 21 differences showing larger risk ratings by the females. However, the gender effect in the expert group was only moderately large. Separate t-tests for each of the 21 aspects were not significant. (There were only 13 members of the group of female experts).

Table 3. Mean ratings of risk aspects, male participants, and results of one-way ANOVAS

	Group			Statistics		
	Experts	Engineers	Public	F	Probability	Eta squared
Little known by those affected	2.17	2.58	2.59	3.329	0.037	0.012
Hard to understand	3.71	3.50	3.43	1.962	0.142	0.007
LIttle known by science	1.33	1.82	2.35	22.714	0.000	0.076
Involuntary	3.60	3.20	3.29	1.718	0.181	0.007
Hard to avoid	3.20	3.86	4.07	13.946	0.000	0.049
Immediate consequences	1.63	2.99	3.80	73.399	0.000	0.213
Irreversible effects	3.17	4.17	4.56	45.499	0.000	0.139
Dread	2.82	3.34	3.61	9.608	0.000	0.033
Hurts children and future generations	1.60	3.23	4.27	120.194	0.000	0.300
Unfair and immoral	0.89	1.96	2.87	59.080	0.000	0.179
Hubris	0.75	1.59	2.77	70.910	0.000	0.206
Cannot be sensed	3.07	3.36	3.27	0.628	0.534	0.002
Sure to kill	1.29	2.49	4.06	159.464	0.000	0.374
Large disasters	1.66	3.34	4.66	163.103	0.000	0.371
Hurts plant life	1.13	2.88	4.32	179.082	0.000	0.394
Hurts animal life	1.26	3.04	4.47	180.589	0.000	0.396
Warning of worse to come	0.83	2.30	3.69	128.670	0.000	0.329
Increases over time	0.97	2.22	3.31	81.214	0.000	0.229
Cancer risk	1.52	2.88	4.12	116.642	0.000	0.301
Kills immediately	0.91	1.52	2.71	54.972	0.000	0.170
Activity against Nature	0.72	1.54	3.49	125.030	0.000	0.318

In conclusion, the analysis of risk perceptions shows that experts rated the level of nuclear risks as much lower than members of the public did. They also rated risk aspects of nuclear waste as much smaller, with the exception of Voluntariness, Dread, and Novel risk. Engineers rated risks similar to experts, but were not as strongly different from the public. Both experts and engineers showed a rhetorical contrast in their ratings, since they expressed more worry about lifestyle risks than the public did, at the same time as they downgraded all nuclear risks. Female experts diverged strongly from women in the public in rating risks as much smaller. They were similar in their ratings to male experts. Possibly, they were somewhat more worried about risks than their male colleagues were.

Risk Perceptions: Structure

The 21 risk aspects of nuclear waste were used to construct indices:

Novel and unknown risk: 3 items, alpha=0.77
Severity of consequences, 10 items, alpha=0.95
Uncontrollable risk, 3 items, alpha=0.61
Immoral and unnatural risk, 3 items, alpha=0.83

Table 4. Mean standardized indices in the three groups, standard deviations, F-value of a one-way ANOVA for each index and effect size

Index	Nuclear waste experts		Engineers		Sample from the public		One-way ANOVA.: F	Effect size (eta squared)
	Mean	Standard deviation	Mean	Standard deviation	Mean	Standard deviation		
Attitude to repository	1.015	0.566	0.467	0.778	-0.329	0.923	166.772***	0.269
General risk sensitivity	-0.149	0.838	-0.188	0.820	0.074	1.059	5.756**	0.013
Novelty of risk	-0.187	1.082	-0.064	0.992	0.058	0.978	3.623*	0.008
Dread	-0.570	1.083	-0.178	0.973	0.164	0.934	34.576**	0.073
Social trust	0.754	0.634	0.273	0.827	-0.231	1.004	70.757***	0.138
Nuclear waste risk, personal	-0.884	0.458	-0.438	0.681	0.293	0.997	118.221***	0.209
Nuclear waste risk, general	-0.908	0.451	-0.422	0.692	0.298	0.990	123.082***	0.219
Epistemic trust	1.099	0.741	0.379	0.902	-0.329	0.856	181.205***	0.285
Attitude to nuclear power	0.908	0.621	0.471	0.868	-0.306	0.931	133.181***	0.227

*p<0.05
** p<0.01
***p<0.0005

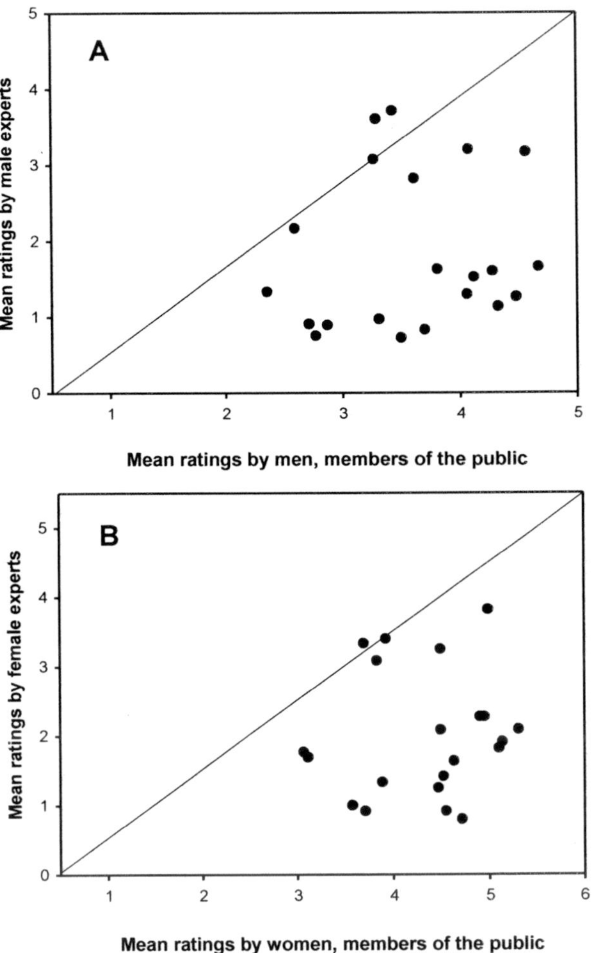

Figure 3. Differences between risk aspects judged by experts and the public; for men (A) and women (B).

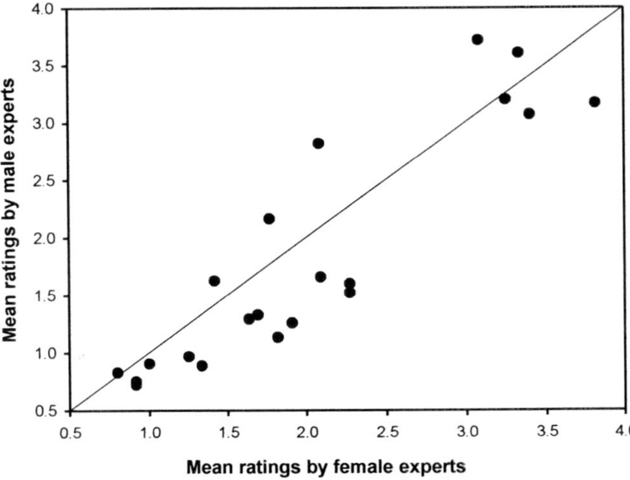

Figure 4. Risk aspects judged by male experts plotted against corresponding data from female experts.

In addition, dread and catastrophic risk were measured with 1 item each.

Personal and general risks of nuclear waste were combined to form a dependent variable. The means and standard deviations of this risk measure varied as expected. It should be noted, in particular, that the variability was much smaller (less than half) in the group of experts compared to the public. For experts, engineers and the public, the standard deviations were 0.60, 0.94, and 1.35, respectively. (The means were 0.61, 1.26, and 2.26).

A separate regression analysis was carried out with each of the three groups, using demographics and the six risk aspect dimensions as explanatory variables. The adjusted squared multiple correlations were, for the experts, engineers and the public 0.190, 0.345, and 0.359. The deviating value for the experts probably reflects the low level of variation in that group. The regression coefficients (unstandardized) are plotted in Figure 5. The figure shows that the models for experts and engineers were quite similar to the model fitted to data from the public.

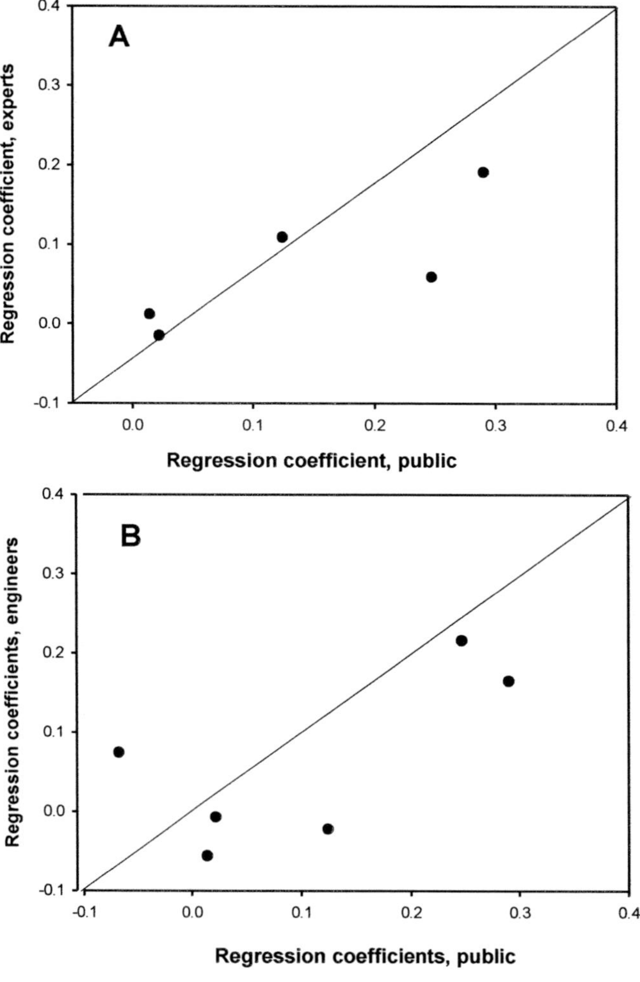

Figure 5. Regression coefficients in model of perceived nuclear waste risk, results from analyzing data from experts (A) and engineers (B) plotted against results from analyzing data from the public.

Large regression coefficients were found for severity of consequences, novel risk, and immoral and unnatural risk. Uncontrollable risk, dread and catastrophic risk did not contribute to the explanatory power of the models.

Analysis of Attitude to a Repository

All indices were standardized to z-scores. Table 4 gives means, standard deviations, and results of one-way ANOVA tests for group differences with regard to the constructed indices. The table shows considerable group differences for all indices except Dread and Novelty, and General risk sensitivity. All differences were statistically significant. Note the smaller standard deviations in groups of experts and engineers.

In the next stage of analysis, we carried out two hierarchical regression analyses of attitude to a repository, one for data from the public and the other from nuclear waste experts. The results are given in Tables 5-8.

Table 5. Hierarchical regression analysis of attitude to a repository data from the public

Explanatory variablesand models	R^2_{adj}	ΔR^2_{adj}	Significance of Δ
A. Demographics: gender, age, level of education, general risk sensitivity	0.148		
B. Dread and Novelty of risk	0.176		
C.Trust, social and epistemic	0.371		
D. Perceived personal and general risk of nuclear waste	0.338		
E. Attitude to nuclear power	0.419		
A+B	0.233	0.093	p<0.0005
A+B+C	0.426	0.193	p<0.0005
A+B+C+D	0.504	0.079	p<0.0005
A+B+C+D+E	0.556	0.052	p<0.0005

Table 6. Hierarchical regression analysis of attitude to a repository, data from experts

Explanaory variable and models	R^2_{adj}	ΔR^2_{adj}	Significance of Δ
A. Demographics: gender, age, level of education, general risk sensitivity	0.017		
B. Dread and Novelty of risk	0.045		
C. Trust, social and epistemic	0.339		
D. Perceived personal and general risk of nuclear waste	0.118		
E. Attitude to nuclear power	0.315		
A+B	0.019	0.044	p < 0.05
A+B+C	0.308	0.287	P < 0.0005
A+B+C+D	0.325	0.025	Not significant
A+B+C+D+E	0.408	0.082	P < 0.0005

Table 7. Zero-order correlations with attitude to a repository and beta weights data from the public

Explanatory variable	Zero-order correlation	Beta	Significance level of beta
Gender	-0.335	-0.091	0.002
Age	-0.073	-0.008	0.775
Educational level	0.102	0.019	0.487
General risk sensitivity	-0.204	0.077	0.021
Dread	-0.388	-0.035	0.268
Novel risk	-0.260	-0.010	0.739
Social trust	0.476	0.154	0.000
Epistemic trust	0.553	0.157	0.000
Nuclear waste risk, personal	-0.560	-0.114	0.055
Nuclear waste risk, general	-0.569	-0.186	0.002
Attitude to nuclear powwer	0.647	0.310	0.000

Table 8. Zero-order correlations attitude with to a repository and beta weights data from experts

Explanatory variable	Zero-order correlation	Beta	Significance level of beta
Gender	-0.042	-0.075	0.270
Age	0.053	0.008	0.905
Educational level	-0.059	0.042	0.532
General risk sensitivity	-0.099	0.057	0.437
Dread	-0.054	0.015	0.825
Novel risk	-0.212	0.066	0.386
Social trust	0.518	0.269	0.002
Epistemic trust	0.512	0.189	0.034
Nuclear waste risk, personal	-0.341	-0.178	0.108
Nuclear waste risk, general	-0.298	0.079	0.477
Attitude to nuclear powwer	0.562	0.358	0.000

It should be noted that the level of explained variance was high: 56 percent for the public and 41 percent for the experts. The experts has a smaller standard deviation in most of their data, see Table 4, which may explain this difference. The most important explanatory variables were attitude to nuclear power, trust (both social and epistemic), and perceived risk. However, perceived risk was of somewhat marginal importance.

Analysis of Inter-Group Variation

Attitude to a repository was analyzed, ANCOVA, across all three groups, and first with gender, age, educational level, and general risk sensitivity as covariates. The unadjusted effect of Group was 0.269. It dropped to 0.121 after controlling for demographics. The Group effect was still substantial and statistically significant. However, the model fit was quite modest, $R^2_{adj} = 0.358$.

In the final analysis, all explanatory variables were entered as covariates, in addition to demographics. The effect of Group vanished, dropping to 0.006. It was no longer significant: $F (2, 783) = 2.490$, $p=0.084$. The adjusted means of attitude to a repository varied only little, from 0.090 (experts) to -0.033 (public). The model fitted quite well with an R^2_{adj} of 0.675. The main explanatory factors were epistemic and social trust, as well as attitude to nuclear power. Perceived risk of nuclear waste, especially general risk, had some effect. Neither Dread nor Novel Risk added to the explanatory power of the model. The explanatory powers of the various independent variables were quite similar for intra- and intergroup analyses.

Comparison with 2006 Data

Some of the questions were repeated in a study of a random sample from the public in 2006; see Sjöberg (2006a) for details. It is well known that the attitude towards nuclear power has become more positive in Sweden since the end of the 1980's. The data show such a trend, with a standardized difference of 0.18, t(1117)=2.917, p=0.004. Thus, the difference is small, albeit statistically significant due to the large sample sizes.

Intention to vote in a local referendum about a siting proposal had not changed from 1992 to 2006, see Table 9. The chi square test gave no significant difference between the two samples.

Table 9. Response distributions regarding the intention to vote question, 1992 and 2006, random samples from the public, percent of the respondents

Response alternative	1992 data	2006 data
Certainly pro	7.8	6.8
Probably pro	11.7	13.0
Uncertain	16.4	16.9
Probably con	20.5	21.6
Certainly con	43.5	41.6

CONCLUSION

It was found that experts, and to some extent engineers, had risk perceptions of nuclear technology which strongly diverged from those of the public. This was an expected finding, but there were some details, which throw new light on the issue of experts vs the public.

First, there was a rhetorical contrast in the data. Experts judged some risks as larger than the public did, viz. lifestyle risks. This finding can be interpreted by the promoter/protector dimension (Sjöberg, 1991, Fromm, 2006 #8797). Protectors are experts who worry about threats to the health of the public, within their area of expertise. A typical example would be medical doctors. Promoters are worried that people exaggerate risks of a technology and bring forward other hazards in rhetorical comparisons. The experts argue, or so it seems, that lifestyle risks are worse than the nuclear risks, implying that people should be willing to accept the nuclear risks. They argue that there are worse things that are already widely accepted[1].

Second, the risk aspects of an extended Psychometric Model showed large differences in most cases, but not with the traditional dimensions of Novel risk and Dread. It is also interesting to note that voluntariness had little power in differentiating the groups. This is the dimension, which was first suggested as an explanatory concept in accounting for societal risk acceptance (Starr, 1969).

Third, expert groups typically have an over-representation of men. Since men usually are less worried than women are about risks, this fact alone could explain the gap between experts and the public. In the present data, we found a modest gender difference in the expert group but the main thrust of the findings is that both male and female experts had risk perceptions, which diverged from the public.

Some of the possible reasons for different risk perceptions by experts and the public, which were mentioned in the introduction, can now be rejected, while others remain as possible and plausible causes:

- *Demographics.* General education in technology at an advanced level did not explain the whole difference, since graduate engineers had risk perceptions in-between those of the experts and the public. Demographics, including also gender, age, and General Risk Sensitivity accounted for only 1/3 of the intergroup variance.
- *Realism.* This is a possible factor in the present case, but it should be noted that nobody fully knows how large the risk of nuclear waste is[2], and that experts varied in their risk assessments. In addition, the structure of risk perception of nuclear waste was similar in the three groups. The notion that the gap between lay people and experts is the result of *qualitative* differences of risk perception is not compatible with the present results – but of course there are large differences in *level* of perceived risk
- *Different risk definitions.* This is not a likely explanation since we found differences between experts and the public not only in risk ratings but also with other types of response formats, such as policy attitude questions.
- *Self-selection.* This is a possibility, but we do not have the longitudinal data necessary to assess it.
- *Socialization of values and risk perception in professional training and work, and social validation.* We find this explanation highly plausible, based on general social psychological knowledge (Cialdini and Trost, 1998). It is a promising basis for further research.
- *Perceived control and familiarity.* This, too, is a possible starting-point for further research.
- *Professional role.* There were indications that the protector/promoter concept would throw some light on the belief dynamics of the type of experts investigated here. Figs. 1 and 2 show that experts not only judged nuclear risks as smaller than the public, they also judged lifestyle risks (and a few others) as larger. In the latter case,

---

[1] We do not wish to assert that all experts reason in this way, but it has been a common argument in the debates, starting with Sowby's risk comparison paper (Sowby, 1965).

[2] Consider such factors as the extremely long time range during which a repository is required to be safe (100,000 years), and risks associated with war and terrorism. Such risks cannot be scientifically assessed.

the group differences were smaller, but still clear, systematic, and statistically significant. These findings may be interpreted as a reflection of a promoter/protector rhetoric

- *General political ideology.* We find it unlikely that there are extreme differences as to political preferences between experts and the public.
- *General tendency to dismiss risks.* The data did not support this type of explanation.

An important factor distinguishing experts and the public was found to be trust, both social and epistemic. Recent research has documented the importance of epistemic trust, both for the public (Sjöberg, in press-a) and for experts (Sjöberg, 2004b). Attitude to nuclear power was also very important. There is prior research testifying to a link between attitudes and risk perception (Sjöberg, 1992a).

The consistently higher ratings in the public regarding nuclear risks were in line with their higher ratings of environmental risks. It is worth noting that the public seemed more concerned about industrial pollution, the ozone layer, and climate change than engineers and nuclear experts – a result that invites further research.

The reasons for different levels of trust, and attitude to nuclear power cannot be determined in the present study. Explanations may have to do both with self-selection into a profession, and socialization of values in the work environment.

It should be noted that the basic dimensions of the Psychometric Model, Dread and Novelty, did not contribute to accounting for the differences between experts and the public. It has been found repeatedly that these dimensions, albeit historically important in risk perception research, play only a minor role in risk related attitudes. In addition, experts appear to make risk ratings in much the same manner as the public, only lower whenever the hazards rated refer to their own area of responsibility (Sjöberg, 2002), see the analysis of the structure of risk perception reported in the present article.

The small effect of Dread may have come as a surprise since it is one of the major considerations behind the "risk as feeling" hypothesis (Loewenstein, Weber, Hsee, and Welch, 2001). There could be two explanations for the current findings (and several others where Dread did not have a strong effect). First, it is possible that emotions have little or no correlations with risk perception. This hypothesis can be rejected with reference to recent research where a number of different emotions all correlated strongly with risk related attitudes (Sjöberg, 2003b, in press-b). A second hypothesis is related to the way people are instructed to rate dread. It is typically left unspecified *whose* dread they should rate. When people rate the expected emotional reactions of others, little explanatory power is generated. The instructions must specify that it is their own emotional reaction, which is of interest (Sjöberg, 2003b). In future research, it would be important to investigate differences between experts and the public with regard to emotional reactions to the hazards under study.

As noted in the introduction, the present data were collected in 1992 and attitudes have of course changed since then. Are the results still valid? We found that the attitude to nuclear power had become somewhat more positive, a finding in line with other research (Forskningsgruppen för Samhälls- och Informationsstudier, 2005). On the other hand, the policy attitude towards a nuclear waste repository, measured by voting intention, remained the same. The over-all structure of the data probably is not misleading with regard to current risk perceptions and attitudes.

How can the differences between experts and the public be resolved? The field of risk communication is devoted to that question (Renn, 1992) and one suggestion is that of so-called citizen panels (Renn, Webler, and Johnson, 1991; Renn, Webler, Rakel, Dienel, and Johnson, 1993). Communication only is a too limited solution; participation is more and more often seen as the most viable approach (Laird, 1993). In general, empowerment of the public may well turn out to be the only feasible solution (Sjöberg, 1992a, 1992b; Slovic, 1993). The notion that conflicts can be resolved by means of economic compensation (Inhaber, 1992) has largely failed.

However, the hope of a consensus between experts and the public may rest on an unrealistic assumption, *viz.* that experts can assess the risk with sufficient precision, and that they can agree on it. Kraus, Malmfors and Slovic (Kraus et al., 1993), in their work on intuitive toxicology, pointed out that

> "controversies over chemical risks may be fueled as much by limitations of the science of risk assessment and disagreement among experts as by public misconceptions" (p. 441).

In the field of nuclear waste, controversies abound. Researchers have pointed to many uncertainties, in particular with regard to very long-term predictions about events in a repository (Kraus et al., 1993). Shrader-Frechette pointed to many questions, which have to be decided based on expert opinion, rather than established scientific facts (Shrader-Frechette, 1993).

Recommendations to experts to use appeals to "feelings" (Johnson, 2001) are questionable both on factual and ethical grounds. A different, and probably more useful, approach is to acknowledge the limits of expert knowledge and science and the role of ideology (Sjöberg, 2001a; Sjöberg and Wester-Herber, in press) and take the arguments raised by the public seriously (Halfacre, Matheny, and Rosenbaum, 2000; Sohn, Yang, and Kang, 2001; Thomas, 2001).

It could also be the case that discrepancies in opinion and judgments between experts on the one hand, and between experts and the public on the other, are important prerequisites for a continuous focus on safety (Sjöberg, 2006b). In the field of nuclear wastes and siting policy, however, the different risk judgments and attitudes sometimes hinder constructive discussions, which could jeopardize constructive developments. The present paper shows that the *structure* of perceived risk was similar for experts and the public but that the risk *levels* and the *policy attitudes* differed. We have here pointed social validation of risk perception and related attitudes as a factor of interest in further analyses of the differences.

REFERENCES

Allen, M. J., and Yen, W. M. (2003). *Introduction to measurement theory.* Long Grove, IL: Waveland Press.

Anshelm, J. (2006). Kärnavfallet - från energireserv till kvittblivningsproblem In K. Vikström (Ed.), *Samhällsforskning 2006. Betydelsen för människorna, hembygden och regionen av ett slutförvar för använt kärnbränsle* (pp. 135-157). Stockholm: SKB.

Barke, R. P., Jenkins-Smith, H., and Slovic, P. (1997). Risk perceptions of men and women scientists. *Social Science Quarterly, 78,* 167-176.

Barke, R. P., and Jenkins-Smith, H. C. (1993). Politics and scientific expertise: Scientists, risk perception, and nuclear waste policy. *Risk Analysis, 13*, 425-439.

Barny, M.-H., Brenot, J., Dos Santos, J., and Pages, J.-P. (1990). *Perception des risques majeurs dans la population bordelaise et chez les experts* (Note SEGP/LSEES No. 90/17): Centre d'etudes nucleaires de Fontenay-aux-roses.

Becker, K. (2001). Reflections on public acceptance of nuclear energy and the low dose issue. *Atw-Internationale Zeitschrift Fur Kernenergie, 46*, 54-+.

Becker, K. (2004). Causes, consequences, and therapy of the radiophobia syndrom. *Atw-International Journal for Nuclear Power, 49*, 177-+.

Bella, D. A., Mosher, C. D., and Calvo, S. N. (1988). Technocracy and trust: Nuclear waste controversy. *Journal of Professional Issues in Engineering, 114*, 27-39.

Carle, B., Charron, S., Milochevitch, A., and Hardeman, F. (2004). An inquiry of the opinions of the french and belgian populations as regards risk. *Journal of Hazardous Materials, 111*, 21-27.

Cialdini, R. B., and Trost, M. R. (1998). Social influence: Social norms, social conformity, and compliance. In D. T. Gilbert, S. T. Fiske and G. Lindzey (Eds.), *The handbook of social psychology. Vol. Ii* (pp. 151-192). Boston: McGraw-Hill.

Davidson, D. J., and Freudenburg, W. R. (1996). Gender and environmental risk concerns: A review and analysis of available research. *Environment and Behavior, 28*, 302-339.

Drottz-Sjöberg, B.-M. (1991). *Perception of risk. Studies of risk attitudes, perceptions and definitions* (Vol. 1). Stockholm: Stockholm School of Economics, Center for Risk Research.

Drottz-Sjöberg, B.-M. (1993). *Risk perceptions related to varied frames of reference.* Paper presented at the SRA Europe Third Conference. Risk analysis: Underlying rationales, Paris.

Drottz-Sjöberg, B.-M., and Persson, L. (1993). Public reaction to radiation: Fear, anxiety or phobia? *Health Physics, 64*, 223-231.

Drottz-Sjöberg, B.-M., and Sjöberg, L. (1991). Attitudes and conceptions of adolescents with regard to nuclear power and radioactive wastes. *Journal of Applied Social Psychology, 21*, 2007-2035.

Fischhoff, B., Slovic, P., Lichtenstein, S., Read, S., and Combs, B. (1978). How safe is safe enough? A psychometric study of attitudes towards technological risks and benefits. *Policy Sciences, 9*, 127-152.

Flynn, J., Slovic, P., and Mertz, C. K. (1993). Decidedly different: Expert and public views of risks from a radioactive waste repository. *Risk Analysis, 13*, 643-648.

Forskningsgruppen för Samhälls- och Informationsstudier. (2005). *Uppdaterad version av fsi05:V02kk* (Release ur Kajsa No. v02kkb). Stockholm: Forskningsgruppen för Samhälls- och Informationsstudier.

Frey, J. H. (1993). Risk perceptions associated with a high-level nuclear waste repository. *Sociological Spectrum, 13*, 139-151.

Fromm, J. (2006). Experts' views on societal risk attention. *Journal of Risk Research, 9*, 243-264.

Fulk, J. (1993). Social construction of communication technology. *Academy of Management Journal, 36*, 921-950.

Halfacre, A. C., Matheny, A. R., and Rosenbaum, W. A. (2000). Regulating contested local hazards: Is constructive dialogue possible among participants in community risk management? *Policy Studies Journal, 28*, 648-667.

Inhaber, H. (1992). Yard sale. Society should bid for the right to site its prisons and dumps. *The Sciences, 32*, 16-21.

Jacob, G. (1990). *Site unseen: The politics of siting a nuclear waste repository.* Pittsburgh, PA: University of Pittsburgh Press.

Johnson, R. H. (2001). The role of the radiation safety specialist as witness: Risk communication with attorneys, judges, and jurors. *Health Physics, 81*, 661-669.

Kraus, N., Malmfors, T., and Slovic, P. (1992). Intuitive toxicology: Expert and lay judgments of chemical risks. *Risk Analysis, 12*, 215-232.

Kraus, N., Malmfors, T., and Slovic, P. (1993). Intuitive toxicology: Expert and lay judgments of chemical risks. *Comments on Toxicology, 4*, 441-484.

Krosnick, J. A., Holbrook, A. L., Berent, M. K., Carson, R. T., Hanemann, W. M., Kopp, R. J., et al. (2002). The impact of "No opinion" Response options on data quality: Non-attitude reduction or an invitation to satisfice? *Public Opinion Quarterly, 66*, 371-403.

Laird, F. N. (1993). Participatory analysis. Democracy, and technological decision making. *Science, Technology, and Human Values, 18*, 341-361.

Lemons, J., and Malone, C. (1991). Scientific, public policy, and ethical implications of the nuclear waste policy act and its amendments. *Research in Philosophy and Technology, 11*, 5-32.

Loewenstein, G. F., Weber, E. U., Hsee, C. K., and Welch, N. (2001). Risk as feelings. *Psychological Bulletin, 127*, 267-286.

Lynn, F. M. (1986). The interplay between science and values in regulating environmental risks. *Science, Technology and Human Values, 11*, 40-50.

Marshall, B. K., Picou, J. S., and Nicholls, K. (2006). Environmental risk perceptions and the white male effect: Pollution concerns among deep-south coastal residents. *Journal of Applied Sociology/Sociological Practice, 23*, 31-49.

Openshaw, S., Carver, S., and Fernie, J. (1989). *Britain's nuclear waste: Safety and siting.* London: Belhaven Press.

Renn, O. (1992). Risk communication: Towards a rational discourse with the public. *Journal of Hazardous Materials, 29*, 465-519.

Renn, O., Webler, T., and Johnson, B. B. (1991). Public participation in hazard management: The use of citizen panels in the u.S. *Risk - Issues in Health and Safety, 2*, 197-226.

Renn, O., Webler, T., Rakel, H., Dienel, P., and Johnson, B. (1993). Public participation in decision making: A three-step procedure. *Policy Sciences, 26*, 189-214.

Rothman, S., and Lichter, S. R. (1987). Elite ideology and risk perception in nuclear energy policy. *American Political Science Review, 81*, 383-404.

Rowe, G., and Wright, G. (2001). Differences in expert and lay judgments of risk: Myth or reality? *Risk Analysis, 21*, 341-356.

Savadori, L., Savio, S., Nicotra, E., Rumiati, R., Finucane, M. L., and Slovic, P. (2004). Expert and public perception of risks from biotechnology. *Risk Analysis, 24*, 1289-1299.

Shrader-Frechette, K. S. (1992). Risk estimation and expert judgment. The case of yucca mountain. *Risk - Issues in Health and Safety, 3*, 283-315.

Shrader-Frechette, K. S. (1993). *Burying uncertainty: Risk and the case against geological disposal of nuclear waste.* Berkeley, CA: University of California Press.

Sjöberg, L. (1980). The risks of risk analysis. *Acta Psychologica, 45,* 301-321.

Sjöberg, L. (1991). *Risk perception by experts and the public* (Rhizikon: Risk Research Report No. 4). Stockholm: Center for Risk Research.

Sjöberg, L. (1992a). Psychological reactions to a nuclear accident. In J. Baarli (Ed.), *Conference on the radiological and radiation protection problems in nordic regions, tromsö 21-22 november, 1991* (pp. Paper 12). Oslo: Nordic Society for Radiation Protection - see http://www.dynam-it.com/lennart/ for downloading.

Sjöberg, L. (1992b). *Risk perception and credibility of risk communication* (RHIZIKON: Risk Research Report No. 9): Center for Risk Research, Stockholm School of Economics.

Sjöberg, L. (2000). Specifying factors in radiation risk perception. *Scandinavian Journal of Psychology, 41,* 169-174.

Sjöberg, L. (2001a). Limits of knowledge and the limited importance of trust. *Risk Analysis, 21,* 189-198.

Sjöberg, L. (2001b). Political decisions and public risk perception. *Reliability Engineering and Systems Safety, 72,* 115-124.

Sjöberg, L. (2001c). Whose risk perception should influence decisions? *Reliability Engineering and Systems Safety, 72,* 149-152.

Sjöberg, L. (2002). The allegedly simple structure of experts' risk perception: An urban legend in risk research. *Science Technology and Human Values, 27,* 443-459.

Sjöberg, L. (2003a). The different dynamics of personal and general risk. *Risk Management: An International Journal, 5,* 19-34.

Sjöberg, L. (2003b). Risk perception, emotion, and policy: The case of nuclear technology. *European Review, 11,* 109-128.

Sjöberg, L. (2004a). Explaining individual risk perception: The case of nuclear waste. *Risk Management: An International Journal, 6,* 51-64.

Sjöberg, L. (2004b). *Gene technology in the eyes of the public and experts. Moral opinions, attitudes and risk perceptions* (SSE/EFI Working Paper Series in Business Administration No. 2004:7). Stockholm: Stockholm School of Economics.

Sjöberg, L. (2004c). Local acceptance of a high-level nuclear waste repository. *Risk Analysis, 24,* 739-751.

Sjöberg, L. (2006a). *Opinion och attityder till förvaring av använt kärnbränsle. (Opinion and attitudes to a repository for spent nuclear fuel)* (Research Report No. R-06-97). Stockholm: SKB. Svensk Kärnbränslehantering AB.

Sjöberg, L. (2006b). Rational risk perception: Utopia or dystopia? *Journal of Risk Research, 9,* 683-696.

Sjöberg, L. (in press-a). Antagonism, trust and perceived risk. *Risk Management: An International Journal.*

Sjöberg, L. (in press-b). Emotions and risk perception. *Risk Management: An International Journal.*

Sjöberg, L., and Drottz-Sjöberg, B.-M. (1993). *Attitudes to nuclear waste* (Rhizikon: Risk Research Report No. 12). Stockholm: Center for Risk Research.

Sjöberg, L., and Drottz-Sjöberg, B.-M. (1994). *Risk perception of nuclear waste: Experts and the public* (Rhizikon: Risk Research Report No. 16): Center for Risk Research, Stockholm School of Economics.

Sjöberg, L., and Drottz-Sjöberg, B.-M. (1997). Physical and managed risk of nuclear waste. *Risk - Health, Safety and Environment, 8*, 115-122.

Sjöberg, L., and Drottz-Sjöberg, B.-M. (in press). Public risk perception of nuclear waste. *International Journal of Risk Assessment and Management.*

Sjöberg, L., and Wester-Herber, M. (in press). Too much trust in (social) trust? The importance of epistemic concerns and perceived antagonism. *International Journal of Global Environmental Isssues.*

Slovic, P. (1987). Perception of risk. *Science, 236*, 280-285.

Slovic, P. (1993). Perceived risk, trust, and democracy. *Risk Analysis, 13*, 675-682.

Slovic, P. (1996). Perception of risk from radiation. *Radiation Protection Dosimetry, 68*, 165-180.

Slovic, P., Fischhoff, B., and Lichtenstein, S. (1979). Rating the risks. *Environment, 21*, 14-20,36-39.

Slovic, P., Flynn, J. H., and Layman, M. (1991). Perceived risk, trust, and the politics of nuclear waste. *Science, 254*, 1603-1607.

Sohn, K. Y., Yang, J. W., and Kang, C. S. (2001). Assimilation of public opinions in nuclear decision-making using risk perception. *Annals of Nuclear Energy, 28*, 553-563.

Sowby, F. D. (1965). Radiation and other risks. *Health Physics, 11*, 879-887.

Starr, C. (1969). Social benefit versus technological risk. *Science, 165*, 1232-1238.

Sundqvist, G. (2002). *The bedrock of opinion. Science, technology and society in the siting of high-level nuclear waste.* Dordrecht, the Netherlands: Kluwer.

Thomas, J. P. (2001). Rebuilding trust in established institutions. *Health Physics, 80*, 379-383.

Vishwanath, A. (2006). The effect of the number of opinion seekers and leaders on technology attitudes and choices. *Human Communication Research, 32*, 322-350.

Wright, G., Pearman, A., and Yardley, K. (2000). Risk perception in the UK oil and gas production industry: Are expert loss-prevention managers' perceptions different from those of members of the public? *Risk Analysis, 20*, 681-690.

In: Nuclear Waste Research: Siting, Technology and Treatment   ISBN 978-1-60456-184-5
Editor: Arnold P. Lattefer, pp. 75-105   © 2008 Nova Science Publishers, Inc.

Chapter 2

MILLIMETER-WAVE MONITORING FOR NUCLEAR WASTE VITRIFICATION

P. P. Woskov[1], S. K. Sundaram[2] and W. E. Daniel[3]
[1] Plasma Science and Fusion Center
Massachusetts Institute of Technology, Plasma Technology Division
Cambridge, MA
[2] Pacific Northwest National Laboratory
Richland, WA
[3] Savannah River Technology Center
Westinghouse, Aiken, SC

1. INTRODUCTION

Nuclear waste vitrification melters currently lack sophisticated on-line monitoring capability to ensure efficient operation and to guarantee in real time the production of a high quality glass waste product that meets stringent regulatory and disposal requirements. This lack of on-line monitoring capability is of particular concern for the costly multi-decade national effort in the U. S. to clean up cold war legacy waste, where the waste stream is not well characterized and it continues to vary significantly over the long time period it has been produced and stored. Current operations at the Defense Waste Processing Facility (DWPF) at the Savannah River Site and future operations at the Waste Treatment Plant (WTP), now under construction at Hanford, rely on predictive models for vitrification because of the lack of on-line monitoring of most melter processing parameters. This necessitates conservative operation to take into account uncertainties in the modeling and the ability to control actual input parameters. It also makes the operations susceptible to unpredictable anomalies such as foaming, liquidus crystallization, noble metals build up, and salt layer formation. Predictive modeling will become increasingly more difficult in time as the waste glass chemistry evolves with changes in the waste feed compositions. Additionally, the predictive models are limited by experimental validation, thus building more conservatism into the models. Comprehensive on-line real time monitoring of the vitrification process would alleviate the uncertainties, improve processing efficiencies (waste loading and throughput), and safeguard the facilities from anomalies. Advanced high temperature thermal analytical tools would also address the needs identified at the Hanford site and at the Savannah River Site (SRS) for increased waste loading and accelerated cleanup.

Much of the desired on-line monitoring capabilities can be realized through the use of millimeter-wave (MMW) technologies. Electromagnetic radiation in the 10 – 0.3 mm (30 – 1000 GHz) range of the spectrum is ideally suited for remote measurements in harsh, optically unclean, industrial, and unstable processing environments. Millimeter waves are long enough to penetrate optical/infrared obscured viewing paths through dust, smoke, and debris, but short enough to provide spatially resolved point measurements for profile information. Another important advantage is the ability to fabricate efficient MMW melter viewing components from refractory materials because optical or infrared quality is not required in the millimeter-wave wavelength range. The same ceramics and alloys from which the melter is constructed can be used to fabricate MMW waveguide/mirror components that go into the melter for long life survivability.

MMW techniques and technologies therefore make possible new robust diagnostic tools for glass melt monitoring in a high temperature and radioactive melter environment that other technologies can not access. These new tools will also make possible new insights into the dynamics of molten glass science and into nuclear waste glass melt property modeling because of the new way the melt properties are measured. In addition, these new monitoring developments will contribute toward modernizing nuclear waste glass vitrification operations to industrial standards of feed-back control as employed by most industries in manufacturing. The goal of more fully monitoring a melter is illustrated in Figure 1, where a cross-section of the DWPF melter is shown with the desired on-line diagnostics that would improve melter control. Currently, none of these listed measurements are being made in real-time on present production nuclear waste glass melters. However, many of these desired monitoring capabilities have now been demonstrated with the MMW techniques described here. The new MMW monitoring capabilities that have been demonstrated include those for melt viscosity, emissivity, temperature, foaming, salt layer formation, and cold cap temperature mapping. These developments have been accomplished both in the laboratory and in engineering scale test melters. Other monitoring capabilities that have not yet been demonstrated such as liquidus formation and noble metals accumulation have the potential to be developed based on this work. Consequently, the vision of a well-instrumented melter with a feed-back control illustrated in Figure 1 has the potential of becoming a reality with these new developments.

2. IMPROVING MELTER OPERATION AND THROUGHPUT

2.1. Melter Control

The Defense Waste Processing Facility at the Savannah River Site uses a feed-forward system to control the glass properties in the melter to ensure the vitrified waste satisfies waste acceptance protocols for long term storage and disposal as well as process operating constraints. The feed-forward system uses Statistical Process Control techniques where glass properties and their associated processing constraints are predicted using empirical models based on waste feed compositions, mostly validated by laboratory testing with limited research-scale melter testing. Due to this conservative approach, acceptable glasses with higher waste loadings can be rejected because the glass property (liquidus, durability, viscosity) is indeterminate (i.e., the applicability of the model is not certain). This type of feed-forward model predictive control is necessary since there are no reliable on-line glass property measurement devices. With online glass property measurement devices a more traditional feedback control system could be used like shown in Figure 2. Feedback control is the industrial standard employed by more companies due to its ability to maintain control despite unknown disturbances.

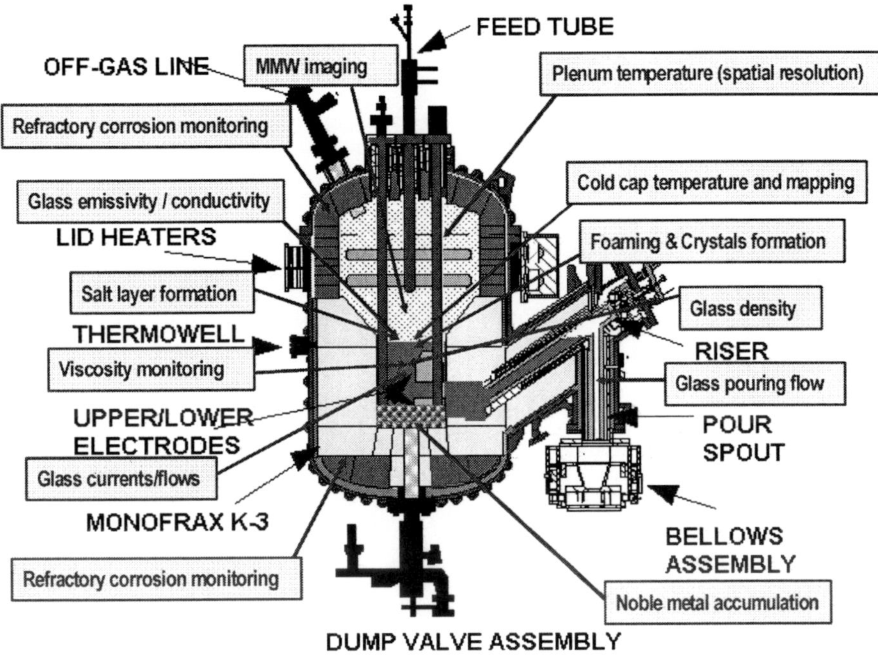

Figure 1. Long range objective to modernize nuclear waste glass melters with on-line monitoring.

For high temperature melters that vitrify radioactive wastes like the ones used at DWPF at SRS and at the West Valley Demonstration Project (WVDP) and the ones planned for the River Protection Project at Hanford, there are no reliable glass properties measuring devices. In addition, the relationship between glass property models and the melter operating conditions have not advanced significantly since early studies conducted at Pacific Northwest National Laboratory (PNNL) and Savannah River Technology Center (now the Savannah River National Laboratory, SRNL) [1-7]. Current melter operations have to deal with cold cap formation and growth in order to keep up production rates. Cold cap formation also affects the melter vapor space temperature and combustion of the melter off gases, which impacts the off gas system as well as melter feed rates. Foaming of the melt also hampers melter operation, which is affected by the redox and the viscosity of the melt [8]. Excessive foaming can slow down production rate by decreasing heat transfer to the top of the melt pool and by limiting feed rates due to maximum tank levels and also potentially lead to catastrophic spills that would put personnel at risk [8, 9].

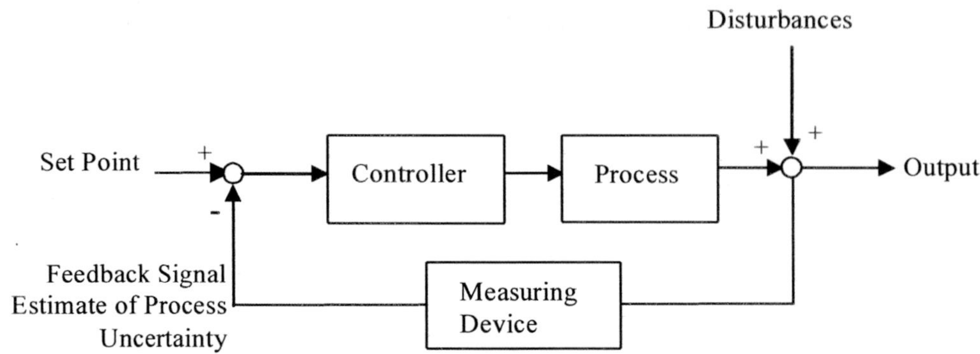

Figure 2. Classic Feedback Control Écheme.

2.2. Melter Monitoring Needs

A basic melter operation problem is not being able to measure liquidus temperature on-line [9]. The liquidus temperature is important due to the formation of crystals that greatly impact the processing of the glass in the melter as well as the durability of the glass product for disposal. Another problem is accounting for the redox (ratio of reducing reactions to oxidizing reactions) of the melt pool when blending a waste batch [8-12]. A higher redox can lead to metals precipitating out and potentially shorting a joule-heated melter. If redox is too low, then foaming can become an issue since there are more oxidation reactions releasing gases. Additionally, the thermal gradients in the melt pool can impact production rates by affecting the density and viscosity of the melt, which in turn affects the ability to mix and properly pour [13]. Having the ability to measure viscosity and density of the melt pool can directly affect the production rate of a melter. To maximize waste loadings, an accurate composition of the melt pool is needed. In current melter operations, only predictions of the melt pool composition are available. These composition predictions are used in property models to predict with the associated inherent modeling errors the glass viscosity, density, redox, and liquidus temperature. The relationship between these anomalies and the measured melter parameters is deemed to be qualitative at best [14]. This lack of on-line measuring devices/technology makes implementing a feedback control loop on glass properties impossible.

With the development of the MMW technology, the situation changes dramatically as it will enable on-line temperature, viscosity, emissivity, and density measurement of the molten glass, simultaneously. It also has the potential for measuring melter vapor space temperature and temperature of the surrounding refractory and even be able to relate measurements back to glass composition. Glass composition is tightly related to glass properties and thus melter controll [1-3, 15, 16]. Using the MMW technology as an on-line glass composition/property tool would be the ultimate goal. With all these MMW on-line glass measurements, a more robust feedback control system could be developed for a vitrification process instead of relying solely on feed-forward statistical process control predictions. In addition, the waste vitrification process has been shown to have non-linear behavior [5, 7, 17] like many processes in the chemical industry. More advanced control techniques like Internal Model Control could be adapted using the MMW on-line measurements. The ability of the MMW techniques to measure on-line glass properties could truly revolutionize the control of glass melters by allowing the implementation of more advanced controls.

The MMW on-line measurements could also be used to improve more than just glass property control. Being able to measure glass properties on-line versus using predictions based on time-delayed analytical data would allow smoother operation at more sustainable melter production rates. On-line monitoring could be used to mitigate problems with cold cap formation, thermal gradients, crystal formation (liquidus temperature), foaming, slow pouring rates, and metal precipitation (redox), which adversely impact melter throughput.

2.3. Cold Cap and Plenum Temperature

In most high temperature joule heated melters, a cold cap forms on top of the melt pool where the waste feed is fed and only partially vaporizes and/or melts. The vaporization and melting adsorb a great deal of melter energy and thus the top of the melter cools off. As the top of the melter cools, a crust begins to form on top of the melt pool and starts to cover the hot molten glass below. In current melter operations, there is no way to measure the extent of the cold cap coverage, which makes it difficult to control the melter temperature and processing. A cold cap also causes the melter vapor space temperature to drop due to loss of radial and convective heat from the molten glass surface. This vapor space temperature drop in turn triggers constraints on minimum temperatures for complete combustion of melter off gas which in turn shuts the feed down until the vapor space temperature goes back above the constraint. When the feed is shut off the melter production rate drops accordingly. MMW diagnostics could be used to profile the temperature on the surface of the melt pool to monitor the status of a cold cap so control of its consistency would be possible by changing the melter blend chemicals, the melter electrode power distribution, and/or the waste feed rate.

An extension of the MMW temperature measurement capabilities would be to measure the melter vapor space temperature. Although this particular measurement technique has not been developed, it is known that MMW or submillimeter-wave radiometry could also provide this valuable measurement. The melter vapor space temperature is important to melter operation because it is the only indicator of poor combustion in the melter vapor space, which directly impacts the safety interlocks on the feed coming into the melter. If the real-time measurement capability could be extended to measuring melter vapor space temperature, a feedback control scheme could be devised to minimize impact on melter throughput. The current control scheme at DWPF uses a conservative predicted value of the melter vapor space temperature to set the feed safety interlocks.

2.4. Improving Throughput

While the vitrification processes at DWPF and those planned for WTP do function well, their ability to increase production and waste loading is limited by the bounds of the operating envelope developed by the feed forward control systems in place [19]. These feed forward control systems are based on first principle and statistical models [1-6, 11, 12, 15-17, 20-22] that are limited by errors in the property composition measurements and databases as well as the understanding of the melting process. MMW measurements would allow relaxation of some of the glass property uncertainties used to define the melter operating envelopes. In turn, it would be possible to expand the operating envelope for acceptable glass blends. The glass viscosity is a key processing constraint, since if the glass viscosity gets too high (much greater than 100 Poise) then it will be difficult to pour the glass melt into the canisters. If the melt viscosity drops below a value of about 20 Poise, then the glass pour will be too fast and concentration gradients could occur in the glass. Melt with low viscosity is known to accelerate refractory and electrode corrosion, with other conditions that remain the same. A low viscosity may also cause problems with foaming. With an on-line glass viscosity measurement, the viscosity dynamics could be monitored real-time and the glass blending chemicals, melter power distribution, and/or feed rates could be adjusted to satisfy a measured value instead of a predicted value.

Traditionally, the "capital expenditure" approach has been used to solve plant bottleneck problems. For example, building new production systems to solve plant capacity problems and adding new facilities to handle quality problems. The U. S. Department of Energy (DOE) of today is limited by budget and stakeholder constraints that limit the available resources to existing operating facilities. The waste problems at Hanford and other sites will tax these limited resources. New sensor technologies like MMW techniques can be developed and applied with existing process control techniques to enhance the operation and control of currently operating and any new planned DOE vitrification facilities. Accurate non-linear modeling techniques already exist in industry [23, 24]. The data to relate melter performance to glass product quality is needed to further develop the current process control models that limit the nuclear waste vitrification processes. Further expansion of these models requires developing a relationship between the melter operating parameters and glass chemistry. MMW diagnostics can bridge this gap.

3. MILLIMETER-WAVE MONITORING TECHNIQUES

3.1. MMW Sensor Configuration

The basic configuration of a millimeter-wave sensor for multiple parameter measurements of the melt pool in a melter is shown in Figure 3 [25]. The main building blocks are the millimeter-wave heterodyne receiver, a beamsplitter in the receiver field-of-view beam, a waveguide/optics transmission line to the melt pool, a window to seal the waveguide, and a thermal return reflection (TRR) mirror aligned with the split signal from the beamsplitter. Millimeter-wave signals are both received from, and transmitted back to the molten glass. With this configuration and appropriate minor modifications, it is possible to monitor all the key melt pool parameters listed in Table 1 using only a single access point into the melter. The physical melt pool effect that is being exploited in the millimeter-wave range to obtain the measured parameter is listed in the second column. The temperature and emissivity of the monitored surface are obtained by measuring the MMW thermal radiation from the surface and using this radiation by the TRR mirror as a probe of the surface reflectivity [26]. The density and viscosity are obtained by measuring the position and flow of the surface in response to a known pressurization of the waveguide. Surface position is measured with submillimeter precision by monitoring the phase of the surface refection of a leaked nonthermal oscillator signal from the receiver. Foaming, salt layer, and other melt pool properties are obtained by combinations of the thermal and nonthermal emission and reflection measurements.

Achieving the sensor configuration as shown in Figure 3 required developing novel millimeter wave components suitable for use in the high temperature and industrial environment of a nuclear waste glass melter. These components are the high temperature waveguide that interfaces with the hot melt pool and the quasi-optical beamsplitter TRR mirror configuration that is both rugged and efficient in the MMW wavelength range. These component developments along with the heterodyne receiver are summarized below after the analytical basis is established for the MMW measurements. Following this section the laboratory/field measurements that have been made to test the performance to this technology are summarized. It should be noted that Table 1 is not comprehensive. It shows only the parameter measurements that have been demonstrated so far by the MMW technology. Additional melt pool parameter measurement capability for liquidus temperature, redox potential, noble metals precipitation, plenum temperature, and others are potentially possible with future development.

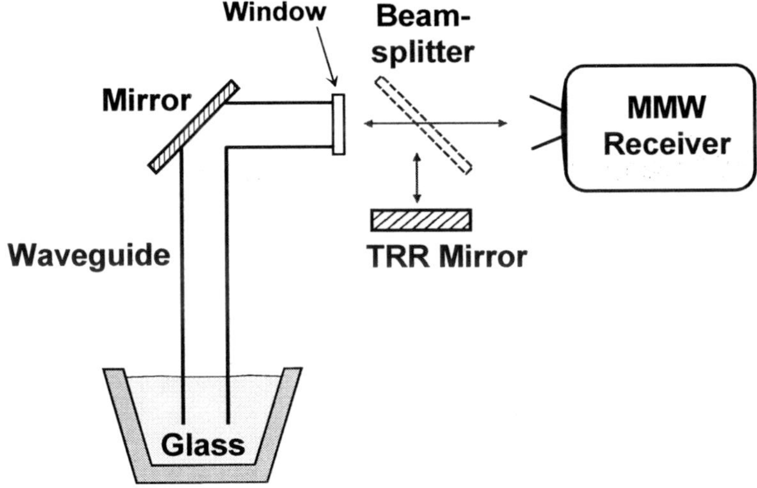

Figure 3. Basic millimeter-wave sensor configuration.

Table 1. MMW Measurements

Melt Parameter	Measured MMW Effect
Temperature (T)	Thermal emission
Emissivity (ε)	Reflection amplitude
Density (ρ)	Reflection phase
Viscosity (η)	Reflection phase rate
Foaming	Surface acceleration/emissivity
Salt Layer	Emissivity/turbulence

3.2. Analytic Basis for Temperature and Emissivity

Thermal emission signals are proportional to the product of the emissivity and temperature (εT) of the material that is viewed, where emissivity (ε) is a factor between 0 and 1 specifying how close the viewed material is to a perfect black body. All current pyrometers require some assumption on emissivity in order to determine temperature. Methods that can resolve the emissivity and temperature in real-time would be of great value to thermal analysis because they would not only improve temperature measurement accuracy, but would also provide information of the properties of the viewed material. The TRR method described here makes it possible to resolve of the temperature and emissivity parts of the thermal emission from a surface that is oriented normal to the viewing direction and is relatively smooth to the viewing wavelength (a distinct advantage for the longer MMW wavelengths over infrared pyrometry). The basic concept is to use the thermal radiation from the viewed source as a probe of its emissivity. Earlier work using a coherent oscillator signal to probe surface reflectivity required taking into account coherent standing wave interference [27]. The use of incoherent thermal radiation as the probe beam eliminates coherent interference and greatly simplifies and improves the reliability of the reflection measurement to determine emissivity.

The analytical basis of the TRR method to monitor temperature and emissivity requires taking into account all sources and losses of the thermal radiation between the source and receiver as viewed by the receiver. The main components of the MMW sensor are shown in Figure 4 along with the thermal parameters that enter into the analysis. The receiver views a sample (S) through a beamsplitter (BS) and waveguide system (WG). Each component has associated with it an emissivity (ε), temperature (T), transmission factor (τ), and/or reflectivity (r). If a component is at room temperature, then its emissivity and temperature will not contribute to the final result since the receiver is calibrated against room temperature. For example, the beamsplitter is at room temperature and its emissivity does not contribute. On the other hand, the waveguide, which is partially inserted into the melter and is on average higher than room temperature, must have its contribution to the thermal signal as seen by the receiver taken into account.

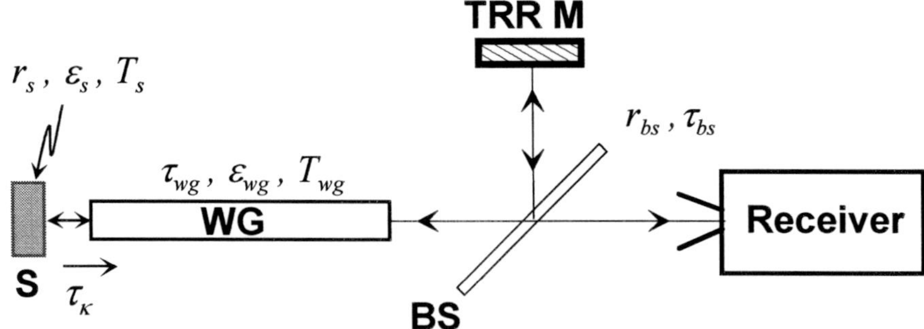

Figure 4. Illustration for defining the terms used in deriving the analytical basis for the TRR method.

Two temperature signal measurements are made with the MMW sensor system. One with the TRR mirror blocked and a second with the TRR mirror unblocked sending a portion of the thermal emission back as a probe of the viewed sample reflectivity. The sample is assumed to be completely opaque with no transmission through it. The analytical results for the measured temperature at the receiver for these two TRR cases are given by [28]:

$$T_{eff} = \varepsilon_{wg} T_{wg} + \tau_{wg} \varepsilon_s T_s + r_s \tau_\kappa \tau_{wg} \varepsilon_{wg} T_{wg} \tag{1}$$

$$T'_{eff} = \frac{T_{eff}}{1 - r_s \tau_\kappa r_{bs}^2 \tau_{wg}^2} \tag{2}$$

where T_{eff} and T'_{eff} are the effective MMW temperatures measured at the receiver without and with the TRR, respectively, T_{wg} and T_s are the waveguide and melt surface temperatures, ε_{wg} and τ_{wg} are the waveguide emissivity and transmission related by $\varepsilon_{wg} = \left(1 - \tau_{wg}\right)$, ε_s and r_s are the emissivity and reflectivity of the viewed surface related by $\varepsilon_s = \left(1 - r_s\right)$, r_{bs} is the beamsplitter reflectivity, and τ_κ is the viewed surface coupling factor. These two equations can be solved to obtain the emissivity and temperature of the viewed sample. The explicit expression for the sample emissivity is given by:

$$\varepsilon_s = 1 - \frac{1}{\tau_k r_{bs}^2 \tau_{wg}^2} \left(1 - \frac{T_{eff}}{T'_{eff}}\right) \tag{3}$$

where the ratio of the two effective MMW temperatures are obtained from the measurements, r_{bs} and τ_{wg} are previously determined instrumentation parameters, and τ_κ can be assumed to be equal to one if the waveguide is very close to the surface (less than the inner radius of the waveguide) and the surface is aligned to perfectly reflect a MMW signal back to the receiver. The result of Equation 3 can then be used with Equation 1 to obtain the sample temperature. If surface roughness is large relative to the wavelength or alignment is not perfect then τ_κ can not be assumed to be one and other means would need to be used to estimate this factor. Assuming it to be equal to one would in this case only give an upper limit to the actual emissivity.

3.3. Viscosity

Viscosity is related to melt glass flow, which is measured by the sensor configuration as shown in Figure 3 by immersing the waveguide into the molten glass and pressurizing the waveguide to displace the glass. In practice, the waveguide is pressurized first to displace the glass melt inside the waveguide and then suddenly the pressure is released by an electrically operated solenoid valve. The rate of flow of the glass melt back into the waveguide under the force of gravity to refill the waveguide is then measured and related to viscosity. This is analogous to the standard room temperature laboratory capillary flow technique for viscosity measurement, but implemented at high temperatures using the refractory waveguide as the "capillary" and the millimeter wave receiver instead of a stopwatch to determine rate of flow.

The actual molten glass surface displacement is determined by monitoring the reflection of the leaked local oscillator (LO) from the MMW receiver. The phase of the coherent LO reflection depends on the path length between the receiver and melt surface. For a path length change of one half wavelength, the phase will change by π or 180°. If the signal is initially in phase (maximum signal), then it would go completely out of phase (minimum or zero signal) for every half wavelength change in path length. Figure 5 illustrates how this signal will appear for an example displacement shown by the top curve and an LO frequency of 137 GHz (λ = 2.2 mm). For every quarter wavelength (¼λ) displacement of the melt level (round trip MMW path length change of ½λ), the video signal will cycle through a minimum–maximum period (fringe) corresponding to a round trip reflection phase shift of π. This is similar to a Michelson interferometer [29] in which the moving mirror against a fixed mirror generates fringe patterns. For a linear displacement the video signal will appear as a rectified cosine function as shown in the beginning of the lower plot in Figure 5. For a non-linear level change, the video signal fluctuations will be distorted from a cosine function as shown in the latter part of the plot in Figure 5. In the case of molten glass refilling the waveguide, the flow observed is typically a gradually slowing rate. Counting the number of fringes and multiplying by ¼λ will give the total distance the melt surface has moved and dividing by the time period of the movement will give the average velocity of the flow.

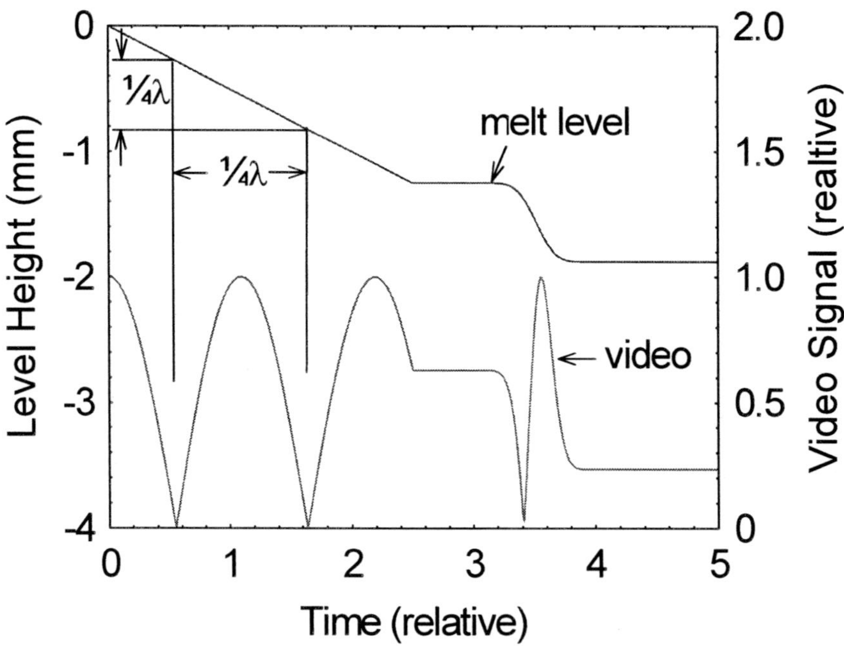

Figure 5. Calculated illustration of the relationship between melt level height change and a millimeter-wave video reflection signal at a frequency of 137 GHz (λ = 2.2 mm).

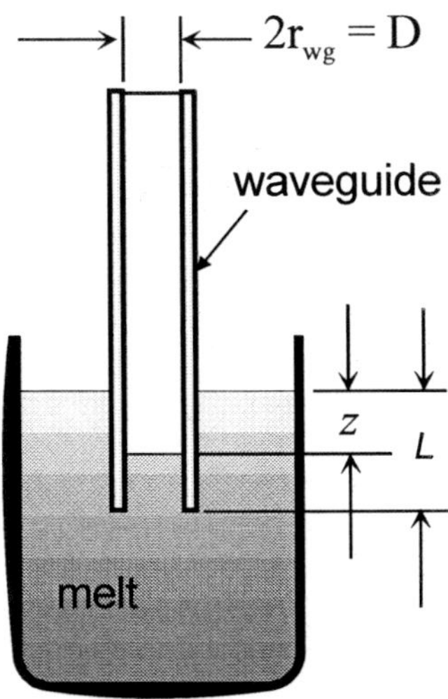

Figure 6. Definition of terms used for derivation of analytic basis of viscosity.

A simplified Navier-Stokes [30] analysis for laminar flow of a Newtonian fluid in a pipe can be used to derive a relationship between the observed measured average waveguide refilling velocity and the molten glass viscosity. With the definition of terms as shown in Figure 6 this equation is given as:

$$v_{ave} = A\rho \frac{z_o}{(L - z_o)\eta} \tag{4}$$

where A is an instrumentation constant, ρ is the molten glass density, z_o is the initial glass displacement, L is the waveguide immersion depth, and η is the glass dynamic viscosity. The instrumentation constant can be determined by making a measurement of a melt with a previously determined viscosity. The melt density can in principle be obtained from the measured fluid displacement for the known waveguide pressurization.

An even simpler form of Equation 4 is possible if the initial glass displacement and waveguide immersion are kept constant for all measurements. In this case the relation between flow rate and viscosity in the waveguide is analogous to the Ubbelohde or Cannon Fenske type capillary viscometers where the relationship between kinematic viscosity and the time for flow, Δt, between two fixed positions in the waveguide is given by [31]:

$$v = C\Delta t \tag{4a}$$

where C is a constant that incorporates the instrumentation constant A as well as the waveguide immersion dept and pressure displacement. The dynamic viscosity is related to the kinematic viscosity as $\eta = v\rho$. The millimeter-wave viscometer measurement is therefore directly analogous to the conventional capillary flow viscometer method where the capillary in this case is the MMW waveguide.

4. MILLIMETER-WAVE MONITORING TECHNOLOGIES

4.1. MMW Receiver

The heart of the MMW sensor configuration for melter monitoring is a MMW receiver system that has the sensitivity to detect the thermal emission from the melter and at the same time leak an oscillator signal for position measurements. It is based on a heterodyne receiver, the main elements of which are shown in Figure 7.

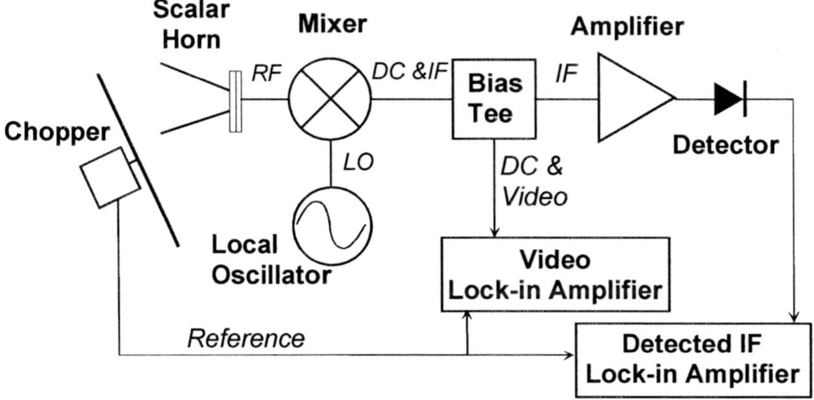

Figure 7. Millimeter-wave heterodyne receiver system for thermal emission and surface position measurements.

A MMW mixer is used with a local oscillator (LO) to down convert the MMW thermal signal to an intermediate frequency (IF) in the microwave frequency range for amplification and detection. A heterodyne receiver naturally leaks a part of its LO out the horn antenna, which is used for melt surface position measurements. A chopper modulates both the thermal signal and the leaked LO to make possible the detection of very weak signals with lock-in amplifiers that are only sensitive to the chopper frequency. A bias tee at the mixer output separates the IF thermal signal from the low chopper frequency position signal (video) to separate channels for amplification and detection.

The chopper also functions as a room temperature reference source for the thermal measurements in the manner as originally described by Dicke [32]. When the chopper bladed blocks the receiver field-of-view the receiver sees room temperature emission off the blade. When the receiver field-of-view is not blocked by the chopper blade the receiver sees the thermal emission from the melt surface. The IF lock-in amplifier detects this difference signal. A thermal difference signal can be produced by viewing either a hot or cold source relative to room temperature, making it convenient to calibrate the receiver with a cryogenic source such as one cooled by readily available boiling liquid nitrogen at a temperature of -196 C. The expression for the receiver measured effective temperature at the position in the view beam where the calibration source is inserted is given by:

$$T_{eff} = \frac{V_s}{V_c}\left(\left|T_c\right|-T_r\right)+T_r \tag{5}$$

where V_s is the detected IF signal when viewing the melt surface, V_c is the IF detected signal when viewing the calibration source at temperature T_c, and T_r is the room temperature. Consequently, the accuracy of the thermal emission measurement is determined by how well the calibration signal and room temperature are known.

Another important point is that the internal noise temperature of the receiver needs to be several times higher than the maximum melter temperature being monitored to insure a linear relationship between the measured signal and temperature. The typical microwave detector diode is not a linear detector. By keeping the internal noise signal on the detector diode large relative to the desired signal, the diode is kept at approximately a constant voltage on its response curve. For example the receiver noise temperature in the measurements presented below was typically in the range of 10,000 to 15,000 K for maximum melter measurements of up to 1,600 °C. A high receiver noise temperature does not mean that it will have poor measurement precision. Wide bandwidth and long signal integration times compensate for high noise temperature. The expression for minimum resolved temperature is given by [33]:

$$\Delta T_{min} = \frac{T_i}{\sqrt{tdf}} \tag{6}$$

where T_i is the receiver noise temperature, t is the signal integration time, and df is the receiver double sideband (DSB) bandwidth. For a receiver with a noise temperature of 15,000 K and a DBS bandwidth of 3 GHz, typical for some of the measurements presented below, the temperature resolution is 0.3 °C for one second integration times. Such temperature precision and time resolution is more than adequate for most melter measurements.

4.2. High Temperature Waveguides

Interfacing the MMW receiver system with the melter requires an efficient millimeter-wave waveguide that can survive temperatures over 1000°C inside the melter and immersion into the melt pool itself for viscosity measurements. Luckily, efficient hollow millimeter-wave waveguides can be fabricated from almost any refractory material such as those used to fabricate melter crucibles. This is accomplished by making the waveguide diameter much larger that the wavelength and using the HE_{11} waveguide mode (transverse magnetic and electric fields) for propagation [34]. The HE_{11} mode is a natural mode inside smooth walled dielectric tubes (the same mode is used inside single mode fiber optic cables at light wavelengths) and can be propagated inside electrically conducting metallic tubes by circumferentially corrugating the inside wall with ¼ λ deep grooves at more than 2.5 grooves per λ of waveguide length [34]. The HE_{11} mode is also ideally suited for monitoring applications because it launches as a free space Gaussian beam, which is optimum for achieving the smallest possible diffraction limited spot sizes for good spatial resolution.

The power transmission factor for a waveguide is given by the exponential relation:

$$\tau_{wg} = e^{-2\alpha z} \tag{7}$$

where α is the waveguide electromagnetic field attenuation coefficient and z is the waveguide length. The electromagnetic field attenuation coefficient has been derived for HE_{11} mode propagation for both the circular smooth walled dielectric and corrugated walled metallic waveguide types. For a hollow circular dielectric waveguide with thick waveguide walls that are opaque to millimeter-wave leakage through the side it is given as [35]:

$$\alpha = 0.0733 \frac{\lambda^2}{a^3} \frac{\left(n^2 + 1\right)}{\sqrt{\left(n^2 - 1\right)}}$$

(8)

where a is the waveguide internal diameter and n is the index of refraction of the dielectric waveguide walls. For circular corrugated metallic wall waveguide with large inside diameter $> 3\lambda$ it is given as [34]:

$$\alpha = 1.94 \times 10^{-4} \frac{\lambda^2}{a^3} R_s$$

(9)

where R_s is the metal wall surface resistivity in ohms. The surface resistivity is given by $R_s = \sqrt{\pi f \mu_o / \sigma}$ where f is the frequency, μ_o is the permeability constant (1.26×10^{-6} henry/m), and σ is the metal conductivity.

It can be seen from Equations 8 and 9 that the transmission losses through the waveguide are proportional to λ^2/a^3. Consequently, they can be made as small as necessary by making a sufficiently large relative to the wavelength. Also, the smooth walled dielectric waveguide needs to be about 10 times larger in diameter than a corrugated metallic waveguide to achieve the same losses. A poor metallic conductor (by microwave standards) with a high surface resistivity can also be used to make an efficient waveguide as was shown with a circular corrugated graphite waveguide used in a non oxidizing furnace [36].

A MMW waveguide configuration developed for use in the oxidizing environment of nuclear waste melters and used for many of the measurements presented below is shown in Figure 8. It consists of a room temperature corrugated brass waveguide section connected by a flat 90° turning miter mirror to a ceramic waveguide section to access the high temperature environment of a melter from the top. Both sections have an internal diameter of 28.6 mm (1 1/8 inch). The MMW receiver frequency used in the tests was 137 GHz ($\lambda=2.19$ mm) and the corrugation dimensions shown for the brass section are optimum for this frequency. According to Equations 7 and 9 the transmission efficiency of the brass waveguide ($R_s = 0.2$ ohm at 137 GHz) is about 10^{-4} dB/m or about a 1% loss for 650 m length. Therefore, this waveguide section could be made as long as necessary to transmit the millimeter-wave signal outside the biological shield of a nuclear waste glass melter facility.

The ceramic waveguide waveguide section used for most measurements was a mullite ($3Al_2O_3 \cdot 2SiO_2$) tube. Mullite was chosen because it was relatively robust to glass melt immersion and has somewhat higher losses for MMW transmission through the sidewalls than the pure alumina or silica materials. Even so, the commercially available wall thickness of 3.2 mm (1/8 inch) was not sufficient to satisfy the condition of Equation 8 for completely MMW opaque side walls.

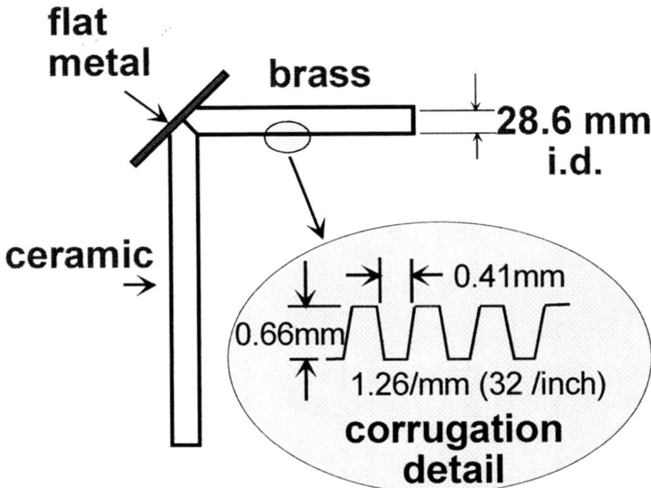

Figure 8. MMW waveguide used for many measurements was a composite of room temperature corrugated brass and high temperature ceramic sections. The corrugation dimensions are shown for a frequency of 137 GHz.

The predicted transmission factor of $\tau_{wg}= 0.7$ for a 50 cm long 28.6 mm diameter mullite tube at 137 GHz ($n \approx 2.5$), in practice, varied over the range 0.6 to 0.8 from tube to tube due to poor tolerance on wall thickness. The higher transmission loss of the ceramic waveguide section necessitated keeping this section as short as possible, typically ≤ 50 cm, only long enough to access the high temperature environment. For some of the laboratory studies a larger diameter brass-mullite waveguide assembly of 41 mm (1 5/8 inch) diameter was used which had a transmission factor of about 0.9 for the 50 cm long mullite section.

Tests were carried out at temperatures of up to 1500°C and with waveguide glass immersions for up to one week in nuclear waste glass melts. Good MMW performance results were achieved. Some penciling (corrosion) of the immersed mullite waveguide was observed, but did not affect the millimeter-wave performance significantly in the tests to date. Advanced refractories could be used to make the waveguide last as long as the melter crucible. This waveguide technology also has important spin off applications to monitoring in nuclear fusion experiments, nuclear power plants, fossil fuel power plants, and manufacturing where diagnostic access to high temperature environments is extremely challenging [37, 38].

4. 3. Compact Quasi-Optical Beamsplitter/TRR Mirror

In addition to the waveguide a MMW signal control component is required to implement the temperature and emissivity monitoring function. At high MMW frequencies optics are the most efficient way of achieving beam splitting and redirection of signal, but MMW optics are large and very sensitive to alignment. In an industrial scale nuclear waste melter environment, it is not desirable to have sensitive optics that take up a lot of space. Therefore, a quasi-optical corrugated waveguide and beamsplitter device have been developed that achieves the TRR function shown in Figure 3 in a compact, rugged component. A cross-section of this device is shown in Figure 9. Crossed corrugated waveguides are machined in an aluminum block with a diagonal spilt as shown to accept a beam splitter. In the experiments carried out so far at 137 GHz, the waveguide block was only 6.4 cm (2.5 inches) square by 3.8 cm (1.5 inches) deep with a quartz beamsplitter.

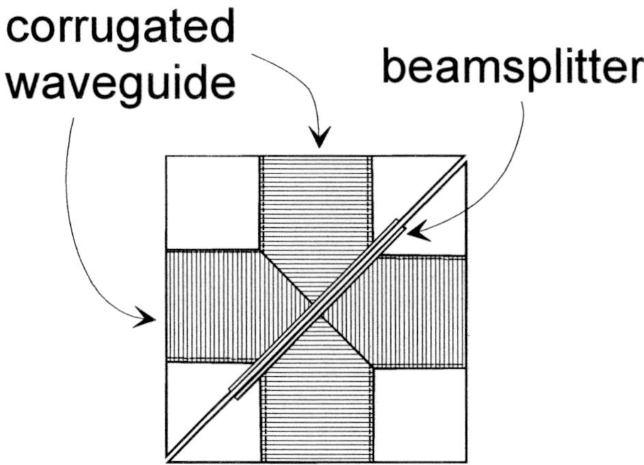

Figure 9. Quasi-optical beamsplitter. corrugated waveguide block.

The MMW signal from the melt and receiver field-of-view transverses the block in one direction and the split signal is directed in an orthogonal direction to a TRR mirror (not shown). Insertion losses were measured to be only 0.5%. This component has been very successful in the Environmental Management Science Program (EMSP) sponsored work and has been copied by an National Institute of Health (NIH) sponsored experiment to monitor a 250 GHz gyrotron beam for dynamic nuclear polarization (DNP) measurements [38]. At frequencies above 100 GHz, this is the most efficient approach to compactly monitor MMW beams.

5. LABORATORY MEASUREMENTS

5.1. Temperature and Emissivity Measurements of Advanced Refractories

The performance of the MMW instrumentation for monitoring temperature and emissivity is illustrated by a laboratory study at 137 GHz of two advanced refractory materials, alumina-zirconia-silica (AZS) and Monofax K3. Horizontally level and flat samples of these refractories were put inside an electric furnace (Deltech model DT-31-RS-12) less than 7 mm from the aperture of the mullite waveguide inserted vertically through the top of the furnace. Measurements were carried out to temperatures of up to 1500°C.

The data for Monofax K3 is shown in Figure 10 where the top trace is the furnace temperature measured by a type S thermocouple and the bottom trace is the MMW thermal emission signal as calibrated just outside the Teflon waveguide window. The effective MMW temperature at the waveguide window is lower than the thermocouple temperature because the emissivity and waveguide transmission losses are not taken into account in this plot. The 3 mm (1/8 inch) thick Teflon window used for these measurements had moth eye antireflection surfaces [39] and was measured to have less than 2% losses.

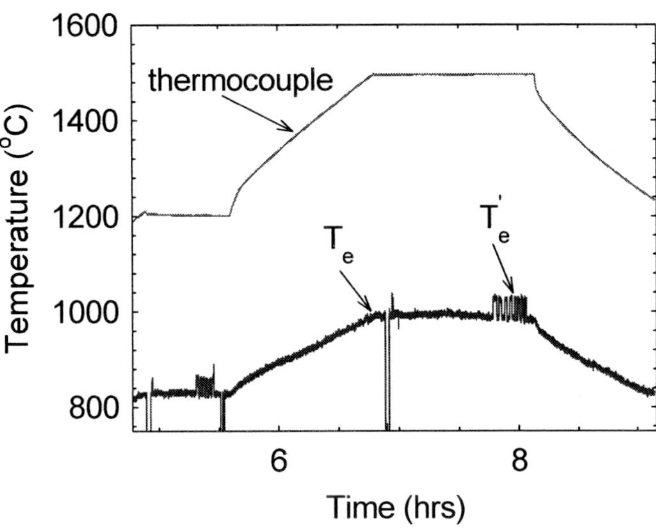

Figure 10. MMW measurements of Monofax K3 refractory.

Over a period of several hours, the Monofax K3 was first heated to 1200°C and then to 1500°C for TRR measurements as indicated by the raised structures on the plot during flat operation at these two temperatures. An expanded view of the TRR measurement at a furnace temperature of 1500°C is shown in Figure 11. The upper temperature is measured with the TRR mirror unblocked; sending a part of the thermal emission back to the refractory for an additional reflection from the viewed surface, and the lower temperature is with the TRR mirror blocked for a completely passive thermal signal measurement. Using the ratio of these two temperatures in Equation 3 the emissivity was found to be about 0.70 at both temperatures showing that Monofax K3 is a relatively stable MMW material at high temperatures.

Similar measurements of AZS, shown in Figure 12, revealed it to be unstable at the highest temperatures studied. Again, the top trace shows the furnace thermocouple temperature and the bottom trace the millimeter-wave temperature as calibrated at the waveguide window.

The AZS was heated to five temperature steps between 1000 and 1500°C. At each step a TRR measurement was made as shown on the millimeter-wave signal by the upward structures. At the temperature transition from 1400 to 1500°C the millimeter-wave signal makes a sudden increase greater than that expected by the thermocouple reading, indicating that the emissivity of the AZS increased anomalously. The TRR data shows that the emissivity gradually decreases from about 0.70 to 0.60 over the temperature range of 1000 to 1400°C and then makes an abrupt increase to about 0.76 in the flat top region at 1500°C, a 25% jump in emissivity.

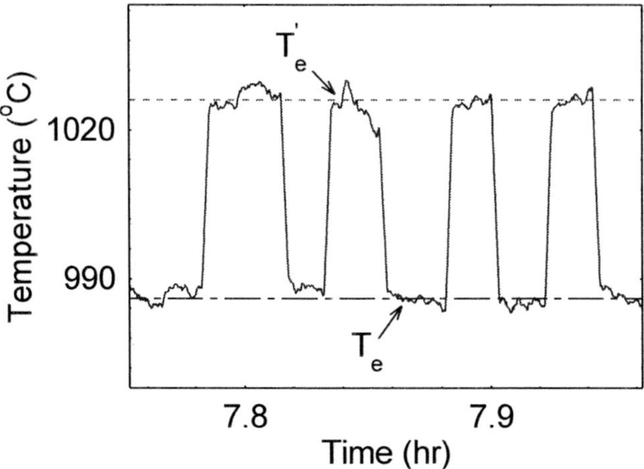

Figure 11. Expanded view of TRR measurement at 1500° C shown in Figure 10.

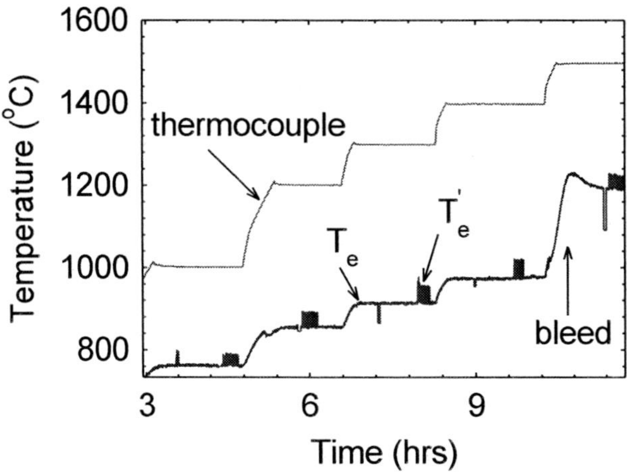

Fgure 12. MMW measurements of alumina-zirconia-silica (AZS) slowing glass bleeding above 1400°C.

Examination of the AZS sample after the furnace cooled down showed that a glass phase from inside the sample had eluded out, changing the surface chemistry and figure. Consequently, the AZS would be unacceptable to use as a MMW waveguide component at high temperatures. Measurements of other refractory materials [31] in addition to Monofax K3 did not reveal unstable MMW behavior. These measurements illustrate the usefulness of millimeter-wave TRR for high temperature thermal analysis of materials [40].

5.2. Viscosity

Millimeter-wave laboratory glass flow data as a function of temperature was taken for a number of glass compositions that included two Hanford glass mixes, DWPF black frit #265, and a commercial E glass. The connection of molten glass flow time in the waveguide with viscosity was empirically established. It was also found that surface tension effects appear to dominate at the end of the flow time when the surface is relaxing to equilibrium and when the displacements are small in the waveguide. The fact that surface tension plays a role in the waveguide melt flow was also seen in the overall reflected signal amplitude variation with changes in pressurization, which effected surface curvature and refection coupling (τ_k) to the waveguide HE_{11} mode. Consequently, the difference in flow time between two pressure displacements was used to subtract out the common starting and ending surface curvature motions to achieve a reliable viscosity measurement minus the surface tension effects. The resulting time difference with this subtraction corresponds only to the laminar flow time in the waveguide.

Representative reflection signals for two pressurizations of Hanford #8 glass (target composition 56.8% SiO_2, 20% Na_2O, 12% Al_2O_3, 9% B_2O_3, and several other trace oxides) at a temperature of 1000°C in 28.6 mm diameter mullite waveguide are shown in Figure 13. The waveguide was immersed about 25 mm into the melt in a 76 mm diameter crucible with at least 50 mm of depth of glass in a Deltech model DT-31-RS-12 furnace. In the top Figure 13a, the waveguide was pressurized with nitrogen to 0.48 inch (12 mm) water pressure above atmosphere and in the lower Figure 13b, the pressurization was to 1.2 inch (30 mm) water. After pressurization and allowing 2 to 3 minutes for the glass melt to reach equilibrium at the higher pressure, a computer controlled valve was opened at time zero as shown on the plots to release the waveguide pressure back to atmosphere. The flow of the melt back into the waveguide pipe is recorded by the fringes in the plots, one fringe corresponding to a displacement of approximately 0.55 mm. Alternate fringes appear strong and weak because of a signal offset in the lock-in amplifier.

The envelope of the fringe pattern varies from a low to high and back to a low level as the surface curvature flexes in response to the sudden flow, changing the focusing of the reflected signal back into the waveguide. Counting the fringes, there are seven for the 12 mm pressurization and eighteen for the 30 mm pressurization corresponding to melt displacements of 3.9 and 9.9 mm, respectively. The ratio of these measured displacements to the pressurization results in a factor of about 3.0, which is not an accurate determination of the specific gravity for this melt. Room temperature measurement of this glass showed that the specific gravity is about 2.5. Even if the finite dimensions of the crucible are taken into account, it is evident that there is additional resistance to the flow in the waveguide. However, this additional resistance factor appears to be a constant over a wide range of melt viscosity.

The measured waveguide flow time for the Hanford #8 glass between 0.5 inch and 1.2 inch pressure displacement is plotted in Figure 14 as open squares against viscosity as determined by Corning Services Laboratory for this glass. Varying the temperature of the glass from 850 to 1200 °C varied the viscosity over the range of 2000 to 20 Poise. The data is plotted on a log-log plot because of the wide dynamic range of the measurements. The line through the experimental points is a linear fit to the experimental points corresponding to Equation 4a with the constant density product given by $C\rho = 58$ Poise/sec. The measured flow time tracks the viscosity perfectly over a 100:1 dynamic range of 20 - 2000 Poise. Accurate viscosity measurement over such a large dynamic range is a unique capability of MMW viscometer in addition to the high temperature operation.

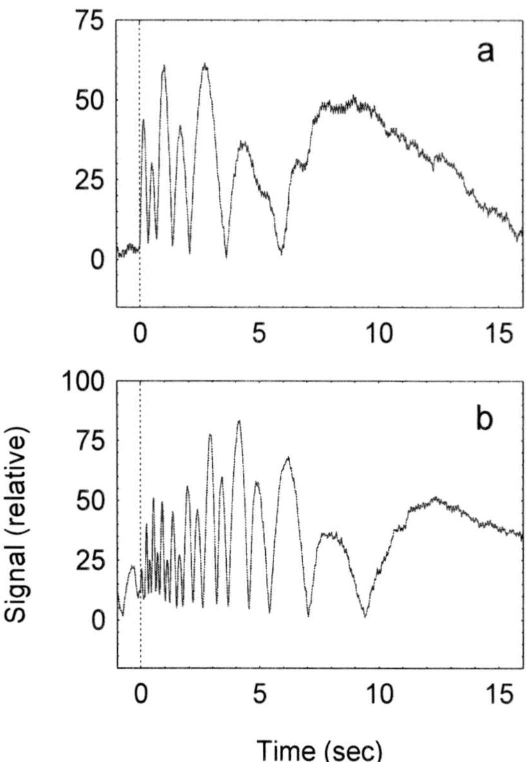

Figure 13. Hanford #8 glass flow signal in 28.6 mm diameter waveguide at 1000°C after pressurization to a) 0.48 inch (12 mm) and b) 1.2 inch (30 mm). Each fringe corresponds to 0.55 mm displacement.

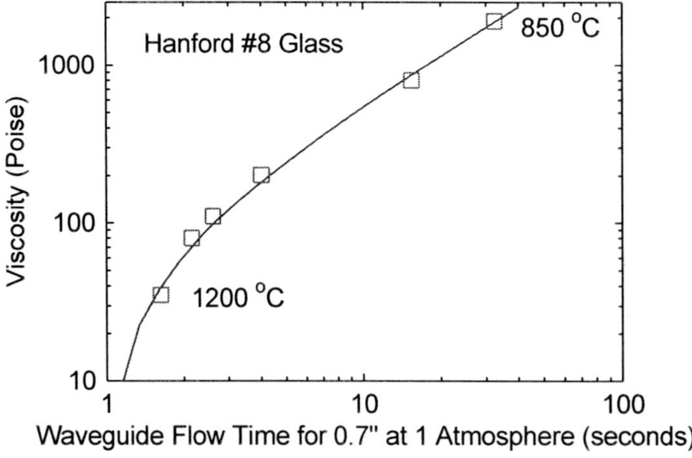

Figure 14. Glass viscosity versus the flow time in a millimeter-wave waveguide.

These laboratory viscosity measurements not only established the viability of MMW viscosity measurements, but also revealed interesting surface dynamics that would be of interest for future study [40]. One of these interesting observations is submillimeter surface turbulence that increases inside the trapped volume of the sealed waveguide versus an unsealed waveguide.

5.3. Foaming

A combination of the MMW monitoring capability for temperature, emissivity, and surface motion can be used to detect a glass melter foaming event. Vitrification operations at DWPF have encountered sudden glass foaming events that seriously disrupt the operations. On-line monitoring that can alert operators to the onset of foaming would be of great value. MMW laboratory measurements with hi-foaming glass have shown that MMW measurements can not only monitor a foaming event, but also can provide a precursor indication long before the foaming gases break the melt pool surface [41].

The data for a 3-inch (76 mm) diameter crucible melt which was formulated to foam (target composition 45.3% SiO_2, 15.3% Na_2O, 8% Al_2O_3, 7% B_2O_3, 0.3% Cr_2O_3, 11.5% Fe_2O_3, 4.5% Li_2O, 6% ZrO_2, and several other trace oxides) is shown in Figure 15. The top trace shows the furnace thermocouple signal, the next trace below shows the MMW thermal signal (εT), and the bottom trace is the surface position video signal. The events of the foaming unfold as follows: 1) at about 65 minutes on the x-axis the surface begins to accelerate as evident by the rapid fringe shifts on the video signal, 2) shortly after the onset of surface acceleration, the MMW thermal signal begins to approach the thermocouple temperature signal (emissivity increasing), it peaks at about 80 minutes on the x-axis, 3) after peaking, the MMW thermal emission decreases, gradually at first and then abruptly at about 100 minutes on the x-axis. When the MMW thermal signal increases, the video reflection signal from the melt surface decreases. At the peak MMW emission, the video signal is approximately zero as would be expected for a surface that has gone to an emissivity of one ($\varepsilon = 1$). At this point, the MMW radiation temperature equals the thermocouple temperature, if the waveguide losses are taken into account. The MMW temperature plot in Figure 15 has been normalized to the thermocouple temperature at this point.

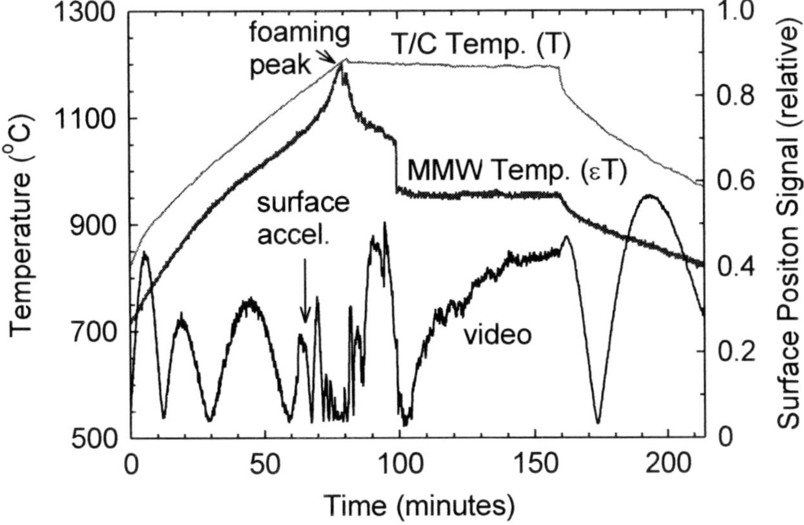

Figure 15. MMW signal record of hi-foaming glass as it is melted down.

A possible interpretation of these measurements is as follows: 1) dissolved gases come out of the melt pool solution and begin pushing on the surface displacing it in the MMW view, 2) it takes about 15 minutes for the gases to reach the melt surface in the 3-inch crucible before they begin to effect the surface emissivity, 3) when the bubbles break the surface, they initially cause a roughening of the surface on MMW scale lengths to prevent reflections and then gradually coalesce to reform a uniform surface, eventually forming a large bubble that pops. If this interpretation is accurate, it will be of great usefulness for nuclear waste glass monitoring. On the much larger scale dimensions of nuclear waste glass melters, it is likely that it would take more time to build up gases below the melt surface before significant foaming occurs on the surface. Monitoring for the onset the MMW surface acceleration and change in surfaces emissivity could therefore give nuclear glass melter operators a longer time than these laboratory measurements before the advent of significant foaming gas releases. Corrective action could then be taken to mitigate the onset of foaming. However, the results in Figure 15 are only preliminary and more research is need to establish this monitoring technique for foaming and to exploit this technique for a better understanding of the physics and chemistry of a melter foaming event.

6. MELTER TESTS

Tests of the MMW monitoring technology have been carried out on engineering pilot scale glass melters. Melter tests achieve two goals: 1) they demonstrate to the melter community that millimeter-wave monitoring technology is not a laboratory curiosity, but fully capable of functioning reliably in a 24 hour, 7 day a week manufacturing environment and 2) they provide access for research with full scale melt pools having continuous top feeding with a cold cap, deep melts with temperature gradients, and glass pour events that can not be easily simulated in the laboratory. Field tests were carried out on the EnVitco (EV-16) joule heated melter at the Clemson Environmental Engineering Technology Laboratory (CETL) and at the Slurry Fed Melt Rate Furnace (SMRF) at Akien County Technology Lab (ACTL) at the Savannah River National Laboratory (SRNL). The EV-16 melter has a melt pool size of 18"x 18"x 14" (46 cm x 46 cm x 36 cm) and is powered typically at 70 kW in two zones by four molybdenum electrodes, one centered in each wall below the melt surface. The SMRF is a smaller melter with an Inconel 690 crucible having dimensions of 8" diameter x 15" deep (20 cm diameter x 38 cm deep) externally heated at 3 kW through the bottom to simulate a joule heated melter. Both these melters are set up for continuous feeding with centrally located bottom pour spouts.

6.1. Cold Cap Measurements

The goal of the first test on EV-16 was to demonstrate two dimensional MMW temperature measurements of the melt surface with and without a cold cap. The cold cap is a solid crust in top fed joule heated melters, playing a key roll in feeding, stabilizing the vitrification process, and maintaining glass production rates. On-line monitoring of the cold cap would result in better process control, increased production, and reduced costs for waste vitrification. For these measurements, the MMW waveguide was modified as shown in Figure 16. An inconel miter mirror cap was attached to the end of a mullite waveguide, which was inserted horizontally into the melter plenum area above the melt pool. The waveguide could be rotated and retracted in and out manually to scan the millimeter-wave receiver field-of-view across the surface of the melt pool.

The MMW receiver could clearly identify hot and cold regions of the cold cap or melt surfaces and their dynamics during processing. Figure 17 shows a photo of the melter cold cap with the movable MMW waveguide entering from the right and an overlay of a surface temperature scan that was taken with the waveguide. The front center of the cold cap above one of the electrodes is identified as being hotter than the back of the cold cap where there are no electrodes. The temperature varies form a high of about 930°C in the front hot spot to a low of about 680°C in the back cold spot.

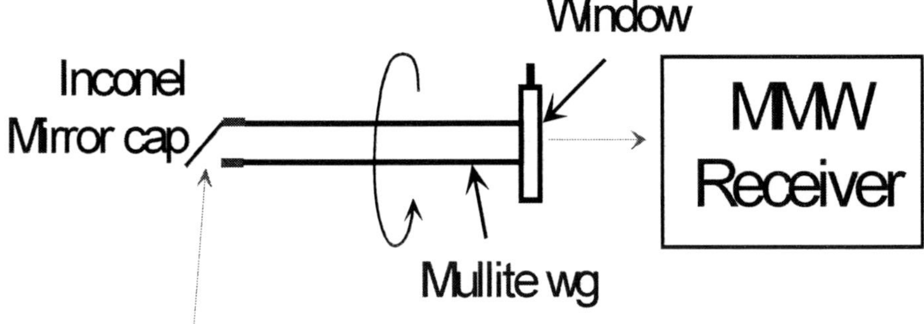

Figure 16. Rotating waveguide that was horizontally inserted into the plenum area above the melt pool in the EV-16 melter.

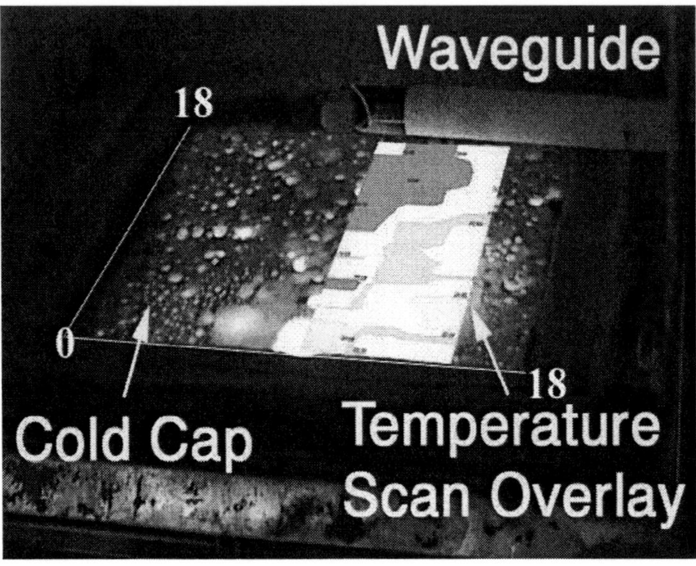

Figure 17. EV-16 melter cold cap and MMW scanned surface temperature. Hotter regions are lighter and yellow in color, and colder regions are darker and green and blue.

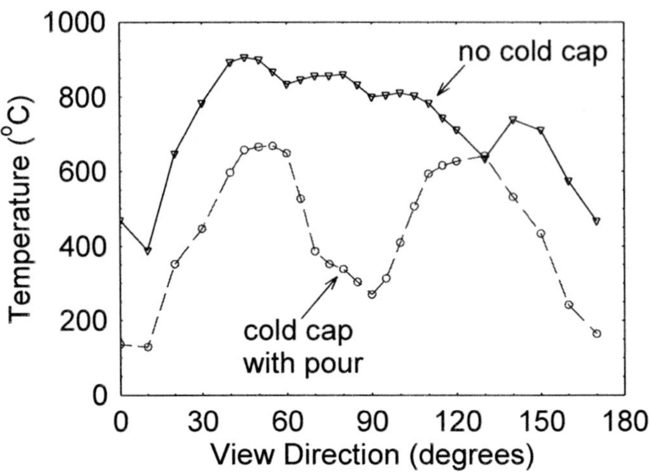

Figure 18. Melt pool surface temperature profiles.

In Figure 18 two temperature surface profiles crossing near the center of the melter are shown. The scans are from front to back across the melter surface. The top trace is with no cold cap and shows a surface temperature gradient that is indicative of how the power was distributed among the electrodes below. The lower trace shows the cold cap temperature profile during glass pour from a spout below the center of the melter. There is a clear drop in surface temperature by more than a factor of two above the location of the pour spout. These measurements demonstrated, for the first time, that on-line monitoring of melter melt pool surface dynamics are possible and that the millimeter-wave instrumentation works reliably in a glass manufacturing scale melter [42].

6.2. Viscosity

In the second test at EV-16, the goals were to demonstrate the feasibility of monitoring viscosity and salt layer formation in an engineering scale melter. For these tests, the MMW waveguide was configured as originally shown in Figures 3 and 8 so that the mullite waveguide could be vertically lowered into the melt pool. Glass flow was measured over a 1″ to 8″ (25 to 203 mm) waveguide immersion range for a series of pressurizations from 0.5″ water (0.9 mm Hg) to a maximum of 12″ water (22.4 mm Hg) pressure at the deepest immersion. The average waveguide refilling glass flow velocity measured for two of the initial waveguide pressurizations of 2″ (2.8 mm Hg) and 4″ (5.6 mm Hg) are plotted as inverted triangles and squares, respectively in Figure 19. Also plotted is a temperature profile measured with a thermocouple. Equation 4 is plotted as the two dashed curves assuming viscosity is inversely proportional to temperature ($\eta \propto 1/T$). Both calculated plots were fit to the same point at 2″ (2.8 mm Hg) pressurization at a depth of 2″ (5 cm) to obtain the instrumentation constant. The general trends of the measured flow velocities with depth and pressurization are reproduced reasonably well by the assumed model. The discrepancies can be explained by effects not modeled such as temperature gradients, drifts in temperature during the course of the measurements, and deviations from the laminar flow assumption, particularly at the shallow depths when immersion depth and displacement are similar. These results demonstrated, for the first time in a glass melter, a capability for not only on-line viscosity monitoring, but also for obtaining viscosity profiles within the melt [43].

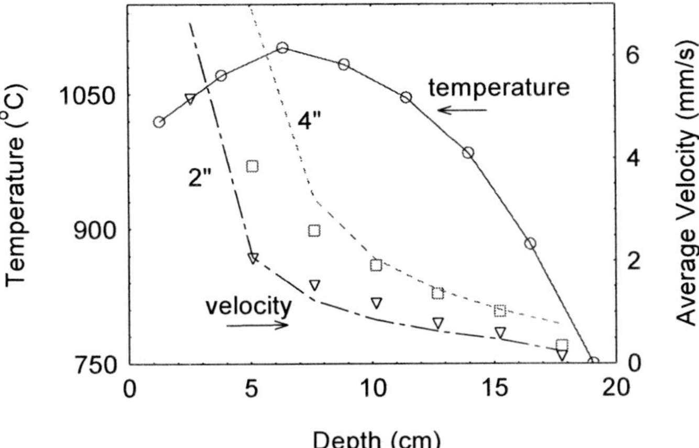

Figure 19. Temperature and melt flow velocity measurements and comparison with simple viscosity theory (dashed curves).

6.3. Salt Layer Formation

Formation of salt pools in nuclear waste glass joule-heated melters is of considerable concern. Supersaturated salt coming out of the glass solution can compromise the integrity of the glass as a permanent nuclear waste storage medium (waste form), increase corrosion in the melter (due to corrosive nature of the molten salt), introduce power instabilities in joule-heated melters, and pose a serious hazard to the melter facility if a continuous layer is formed to short out the electrodes. At the end of the second melter test a total of 4.2 lbs. (1.9 kG) of sodium sulfate ($NaSO_4$) was added to simulate a salt layer in a glass melter [43, 44]. The MMW waveguide was not immersed, but positioned above the melt surface to monitor thermal emission. The dynamics of salt layer formation in a glass melter were observed for the first time. The salt formed small drops first that grew in size until a continuous surface layer was formed. MMW surface fluctuations were coincident with the growth of the molten salt pools on the glass melt.

A quantitative measure of the emissivity of the molten glass and salt at 137 GHz was also obtained with the aid of the TRR method. Figure 20 shows the TRR data that was recorded. Not only is the MMW temperature of the glass melt higher than the molten salt, but also the ratio of the temperature without to that with the TRR is smaller for the glass melt as expected for a material having a higher emissivity. Using equations 1 - 3 and the thermocouple temperature, the emissivity and surface coupling factor were determined to be: $\varepsilon = 0.64 \pm 0.05$, $\tau_K = 0.46 \pm 0.05$ for the DWPF black frit glass and $\varepsilon = 0.44 \pm 0.05$, $\tau_K = 0.60 \pm 0.05$ for the salt. The MMW emissivity of the molten salt was 32% lower than that for the glass. The difference in coupling factor suggested that the molten salt surface was a smoother, flatter surface (due to its low viscosity hence smooth covering for better return coupling) than the glass melt at these wavelengths.

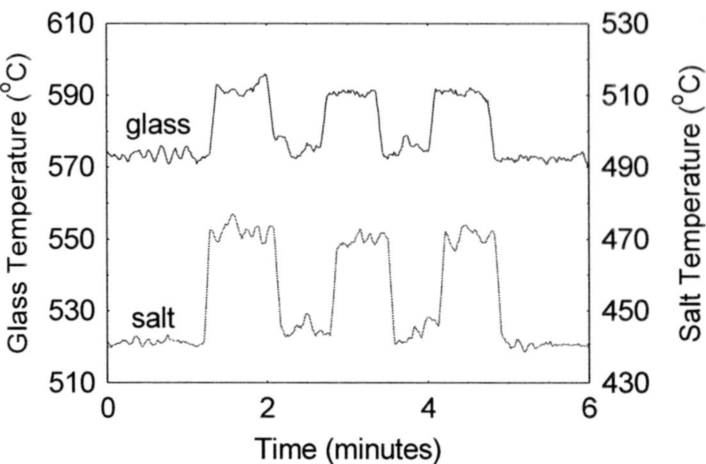

Figure 20. MMW thermal return reflection measurements through an 80% transmitting warm waveguide (~ 360°C) of molten glass and salt at a thermocouple temperature of 950°C.

6.4. Melter Pour Velocity

The capability to measure glass flow velocity in a melter during a pour was demonstrated for the first-time at ACTL [45], as shown in Figure 21 for a SMRF pour of DWPF black glass melt through the bottom spout. The interfreogram shows the receiver video signal, the reflected leaked LO cycling through fringes, as the melt surface drops.

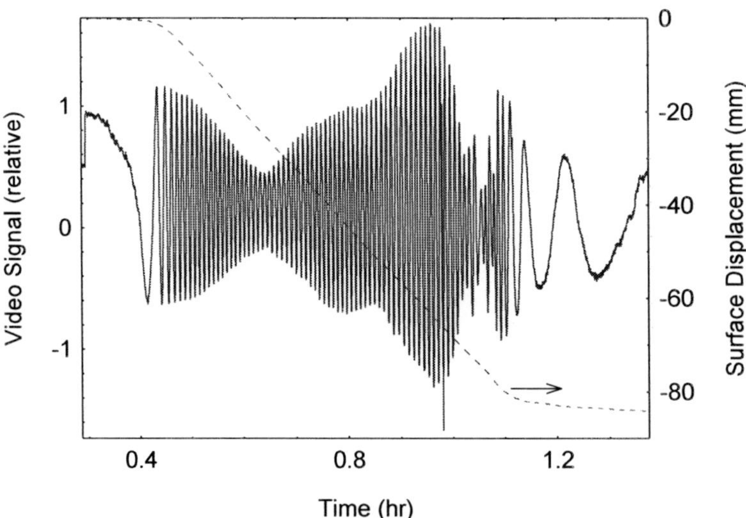

Figure 21. Glass pour interferogram and corresponding surface displacement.

The fringes in this case are not rectified as shown in Figures 5 and 13 so counting the positive peaks is equivalent to a surface displacement of $\frac{1}{2}\lambda$ per cycle. Each cycle therefore corresponds to a surface displacement of 1.1 mm for the 137 GHz monitoring frequency. The overall modulation envelope of the fringes is related to the surface receding and changing curvature, which modulates the coupling factor for reflection of the leaked LO back into the waveguide. The dashed curve plots the corresponding surface position showing the surface displacement to be a linear function of time over most of the pour time, corresponding to a constant flow velocity of 0.035 m/s in the pour spout. This is an interesting result since the pressure on the spout from above is a decreasing function of time as the weight of the melt is reduced by the pour, suggesting that the melt may be behaving like a non Newtonian fluid similar to catchup where the viscosity deceases as the pour progresses. Only at the very end of the pour when the melter is almost empty does the flow velocity finally decrease. The viscosity in the pour spout can in principle be determined from this velocity measurement once the instrumentation constant for the pour spout is established. This is similar to viscosity measurement as described above by inducing a pressurized flow in the waveguide, but not exactly the same because the pressure on the spout during the pour is not a constant function of time. However, this data does show that it may possible to check the melt viscosity in the vicinity of the pour spot without immersing the waveguide, but by simply monitoring surface recession rate during a pour, keeping all other parameters constant. Also interesting new melt pour dynamics are being revealed by this new way of monitoring which may lead to improvements in the way glass is poured.

CONCLUSION

In summary, the main principles of MMW measurements that form the basis of MMW technology for monitoring the vitrification of nuclear wastes have been established. MMWs are ideally suited for monitoring in nuclear waste glass melters because they are tolerant of the imperfections in this environment that would obscure and scatter infrared and optical techniques. Consequently, infrared and optical monitoring techniques have not been reliable in nuclear waste glass melters in the past, but MMW techniques as described here will be. The analytical basis, designs, and components of the MMW technology have been outlined for monitoring temperature, emissivity, viscosity, foaming, salt layer formation, and other melter parameters. All of these melter parameters can be monitored simultaneously with a single instrument with only one penetration into the melter of less than 10 cm (2") diameter. Laboratory scale tests and melter measurements have demonstrated and verified the potential of this technology for comprehensive measurements in nuclear waste vitrification facilities. Overall, the MMW technology represents a significant new development for nuclear waste vitrification monitoring that promises to improve the control, efficiency, and reliability for making a high quality nuclear waste glass storage product. It should finally be possible to advance the nuclear waste glass vitrification process to modern industrial standards for feedback control.

REFERENCES

[1] Jantzen, C.M., "Prediction of Glass Durability as a Function of Environmental Conditions", *Proceedings of the Symposium on Materials Stability and Environmental Degradation*, A. Barkatt et. al. (Eds.), Materials Research Society, Pittsburgh, PA, 1988, 143-159.

[2]   Jantzen, C.M., and Ramsey, W.G., "Prediction of Radioactive Waste Glass Durability by the Hydration Thermodynamic Model: Application to Saturated Repository Environments," *Proceedings of the Symposium on the Scientific Basis for Nuclear Waste Management*, V.M. Oversby and P. Brown (Eds.), Materials Research Society, Pittsburgh, PA, 1990, 217-228.

[3]   Bickford, D.F., Applewhite-Ramsey, A., Jantzen, C.M., and Brown, K.G., "Control of Radioactive Waste Glass Melters: I, Preliminary General Limits at Savannah River," *J. Am. Ceram. Soc*, 1990, *73* [10] 2896-2902.

[4]   Andrews, M.K.; Bibler, N.E.; Jantzen, C.M., and Beam, D.C., "Demonstration of DWPF Process and Product Control Strategy by Vitrification of Actual Radioactive Waste," *Proceedings of the 5th International Symposium on Ceramics in Nuclear Waste Management*, G.G. Wicks, D.F. Bickford, and R. Bunnell (Eds.), American Ceramic Society, Westerville, OH, 1991, 569-576.

[5]   Hutson, N.D.; Jantzen, C.M., and Beam, D.C., "A Pilot Scale Demonstration of the DWPF Process Control Strategy," *Proceedings of the International High-Level Radioactive Waste Management Conference*, Las Vegas, NV, 1992, 525-532.

[6]   Jantzen, C.M., and Brown, K.G., "Statistical Process Control of Glass Manufactured for the Disposal of Nuclear and Other Wastes," *Am. Ceramic Society Bulletin*, 1993, *72*, 55-59.

[7]   Cozzi, A.D.; Jantzen, C.M.; Brown, K.G. and Cicero-Herman, C.A., "Process Control for Simultaneous Vitrification of Two Mixed Waste Streams in the Transportable Vitrification System (TVS)," *Environmental Issues and Waste Management Technologies in the Ceramic and Nuclear Industries, Vol. IV Ceramic Transactions*, J. Marra and D.K. Peeler (Eds.), American Ceramic Society, Westerville, OH, (submitted as WSRC-MS-97-00855), 1997.

[8]   Ryan, J. L., "REDOX Reactions and Foaming in Nuclear Ware Glass Melting," *PNL-10510*, Pacific Northwest Laboratory, Richland, Washington, August 1995.

[9]   Weinberg, M.C., D. R. Uhlmann, G. L. Smith, "Influence of Radiation and Multivalent Cation Additions on Phase Separation and Crystallization of Glass," *DOE FG07-97-ER45670*, University of Arizona, Tucson, AZ, August 2002.

[10] Koopman, D. C.; Lambert, D. P., "Hydrogen Generation and Foaming During Test in the GFPS Simulating DWPF Operations with Tank 42 Sludge and CST," *WSRC-TR-99-00302*, Westinghouse Savannah River Company, Aiken, SC, September 1999.

[11] Jantzen, C.M., "DWPF Glass Redox Determination-Summary of Results from Clemson Subcontract," *U.S. DOE Report 86-745*, 1986.

[12] Jantzen, C.M., and Plodinec, M.J., "Composition and Redox Control of Waste Glasses-Recommendation for Process Control Limit," *U.S. DOE Report 86-773*, 1986.

[13] Calloway, T. B., and Jantzen, C. M., "Analysis of the DWPF Glass Pouring System using Neural Networks(U), *Proceeding of Waste Management 1998*, Tucson, AZ. March 3-4, 1998.

[14] Rawlings et. al., "Non-linear Model Predictive Control: A Tutorial and Survey", *Proceedings of ADCHEM '94*, Kyoto, Japan, 1994.

[15] Jantzen, C.M. "Characterization of Off-Gas System Pluggages, Significance for DWPF and Suggested Remediation," *U.S. DOE Report WSRC-TR-90-205*, 1991.

[16] Jantzen, C.M., "Relationship of Glass Composition to Glass Viscosity, Resistivity, Liquidus Temperature, and Durability: First Principles Process-Product Models for

Vitrification of Nuclear Waste," *Proceedings of the 5th International Symposium on Ceramics in Nuclear Waste Management*, G.G. Wicks, D.F. Bickford, and R. Bunnell (Eds.), American Ceramic Society, Westerville, OH, 37-51, 1991.

[17] Jantzen, C.M., Pickett, J.B., Brown, K.G., Edwards, T.B., and Beam, D.C., "Process/Product Models for the Defense Waste Processing Facility (DWPF): Part I. Predicting Glass Durability from Composition Using a Thermodynamic Hydration Energy Reaction MOdel (THERMO)," *US DOE Report WSRC-TR-93-0672*, September, 1995.

[18] Rivera, D. E.; M. Morari, and S. Skogestead, "Internal Model Control, 4. PIC Controller Design", *Industrial and Engineering Chemistry Process Design and Development*, Vol. 25, 1986.

[19] Cozzi, A.D.; Jantzen, C.M.; Brown K.G., and Cicero-Herman, C.A., "Process Control for Simultaneous Vitrification of Two Mixed Waste Streams in the Transportable Vitrification System (TVS)," *Environmental Issues and Waste Management Technologies in the Ceramic and Nuclear Industries, Vol. IV Ceramic Transactions*, American Ceramic Society, Westerville, OH, 1997.

[20] Morari, M., and Zafiriou, E., *Robust Process Control*, Prentice Hall, Engle Cliffs, New Jersey 07632, 1991.

[21] Jantzen , C.M., "Prediction of Glass Durability as a Function of Glass Composition and Test Conditions: Thermodynamics and Kinetics", *Proceedings of the First Intl. Conference on Advances in the Fusion of Glass*, American Ceramic Society, Westerville, OH , 1988, 24.1-24.17.

[22] Ramsey, W.G., Jantzen, C.M., and Taylor, T.D. "Predictive Modeling of Simulated High-Level Waste Glass pH," *Proceedings of the 5th International Symposium on Ceramics in Nuclear Waste Management*, G.G. Wicks, D.F. Bickford, and R. Bunnell (Eds.), American Ceramic Society, Westerville, OH, 1991, 105-114.

[23] Picton, Phil D. et.al., "Identifying Chemical Species in a Plasma Using Neural Networks", *Proceedings of the 9th International Conference on Ind. Amd Eng. Application of Artificial Intelligence and Expert Systems*, Fukuoka, Japan, June 4-7, 1996.

[24] Shin, Jooho et. al., Design of a Composition Estimator for Inferential Control of High-Purity Distillation Columns", *Proceedings of the 5th International Conference Chemical Process Control*, 1997, AICHE Symposium Series, *93*.

[25] Woskov, P.P.; Sundaram, S. K.; Daniel, G.; Machuzak, J. S.; Thomas, P., "Millimeter-Wave Monitoring of Nuclear Waste Glass Melts – An Overview", *Environmental Issues and Waste Management Technologies VII*, Ceramic Transactions, 2002, *132*, 189-201.

[26] Woskov, P.; Hadidi, K.; Sundaram, S. K., and Daniel, Jr., W. E., "Millimeter-Wave Radiometer Measurement of Emissivity and Temperature by Thermal Return Reflection", *Inter. Conf. on Infrared and Millimeter-Waves*, San Diego, Sept. 22-26, 2002, *IEEE 02EX561*, 211-212,.

[27] Woskov, P. P.; Cohn, D. R.; Rhee, D. Y.; Thomas, P.; Titus, C. H., and Surma, J. E., "Active Millimeter-Wave Pyrometer", *Rev. Sci. Instrum.*, 1995, 66, 4241-4248.

[28] Woskov, P.P., and Sundaram, S. K. "Thermal return reflection method for resolving emissivity and temperature in radiometric methods", *J. Appl. Phys.*, 2002, *92*, 6302-6310.

[29] Goldsmith, P. F. *Quasioptical Systems*; IEEE Press, New York, NY 1998; pp. 256-259.

[30] Granger, R. A. *Fluid Mechanics*; Dover Publishers, New York, NY 1985; *Haake Viscometers*; Gebrüber HAAKE GmbH, D-7500 Karlsruhe, Germany, 1981.

[31] Dicke, R. H., *Rev. Sci. Instr.* 1946, *17*, 268-75.

[32] Tiuri, M. E. In *Radio Astronomy*, by J. D. Kraus, Cygnus-Quasar Books, Powell, OH, 1986, Chap. 7.

[33] Doane, J. L., "Propagation and Mode Coupling in Corrugated and Smooth-Walled Circular Waveguides", In *Infrared and Millimeter Waves*, Button, K. J., Ed., Academic Press, NY, 1985, Vol. 13, Chap. 5.

[34] Marcatili, E. A. J., and Schmeltzer, "Hollow Metallic and Dielectric Waveguides for Long Distance Optical Transmission and Lasers", *The Bell System Technical Journal*, 1964, *43*, 1783-1808.

[35] Woskov, P. P., and Titus, C. H., "Graphite Millimeter-Wave Waveguide and Mirror for High Temperature Environments", *IEEE Transactions on Microwave Theory and Techniques*, 1995, *43*, 2684-2688.

[36] Woskov P. P.; J. Machuzak, P. Thomas, S. K, Sundaram, and G. Daniel, "High temperature refractory material waveguides for millimeter-wave diagnostics", *Proceedings 10th Intern. Symp. On Laser Aided Plasma Diagnostics*, 6 pages, Fukuoka, Japan, , Sept. 24-28, 2001.

[37] Woskov, Paul P.; Bajaj, Vikram S. ; Hornstein, Melissa K.; Temkin, Richard J., and Griffin, Robert G., "Corrugated waveguide and directional coupler for CW 250 GHz gyrotron DNP experiments",*IEEE Trans. of Microwave Theory and Techniques*, 2005, *53*, 1863-1869.

[38] Ma, J. Y. L., and Robinson, L. C., " Night moth eye window for the millimetre and sub-millimetre wave region", Optica Acta, 1983, *30*, 1685-1695.

[39] Woskov, P. P.; Sundaram, S. K.; Daniel, Jr., W. E., "Waste Glass Melter Process Monitoring with Millimeter Waves", Spectrum 2002, *9th Biennial Conference on Nuclear and Hazardous Waste Management*, American Nuclear Society, Reno, NV, August 4-8, 2002, 5 pages.

[40] Woskov, P. P.; Hadidi, K.; Bromberg, L.; Sundaram, S. K.; Rodgers, L. A.; Daniel, Gene, and Miller, Don, "Glass Melt Emissivity, Viscosity, and Foaming Monitoring with Millimeter-Waves", *226th American Chemical Society Meeting CD*, Division of Nuclear Chemistry and Technology, New York, Sept. 7-11, 2003, 81.

[41] Sundaram, S. K.; Woskov, P.P.; Machuzak, J.S., and Daniel, Jr., W.E., "Cold Cap Monitoring using Millimeter Wave Technology", In *Environmental Issues and Waste Management Technologies VII*, Editors: G. L. Smith, S. K. Sundaram, and D. R. Spearing, Ceramic Transactions, *132*, 2002, 203-213.

[42] Woskov P. P.; Sundaram, S. K.; Daniel, Jr., W. E.; Miller, D.; Harden, J., "Millimeter-wave measurements at137 GHz of DWPF black frit glass flow and salt layer pooling in a pilot scale melter" *227th American Chemical Society Meeting CD*, Division of Environmental Chemistry, March 28 – April 1, 2004, 63, psfc.mit.edu/library/04JA001.

[43] Woskov, P. P.; Sundaram, S. K; Daniel, W. E.; Jr., Miller, D., "Molten salt dynamics on glass melt using millimeter-wave emissivity measurements", *J. of Non Crystalline Solids*, *341/1-3*, 2004, 21-25.

[44] Woskov, P. P.; Sundaram, S. K; Daniel, W. E.; Jr., Miller, D.,"Millimeter-Wave Measurements of Nuclear Waste Glass Melts", *Proceedings IRMMW-THZ 2005*, IEEE Cat. No. 05EX1150C, September, 2005, 223-224.

In: Nuclear Waste Research: Siting, Technology and Treatment   ISBN 978-1-60456-184-5
Editor: Arnold P. Lattefer, pp. 107-141   © 2008 Nova Science Publishers, Inc.

Chapter 3

ON STATISTICALLY OPTIMAL ALGORITHMS OF REGULATION OF A PULSED REACTOR

Alexander K. Popov[*] *and Alexey A. Marachev*[**]

[*]Joint Institute for Nuclear Research, Russia

[**]Moscow Center of Continuous Mathematical Education, Russia

ABSTRACT

The principle of work and feature of a design of a pulsed reactor are the reason for essential fluctuations of power pulse amplitudes generated by a reactor. These fluctuations reaching 40% are caused by high sensitivity of a pulsed reactor to changes of reactivity. The general problem of definition of optimal (in statistical sense) automatic control algorithms is considered for various operating modes of a pulsed reactor. A minimum of the expected mean-square deviation of a control parameter of the future power pulse from base value of this parameter is accepted as the optimality criterion. To define optimal algorithm elements of the theory of optimal systems are used. The optimal algorithm is received in the assumption, that in a regulator any information on the previous pulses can be used. A feature of the definition of algorithm is the use of an added concept of a degree of ageing of the information. This concept has a clear enough physical sense. The decision of a problem in such a general statement gives the basis to simplify the algorithm and to lead it to the kind more convenient for realization (when in formation of control action the parameters accessible to direct measurement are used only). The algorithms achieved as a result of physically proved simplifications essentially do not miss traditional algorithms. On the one hand, it specifies a correctness of the considered method of a choice of algorithm, and with another, once again confirms efficiency of the engineering intuitive approach demanded from a regulator certain inertia. In the model of the reactor modeling transients are realized at random and regular disturbances of reactivity for the mode of power stabilization and for the mode of reactor going up on the set period. The estimation of influence of the parameter of the regulator on transients is given. The expediency of a compromise choice of a degree of regulator inertance is confirmed.

---

[*] E-mail: popov_ak@nf.jinr.ru

INTRODUCTION

Pulsed reactors on fast neutrons of periodic action have almost fifty years' history. The first such reactor IBR had started up in operating in the Joint Institute for Nuclear Research (Dubna, Russia) in 1960. Originally its mean power had made 1 kW, then it had been raised to 6 kW. After reconstruction in 1969 the reactor began to refer to IBR-30, its mean power had been finished up to 25 kW.

The operational experience of these reactors had allowed the design of the pulsed reactor IBR-2 with mean power 2 MW. The reactor started operating in 1984.A feature of each pulsed reactor is the presence in it of a reactivity modulator. The reactivity modulator gives a pulse character to the coefficient of multiplication. During these pulses the coefficient of multiplication of reactor IBR-2 changes approximately on 3 %. Use of such a reactivity modulator allows one to introduce periodically the reactor for short time in *the supercritical condition on prompt neutrons*. In intervals between pulses the reactor remains deeply subcritical. As a result the reactor generates powerful neutron pulses (power pulses). Such a principle of work and feature of a design of a pulse reactor are the reason for essential fluctuations of reactivity and, hence, amplitudes of power pulses.

Presence of essential fluctuations of reactivity was always considered by design of automatic regulators of pulsed reactors. The choice of structure of a regulator and its algorithm was based on engineering intuition. It specified a correct direction of the decision of this problem, namely, demanded certain inertia of an automatic regulator.

At a design stage of the IBR-2 reactor the method of the decision of this problem, based on minimization of beforehand chosen criterion of a statistical optimality [1] has been offered. The opportunity of application of this method to rather specific operating mode of a pulse reactor [2] gives the basis to consider this method universal enough.

In this chapter the detailed statement of the method of a choice of statistically optimal regulator is given. In the chapter a) the detailed substantiation of a method is given, b) for various operating modes of a pulsed reactor algorithms of optimal regulation are deduced, c) as a result of the proved simplifications of algorithms structures of the regulators realizing simplified algorithms of regulation of a pulsed reactor are certain, d) results of modeling of the work of a reactor with these regulators are presented.

BRIEF DESCRIPTION OF THE IBR-2 REACTOR

The pulsed reactor of periodic action on fast neutrons IBR-2 is characterized by very high (in comparison with an average level) neutron flux in a pulse. Amplitude of the neutron flux pulse is on three orders above, than an average level of flux over the period of following the pulses. Reactor IBR-2 (owing to this feature) is very convenient for many physical experiments in which the main thing is the greater power generated during a small time interval, necessary for an experiment, but not the greater power generated continuously.

The core of the reactor has the shape of an irregular hexahedral prism with a vertical axis. Heat removal from the core is carried out by liquid sodium. Two rotating mobile reflectors (main and auxiliary), representing the reactivity modulator, pass beside one of the prism faces. The main and auxiliary mobile reflectors are coaxially positioned and kinematically connected to one another and rotate with different speeds. The case in which the mobile reflectors rotate, is filled by helium.

At rotation mobile reflectors create periodic reactivity pulses in the IBR-2 reactor. The greatest reactivity pulses are created at simultaneous passage by the core of the main and auxiliary mobile reflectors. From 1984 through 2004 the reactivity modulators in which speeds of the main and auxiliary mobile reflectors made 1500 and 300 rpm accordingly were used. Thus each fifth pulse of reactivity had the greatest amplitude – more than 0.03 absolute units of reactivity (Figure 1).

The non-automatic reactor controls were set in such a position that the reactor reactivity caused by every fifth pulse of the reactivity modulator periodically became prompt-neutron positive for a short time (~400 μs). As a result, the reactor generated powerful neutron pulses (power pulses) with a frequency of 5 Hz (Figure 2).

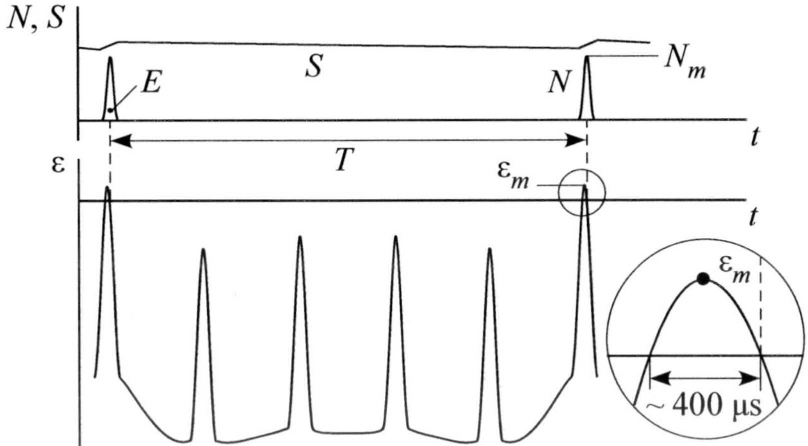

Figure 1. Prompt neutron reactivity ε, power N and the intensity of sources of delay neutrons S (E – energy of the power pulse, t – the time, T – the power pulse period).

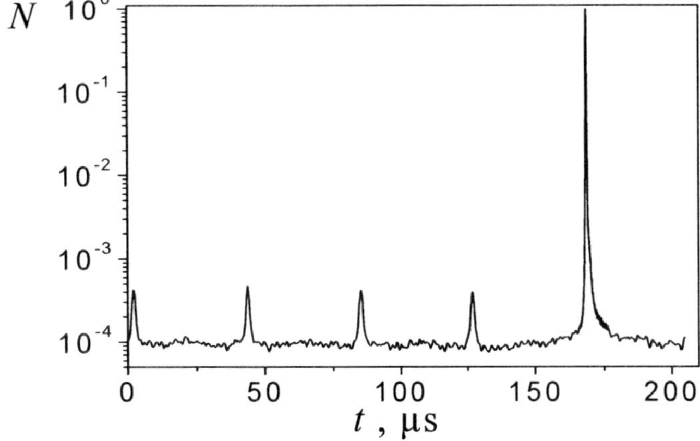

Figure 2. The reactor power (relative units) for the period at the mean reactor power 2 MW.

The width of power pulses (pulses of fast neutrons) on half of height made 215 μ s. In intervals between power pulses when the reactor was deeply subcritical 7 % of total energy generated in the reactor was released only. Almost the total energy was released during power pulses (93 %). Up to 1997 the reactor worked at mean power 2 MW, and then – 1.5 MW.

In 2004 replacement of the reactivity modulator (PO-3 instead of PO-2M) is made. The new form of mobile reflectors has allowed to lower speed of the main mobile reflector up to 600 rpm. Speed of the auxiliary mobile reflector has remained former (300 rpm) therefore the frequency of power pulses has remained former (5 Hz). At the new reactivity modulator the power pulse corresponds to each second pulse of reactivity instead of everyone the fifth (Figure 3). The width of the power pulse on half of height has made 245 μ s. The share of the energy released in power pulses has made 92 %.

Duration of the power pulses is very small. It almost on three orders is less than the period of power pulses and even less than time constants of sources of delay neutrons and time constants of the feedback caused by a warming up of the reactor. Therefore at the analysis of reactor dynamics it is possible to consider power pulses proportional delta-functions.

The reactor periodically becomes prompt neutron-supercritical. As a result, the pulsed reactor is very sensitive to indignations of reactivity which are caused by vibration of the reactivity modulator, trembling of the fuel sub-assemblies in which are grouped fuel elements, fluctuations of temperature and speed of the coolant in the core, etc. Sensitivity of reactor IBR-2 to indignations of reactivity (fluctuations of power pulse energy reach 40 %) approximately in 20 times more, than sensitivity, for example, of stationary reactors with uranium fuel.

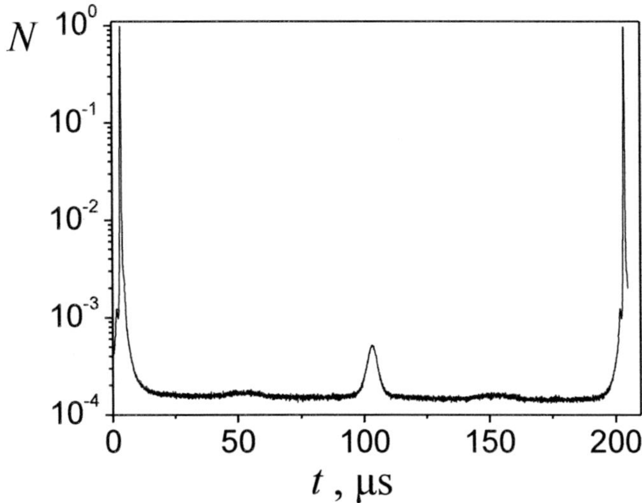

Figure 3. Change of power (relative units) at mean reactor power 1.35 MW.

The law of distribution of probability density for power pulse energy is close to Gauss law (Figure 4).

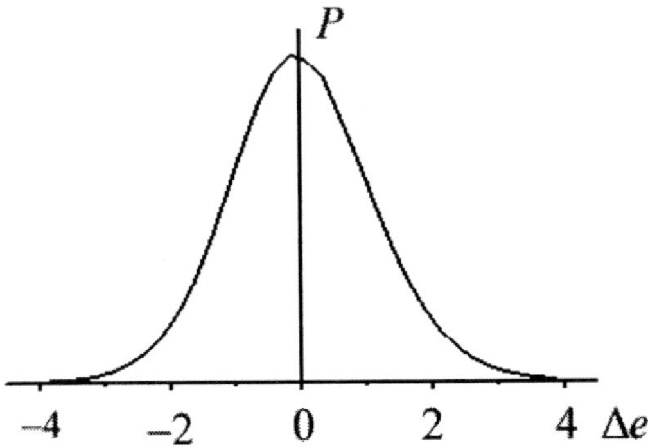

Figure 4. Density of distribution P of the relative deviation $\Delta e = (E - E^0)/E^0$ of power pulse energy E from basic value E^0 at mean power of the reactor 1.5 MW (Δe is expressed in shares of the mean square deviation σ).

The basic operating mode of a pulsed reactor is the mode of stabilization of power. As controlled parameter of the IBR-2 reactor the deviation of amplitude of the power pulse N_m from its set (basic) value N_m^0 in relative units is accepted: $(N_m - N_m^0)/N_m^0$.

As speed of rotation of the main mobile reflector is supported with high accuracy (7×10^{-4}), the relative deviation of amplitude is practically equal to the relative deviation of power pulse energy

$$\Delta e = \frac{E - E^0}{E^0} = \frac{N_m - N_m^0}{N_m^0}, \tag{1}$$

where E, E^0 – power pulse energy and its basic value, respectively.

POWER PULSE ENERGY

The power pulse energy E is convenient to express by the formula [3]

$$E = MS, \tag{2}$$

where S – the intensity of the sources of delay neutrons at the moment corresponding the start of the reactivity pulse, M – the pulse transfer coefficient.

At the description of dynamics of a stationary reactor the reactivity $\delta k = (k-1)/k$ is used, where k – the effective coefficient of multiplication.

At the description of dynamics of a pulsed reactor it is convenient to use the prompt neutron reactivity

$$\varepsilon = \delta k - \beta \tag{3}$$

(where β – the total share of delay neutrons) and the maximal value of reactivity ε_m.

The operational experience of the IBR-2 reactor has shown, that change of power in time is well described by the known equations of the kinetics of a one-point model of a reactor. The parameter M [2] (calculated from the kinetics equations) is shown on Figure 5.

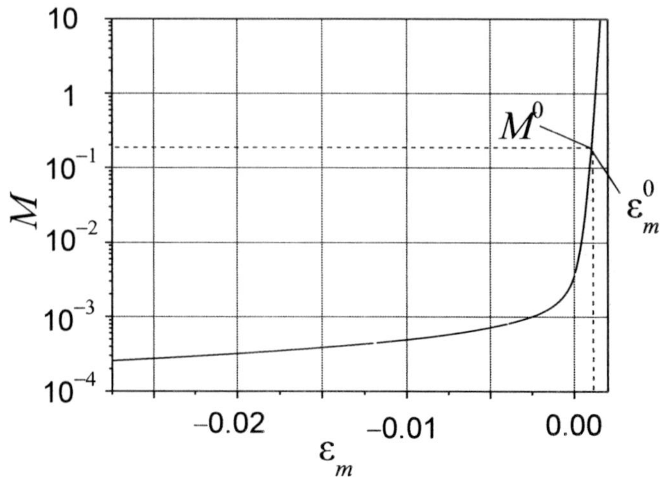

Figure 5. The pulse transfer coefficient (M^0, ε_m^0 – the basic values corresponding to work of the reactor with stable power pulses).

The parameter M is nonlinear function from the maximal value of reactivity ε_m within the reactivity pulse. For reactor modes where the power pulse energy changes not more than 3–4 times as much as its basic value, exponential approximation of pulse transfer coefficient M is possible [4]. In this case the approximation error, for example, in 5 % is reached if M changes rather M^0 in a range from $0.4M^0$ up to $2.9M^0$. The error in 10 % is reached at essential greater changes of M: from $0.3M^0$ up to $4.5M^0$.

At exponential approximation of M the following formula for relative energy of the power pulse corresponds to Eq. (2):

$$\frac{E}{E^0} = \frac{S}{S^0}\exp\left(\frac{\varepsilon_m - \varepsilon_m^0}{\beta_u}\right) = \frac{S}{S^0}\exp\left(\frac{\Delta\varepsilon_m}{\beta_u}\right). \tag{4}$$

In Eq. (4) $\Delta\varepsilon_m = \varepsilon_m - \varepsilon_m^0$ – deviation of the maximal reactivity from its basic value, β_u – the parameter (introduced in the article [5] and named by a pulsed share of delay neutrons) is defined by the formula

$$\beta_u = 1 \bigg/ \frac{d\ln M}{d\varepsilon_m}\bigg|_{\varepsilon_m^0} = M^0 \bigg/ \frac{dM}{d\varepsilon_m}\bigg|_{\varepsilon_m^0}. \tag{5}$$

Reactivity of a stationary reactor is expressed quite often in shares of β. For a pulsed reactor it is convenient to express prompt neutron reactivity not in absolute units and in shares of β_u:

$$\rho = \frac{\varepsilon}{\beta_u}. \tag{6}$$

Here ρ is a prompt neutron reactivity expressed in shares of β_u. The value $\beta_u = 1{,}6 \times 10^{-4}$ corresponded to the reactivity modulator PO=2M, at the modulator PO-3 $\beta_u = 1{,}36 \times 10^{-4}$ [6]. At use of reactivity in shares of β_u the equation (4) will become

$$\frac{E}{E^0} = \frac{S}{S^0}\exp\!\left(\rho_m - \rho_m^0\right) = \frac{S}{S^0}\exp(\Delta\rho_m), \tag{7}$$

where $\Delta\rho_m = \rho_m - \rho_m^0$ is the deviation of the maximal reactivity from its basic value expressed in shares of β_u.

Let's copy the formula (7) in the form of

$$\frac{E}{E^0} = \exp\!\left(\Delta\rho_m + \ln\frac{S}{S^0}\right). \tag{8}$$

In the formula (8) let us expand the logarithm in a Taylor series without taking into account nonlinear components:

$$\frac{E}{E^0} = \exp\!\left(\Delta\rho_m + \frac{\Delta S}{S^0}\right), \tag{9}$$

where $\Delta S = S - S^0$ is the deviation of the intensity of delay neutron sources from its basic value. Let us write down the formula (1) for a relative deviation of the power pulse energy Δe (which is controlled parameter of the reactor IBR-2) using expression (9):

$$\Delta e = \frac{E}{E^0} - 1 = \exp\!\left(\Delta\rho_m + \frac{\Delta S}{dS^0}\right) - 1 \tag{10}$$

STATEMENT OF THE PROBLEM ON STATISTICALLY OPTIMAL ALGORITHM OF REGULATION FOR THE MODE OF STABILIZATION

Let's consider a mode of maintenance of power pulse energy at the set level (the mode of stabilization). Let us define the algorithm of regulation providing a minimum of the mean-square deviation of the relative energy of the future power pulse based on the information obtained from the previous pulses. For a generality let us assume, that for formation of optimal algorithm in a regulator any information which can be obtained from the previous pulses (including the information on reactivity in each pulse) can be used. The decision of a problem in such general statement has essential practical importance. The analysis of the received algorithm will enable to carry out the proved simplifications and to lead algorithm to a kind convenient for realization so that the parameters accessible to direct measurement participated in formation of control action only.

METHOD OF DEFINITION OF OPTIMAL ALGORITHM TAKING INTO ACCOUNT AGEING OF INFORMATION

The deviation of reactance $\Delta\rho_m$ (in the further we shall name it reactivity for brevity) can be presented in as the sum of reactivity of the regulator u, random component of reactivity μ and reactivity of the temperature feedback $\Delta\rho_\theta$:

$$\Delta\rho_m = u + \mu + \Delta\rho_\theta. \tag{11}$$

The reactivity $\Delta\rho_\theta$ caused by a warming up of a reactor, can be presented as the power feedback reactivity and is described by the differential equations of the first order connecting the reactivity not with temperature, but with power. These equations are similar to the equations connecting sources of delay neutrons with power. Considering it, at substitution of the equation (11) in the equation (10) reactivity of the regulator u and noise of reactance μ are allocated separately

$$x = u + \mu. \tag{12}$$

As a result the equation (10) will become

$$\Delta e = \exp\left[x + \left(\Delta\rho_\theta + \frac{\Delta S}{S^0}\right)\right] - 1 \tag{13}$$

As reactivity contains a random component the estimation of this reactivity in the regulator generally should be carried out.

It is supposed, that reactivity x is estimated in a regulator with error η having random character. Thus, control action of the regulator is formed not on the basis of the true value of x, but on the basis of the estimation of reactance y

$$y = x + \eta. \tag{14}$$

The block diagram for definition of algorithm of regulation for a mode of stabilization of power is shown on Figure 6. The index k concerns to the parameters corresponding to the future pulse.

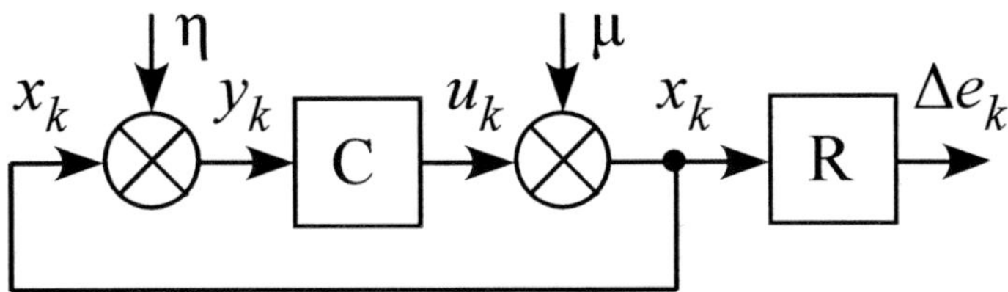

Figure 6. The block diagram of system "the regulator – the reactor" for definition of algorithm of regulation for a mode of stabilization of power (C – regulator, R – reactor).

Specific loss function, i.e. loss function for k-th (future) pulse is entered:

$$Q_k = Q_k(k, x_k). \tag{15}$$

The specific risk R_k representing a mathematical expectation of specific loss function for k-th pulse is considered

$$M(Q_k) = \min. \tag{16}$$

The aspiration to minimize risk for the first future pulse (instead of a combination of risks of several future impulses that would be to integral criterion of quality of transient in automatic systems) is dictated by desire to reduce probability of misoperation of emergency protection of a reactor. For a generality it is considered that the regulator possesses memory. Hence, the output signal of a regulator u_k depends on values y_j and u_j, corresponding the previous pulses ($j < k$).

According to the theory of optimal control systems [7] statistically optimal algorithm is reduced to a choice of value u_k at which the minimum of following function is reached

$$\chi_k = \int_{\Omega(\mu, x_k)} Q(k, x_k) P(x_k | \mu, u_k) P_0(\mu) \left[\prod_{j=0}^{k-1} P(y_j | \mu, u_j) \right] d\Omega(\mu, x_k) = \min \tag{17}$$

Here $P_0(\mu)$ – a priori probability density of μ, $P(x_k|\mu, u_k)$ – a posteriori probability density of x for k-th pulse, $P(y_j|\mu, u_j)$ – a posteriori probability density of y for j-th pulse ($j = 0$ corresponds to a reference mark), Ω – the area including set of possible values μ and x_k.

Elements of the theory of the optimum control systems concerning our case are stated in the Appendix 1.

For our case

$$P(x_k|\mu, u_k) = \delta[x_k - (u_k + \mu)] \tag{18}$$

where δ – delta-function.

It is considered that the a priori probability density of μ is distributed under the normal law:

$$P_0(\mu) = \frac{1}{\sigma_\mu \sqrt{2\pi}} \exp\left(-\frac{(\mu - m_0)^2}{2\sigma_\mu^2}\right), \tag{19}$$

where σ_μ^2 and m_0 – a priori dispersion and mean value of μ, respectively.

It is supposed that the probability density of η also submits to the normal law. The regulator perceives the variable y which is the estimation of the reactivity x measured with the error η. For a generality it is accepted that the received estimations can depend on time. In this connection dependence of probability density of η on a difference of pulse numbers $k - j$ (where $j \leq k$) is entered:

$$P_j(\eta) = \frac{1}{\sigma_{\eta j} \sqrt{2\pi}} \exp\left(-\frac{\eta^2}{2\sigma_{\eta j}^2}\right) = \frac{a_{k-j}}{\sigma_\eta \sqrt{2\pi}} \exp\left(-\frac{a_{k-j}^2 \eta^2}{2\sigma_\eta^2}\right) \tag{20}$$

Here the dispersion in j-th pulse $\sigma_{\eta j}^2$ is expressed through the dispersion in k-th pulse σ_η^2 as follows:

$$\sigma_{\eta j}^2 = \frac{\sigma_\eta^2}{a_{k-j}^2}, \tag{21}$$

where $a_{k-j} \geq 0$ – the coefficient depending on difference of pulse numbers. Let's explain sense of this coefficient.

If $a_{k-j}^2 = 1$ for everything j the probability density of η does not depend on number of the pulse. If $1 = a_0 > a_1 > a_2 > \ldots$, then $\sigma_\eta = \sigma_{\eta k} < \sigma_{\eta k-1} < \sigma_{\eta k-2} < \ldots$. It means that the regulator perceives earlier estimation of reactivity as less exact. Thus, the entered coefficient a_{k-j}^2 characterizes a degree of ageing of the information received by the regulator.

Let's accept as specific loss function a square of a relative deviation of energy of future (k-th) power pulse. In view of expression (13) it will become

$$Q_k(k, x_k) = \Delta e_k^2 = \left\{ \exp\left[x_k + \left(\Delta \rho_{\theta k} + \frac{\Delta S_k}{S^0} \right) \right] - 1 \right\}^2 . \tag{22}$$

After substitution of Eq. (22) in Eq. (17) it is possible to define optimal algorithm of regulation u_k at which the minimum of expression (17) is provided, i.e. the minimum of risk (16) is provided. The conclusion of optimal algorithm is resulted in the Appendix 2. Below the final result is resulted.

Optimal control action u_k (which provides a minimum of the mean-square relative deviation of energy for the future power pulse based on the information obtained from the previous pulses at the account of ageing of the information) is expressed by the formula

$$u_k = -\Delta \rho_{\theta k} - \frac{\Delta S_k}{S^0} - \frac{\dfrac{3}{2}\sigma_\eta^2 + \left(\dfrac{\sigma_\eta}{\sigma_\mu}\right)^2 m_0 + \sum\limits_{j=0}^{k-1} a_{k-j}^2 (y_j - u_j)}{\left(\dfrac{\sigma_\eta}{\sigma_\mu}\right)^2 + \sum\limits_{j=0}^{k-1} a_{k-j}^2} \tag{23}$$

Indexes k and j concerns to the future and to the previous impulses, respectively. Numbers of the previous pulses are 0, 1, ..., $k-1$. The zero index concerns to a reference mark.

The physical sense of the received algorithm is clear enough. The first member in the right part of the formula (23) compensates a deviation (from basic value) of the reactivity caused by a warming up of a reactor. The second member compensates a similar deviation of the intensity of sources of delay neutrons. The third member compensates the estimation (with the error η) of the noise of reactivity μ for k-th pulse, based on measurement of differences $y_j - u_j$ at $j < k$.

The understanding of physical sense of the obtained algorithm allows to make the proved simplifications. For a mode of stabilization as a first approximation it is possible to neglect values $\Delta\rho_{\theta k}$ and ΔS_k because of them inertance. It is possible to assume, that the value σ_η is small enough. The value y_j can be expressed approximately from the equation (13). As a result the algorithm essentially becomes simpler and becomes

$$u_k = \frac{1}{\sum\limits_{j=0}^{k-1} a_{k-j}^2} \sum_{j=0}^{k-1} a_{k-j}^2 (u_j - \Delta e_j).$$

(24)

The algorithm (24) is algorithm of the closed control system as control action u is function of controlled variable Δe.

STRUCTURE OF THE REGULATOR FOR THE MODE OF STABILIZATION

The obtained algorithm essentially depends on a_{k-j}^2. It is represented proved to characterize a degree of ageing of the information exponentially decreasing function

$$a_{k-j}^2 = \exp\left[-\frac{(k-j)T}{T_A} \right],$$

(25)

where T – the period of pulses, T_A – time constant ($T_A > T$). Eq. (25) means, that the smaller weight is given to the information of older origin.

Beginning already from pulse number $k > 5T_A/T$ it is possible to consider that $\sum\limits_{j=0}^{k-1} a_{k-j}^2 = const$. Taking into account it, from the equations (24) and (25) follows, that the algorithm (24) is realized in the form of the output signal of the regulator representing an inertial element with transfer function

$$W_A(s) = \frac{k_A}{T_A s + 1} = \frac{T_A[\exp(T/T_A) - 1]}{T_A s + 1},$$

(26)

where s – Laplace transformation variable, k_A – transfer coefficient , T_A – time constant. In that case as an input signal of a regulator there should be a pulse signal $u_j - \Delta e_j$.

The period of pulses T makes fraction of a second. Therefore from physical reasons the preference should be given to a case, when $T_A \gg T$. Ii that case the regulator becomes simpler turning to the integrator with transfer function

$$W_A(s) = \frac{T}{T_A s}.$$
(27)

In order to the optimal control signal was formed on an output of a regulator at instant corresponding the k-th pulse, the following pulse signal should be given on an input of a regulator

$$\Delta(t) = -\Delta e_j \delta(t - jT),$$
(28)

where t – time, δ – delta-function, $j = 0, 1, ..., k-1$. Let us pay attention that the output signal corresponding to the k-th pulse is caused by the input signals corresponding to the previous pulses.

The considered statistical approach to a choice of structure of a regulator at assumptions and the simplifications based on physical reasons leads to traditional structure of a regulator in the form of the integrator.

ALGORITHM AND STRUCTURE OF THE REGULATOR FOR THE MODE OF REACTOR GOING UP ON THE SET PERIOD

Let's define statistically optimal algorithm for a mode of reactor going up. Instead of the reactor period T_R it is more convenient to use the inverse reactor period $\alpha_R = 1/T_R$. We shall understand as the inverse reactor period at instant corresponding k-th power pulse variable

$$\alpha_{Rk} = \frac{1}{T} \ln \frac{E_k}{E_{k-1}},$$
(29)

where T – the pulse period, E_k, E_{k-1} – energy of the k-th and $(k-1)$-th power pulses, respectively.

Let's take advantage of Eq. (7) for k-th and $(k-1)$-th pulses and result the formula (29) in a following kind

$$\alpha_{Rk} = \frac{1}{T}\left(\rho_{mk} - \rho_{mk-1} + \frac{S_k - S_{k-1}}{S_{k-1}} \right) = \frac{1}{T}(\delta\rho_{mk} + \delta s_k),$$
(30)

where $\delta\rho_{mk} = \rho_{mk} - \rho_{mk-1}$, $\delta s_k = (S_k - S_{k-1})/S_{k-1}$ – increments during T reactivity and relative intensity of sources of delay neutrons, respectively. We shall present the reactivity increment in the form of the sum

$$\delta\rho_{mk} = \delta u_k + \mu + \delta\rho_{\theta k}, \tag{31}$$

where $\delta u_k = u_k - u_{k-1}$ – increment of regulator reactance, μ – random component, $\delta\rho_{\theta k} = \rho_{\theta k} - \rho_{\theta k-1}$ – increment of reactivity of a temperature feedback.

Reactor going up on the period is used in a power range essentially smaller power rating of a reactor. At such powers the feedback caused by a warming up of a reactor practically does not show itself and it is possible to not consider the feedback. Therefore we shall consider only an increment of reactivity of a regulator and noise of reactivity

$$\delta x_k = \delta u_k + \mu. \tag{32}$$

Thus the inverse reactor period will be expressed by the formula

$$\alpha_{Rk} = \frac{1}{T}(\delta x_k + \delta s_k). \tag{33}$$

As well as for a mode of stabilization we shall accept that the random variable μ is distributed under the normal law (19) and that it is estimated by a regulator with the mistake η distributed under the normal law (20). As well as for a mode of stabilization we shall accept that in a regulator the information received from last pulses becomes outdated. As a result the block diagram for calculation of algorithm for a mode of reactor going up can be presented in the form of Figure 7 which is similar Figure 6 for a mode of stabilization.

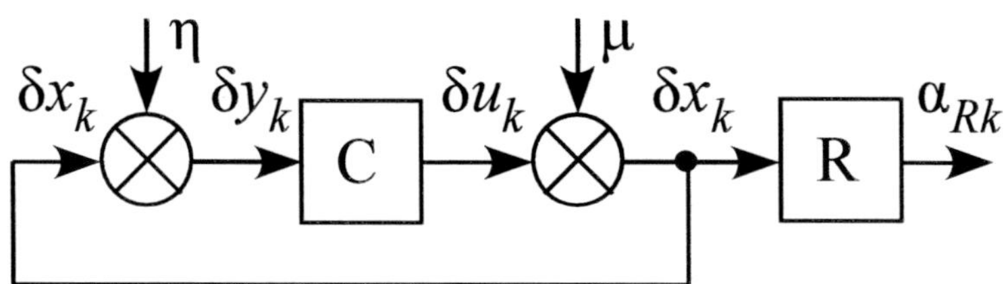

Figure 7. The block diagram for definition of algorithm for a mode of reactor going up (C – regulator, R – reactor).

As specific loss function we shall accept expression

$$Q_k(k, \delta x_k) = [\exp(\alpha_{Rk} - \alpha_R^0) - 1]^2, \tag{34}$$

where α_R^0 — setting value of the inverse reactor period. Chosen the exponential kind of loss function means that excess of value of the inverse period over a setting value is represented to less desirable than reduction equal to it. At small deviations α_{Rk} from α_R^0 expression (34) comes nearer to $(\alpha_{Rk} - \alpha_R^0)^2$.

The reasoning similar to that has been done at a conclusion of algorithm of regulation for a mode of stabilization and the account of that fact that change of intensity of sources of delay neutrons essentially lags behind change of power lead to the following simplified algorithm:

$$\frac{\delta u_k}{T} = \frac{1}{\sum\limits_{j=0}^{k-1} a_{k-j}^2} \sum_{j=0}^{k-1} a_{k-j}^2 \left[\frac{\delta u_j}{T} - (\alpha_{Rj} - \alpha_R^0) \right].$$

(35)

The algorithm (35) is similar to algorithm (24) for a mode of stabilization. However in the algorithm (35) the speed of the regulator reactivity $\delta u_j / T$ participates instead of the regulator reactivity u_j that participates in the algorithm (24). A degree of ageing of the information we shall characterize expression (25) as well as for a mode of stabilization. The reasoning similar to the previous case leads to the conclusion that at $T_A \gg T$ the algorithm (35) is realized by a regulator with transfer function

$$W_A(s) = \frac{T}{T_A s^2}.$$

(36)

The optimal control signal corresponding to the k-th pulse is formed on the output of the regulator at giving the following pulse signal on the input of the regulator

$$\Delta_\alpha(t) = (\alpha_R^0 - \alpha_{Rj})\delta(t - jT),$$

(37)

where δ — delta-function, $j = 0, 1, ..., k-1$.

Transfer function (36) can be presented as product $W_A(s) = W_{A1}(s)W_{A2}(s) = \frac{1}{s}\frac{T}{T_A s}$.

It means that the regulator can be presented as consecutive connection of two elements with transfer functions W_{A1} and W_{A2}.

In this case the element with transfer function $W_{A1}(s) = 1/s$ can be treated as the filter (in the form of the integrator) on which input the signal (37) gives.

Element with transfer function $W_{A2}(s) = \dfrac{T}{T_A s}$ it is possible to treat as the proper regulator which transfer function coincides with transfer function of the regulator (27) for a mode of stabilization. It is necessary to note that at giving on an input of the filter of the signal (37) in the form of sequence of ideal pulses (i.e. the pulses proportional the delta-functions) on an output of the filter will be generated a continuous signal of the step form.

ALGORITHM OF REGULATION FOR THE POSSIBLE OPERATING MODE OF THE REACTOR WITH THE INJECTION AS THE REGULATOR

Many years pulse reactors IBR and IBR-30 operated not only in a reactor mode but also in a booster mode.In a booster mode the reactivity modulator created as before pulses of reactivity but controls were in such position that the reactor was always subcritical. At the moments of the least subcriticality in a target located in a core of a reactor acted short (shares-units of micro second) pulses from the electron accelerator (injector). Under their action the target generated the same short pulses of neutrons which were multiplied by the core of a reactor. As a result the reactor generated power pulses which duration was a little bit greater than duration of electronic pulses but many times over smaller than duration of power pulses in a reactor mode. In a booster mode of a pulse reactor the gain of the neutrons generated by a target much more than the gain of usual subcritical assembly. It is reached because during the moment of injection the reactor subcriticality (because of presence of the reactivity modulator) much less than subcriticality of the usual assembly. The project of reactor IBR-2 also provided opportunity of a booster mode with the powerful electron accelerator as the injector.

At a design stage of reactor IBR-2 the intermediate operating mode of a pulse reactor together with an injector (a mode of correction) [8] has been offered. The attention has been again attracted to this mode in work [9]. This possible mode consists in the following. The reactor is periodically gone in prompt-neutron supercriticality but the maximal value of this supercriticality ε_m less than the equilibrium level corresponding to a reactor mode. This mode is illustrated on Figure 8.

At instant t_I when the reactor is supercritical pulses of the accelerator are injected in the target of the reactor. Borders of a range of injection are the instants t^+ and t^-. As a result there is an opportunity to correct energy of each power pulse separately by necessary displacement of the moment of injection. The more close the moment of injection t_I to t^+ the more (at the same value ε_m) energy of pulse E is (Figure 8). Thus, the injector plays a role of a regulator and the moment of injection is control action.

The mode of correction of each power pulse entails difficulties of definition of the moment of injection. The moment of injection depends on an estimation of ε_m. Value ε_m can be calculated on the portion of energy of the power pulse E' measured during enough short interval of time before the left border of the range of injection (Figure 8).

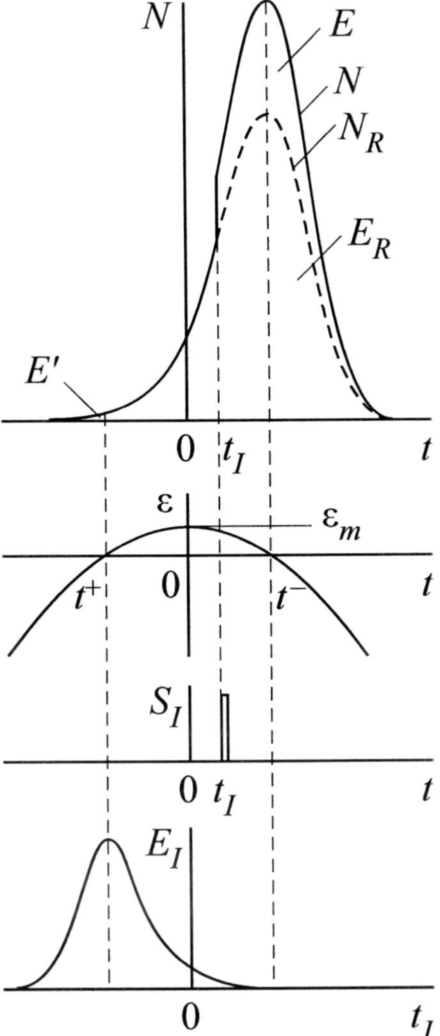

Figure 8. An illustration of an operating mode of a reactor together with an injector: t – time, ε – reactivity, S_I – intensity of the neutrons generated by a target under the action of injector, N, N_R – reactor power taking into account injection and its portion without taking into account injection, respectively, E – energy of a power pulse, E_R – the portion of energy E caused by a proper reactor (without taking into account effect of injection), E_I – the portion of energy E caused by injector at fixed value ε_m as function of the moment of injection t_I, E' – portion of energy E in the beginning of development of a power pulse till the moment t^+.

The more powerful the injector is the more to the right from the moment t^+ it is possible to establish the left border of the range of injection. At reduction of injector power the left border of the range should be shifted to the left (in a limit till the moment t^+). In this case it is necessary to measure energy E' till the moment of time t^+ when power of a developing pulse is small enough (at instant t^{+} it is on two orders less than peak value). In this connection it is necessary to expect mistakes in calculation ε_m and, hence, in calculation of the moments of injection. Considering it we shall consider a mode of correction from the point of view of statistical optimization of the moment of injection.

As energy of a power pulse cannot be negative we shall present it in the form of exponential function with linearized exponent. Energy of k-th pulse E_k we shall write down in the form of the following equation

$$\frac{E_k}{E^0} = \exp(Au_k + B\mu) . \tag{38}$$

In the equation (38) u − deviation of the moment of injection from basic (mean) value u^0, μ − noise of reactivity, A and B − coefficients of Taylor expansion concerning basic values of parameters u^0 and ε_m^0: $A = \dfrac{\partial}{\partial\varepsilon}\ln\dfrac{E}{E^0}\bigg|_{\varepsilon_m^0,u^0}$, $B = \dfrac{\partial}{\partial u}\ln\dfrac{E}{E^0}\bigg|_{\varepsilon_m^0,u^0}$, energy E^0 corresponds to basic values of parameters ε_m^0 and u^0.

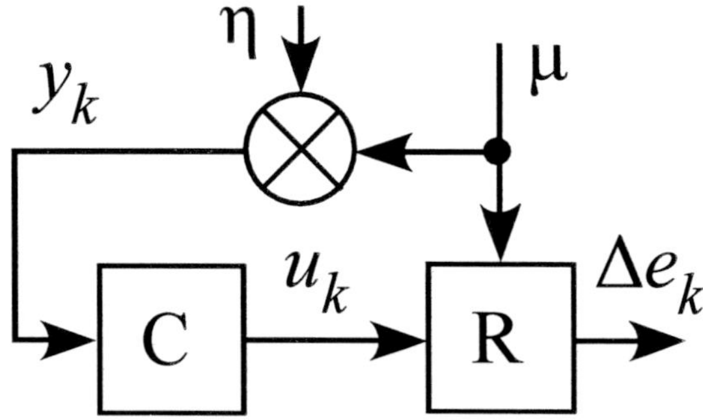

Figure 9. The block diagram for definition of algorithm for a mode of correction (C − regulator, R − reactor).

In Figure 9 the block diagram for definition of optimal algorithm of the moment of injection is shown. We consider that the regulator estimates noise μ with error η. As a result the regulator receives the information on noise of reactivity in the form of $y = \mu + \eta$. As well as for the considered mode of stabilization it is accepted a relative deviation of energy of the k-th power pulse $\Delta e_k = \dfrac{E_k}{E^0} - 1$ as controlled parameter. At that in view of Eq. (38) specific loss function will become

$$Q_k(u_k,\mu) = \Delta e_k^2 = [\exp(Au_k + B\mu) - 1]^2 . \tag{39}$$

As before we consider that probability density of μ and η are defined by the equations (19) and (20), and the degree of ageing of the information is characterized by the equation (21). Minimizing function χ_k (17) on u_k we shall receive the formula for statistically optimal algorithm:

$$u_k = -\frac{B}{A} \cdot \frac{\dfrac{3}{2} B\sigma_\eta^2 + \left(\dfrac{\sigma_\eta}{\sigma_\mu}\right)^2 m_0 + \displaystyle\sum_{j=0}^{k-1} a_{k-j}^2 y_j}{\left(\dfrac{\sigma_\eta}{\sigma_\mu}\right)^2 + \displaystyle\sum_{j=0}^{k-1} a_{k-j}^2}. \tag{40}$$

After the same simplifying assumptions as for a mode of stabilization algorithm (40) we shall lead to a simple kind:

$$u_k = \frac{1}{\displaystyle\sum_{j=0}^{k-1} a_{k-j}^2} \sum_{j=0}^{k-1} a_{k-j}^2 \left(u_j - \frac{1}{A}\Delta e_j\right). \tag{41}$$

From the formula (41) follows that at use of the information received in the previous pulses the moment of injection is formed as the output signal of the integrator which transfer function is defined by expression (27): $W_A(s) = T/(T_A s)$. At that it is necessary to give the

pulse signal $\Delta_I(t) = -\dfrac{1}{A}\Delta e_j \delta(t - jT)$ on the input of the integrator, where δ – delta-

function, $j = 0,1,...,k-1$.

Once again we shall note, that in an ideal the moment of injection can be defined using only the information received in the beginning of developing (k-th) power pulse (namely, using energy E') [8,9]. The control system operating by such principle (it can be named ideal) is opened. The moment of injection of ideal system we shall designate u_{Ik}. The

system which forms the moment of injection u_k (41) on the basis of the information received in the previous pulses is inertial and closed. It is possible to imagine a compromise variant of formation of the moment of injection $\tilde{u}_k$ according to the formula

$$\tilde{u}_k = u_{Ik} + c(u_k - u_{Ik}),$$

where c – some factor (its value gets out of a range from 0 up to 1). The factor c reflects weight of the moment of injection u_k in comparison with the ideal moment of injection u_{Ik} in formation of the compromise moment of injection $\tilde{u}_k$.

RESULTS OF MODELLING OF TRANSIENTS IN THE IBR-2 REACTOR

For the mode of stabilization and the mode of reactor going up on the set period influence of parameters of automatic regulators with transfer functions (27) and (36) on power transients is estimated. In model of reactor IBR-2 the feedback caused by a warming up of a reactor is considered. The structure and parameters of this feedback are received as a result of mathematical treatment of the registered test transients caused by deliberate changes of reactivity.

Transients were modeled both at random disturbance of reactivity and at regular (jump of reactivity). It was considered that random disturbance of reactivity are distributed under Gauss law with the dispersion $\sigma_\rho^2 = 0,05^2$. Real disturbance of reactivity have, naturally, more complex character, and character of disturbance depends on mean reactor power [10]. By way of illustration in Figure 10 the sequence of energy of power pulses (registered in January, 2002 when reactor IBR-2 worked at mean power 1.435 MW in a mode of self-regulation, i.e. at the disconnected automatic regulator) is shown. Apparently, the registered process is characterized by essential fluctuations.

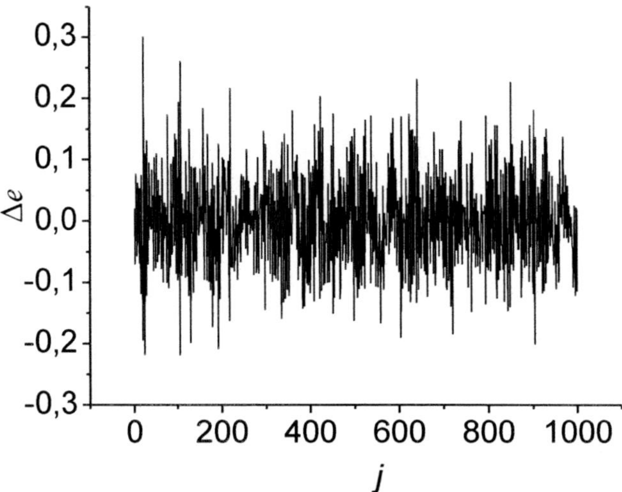

Figure 10. The registered sequence of energy of power pulses.

In Figure 11 the calculated frequency spectrum of reactivity corresponding to these fluctuations of power is shown. Sharp enough peaks of spectral density S_ρ in Figure 11 indicate that regular components (close to sinusoidal components) are present at noise of reactivity.

For a mode of stabilization on the IBR-2 reactor model of power transients are calculated at random disturbance of reactivity. The quantity of the calculated power pulses is accepted equal to $K=1000$ (at that duration of transients is equal to 208 s). In Table 1 the mean square deviations of energy of power pulses $\sigma = \sqrt{\dfrac{1}{K}\sum_{j=1}^{K}\Delta e_j^2}$ are resulted for various levels of the

reactor mean power $\overline{N}$ at various values of parameter of the automatic regulator T_A.

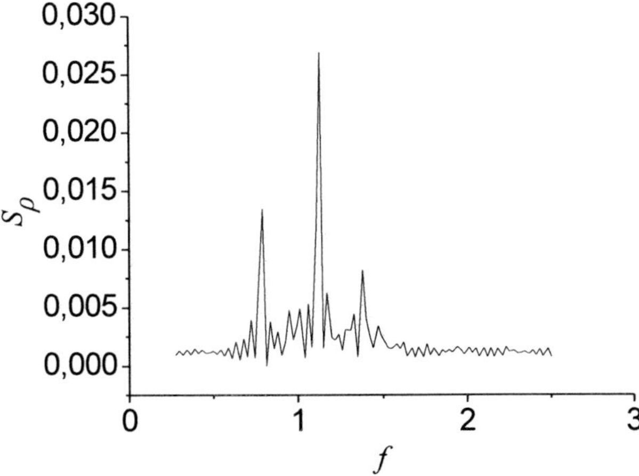

Figure 11. The calculated frequency spectrum of the reactivity corresponding registered process shown in Figure 10.

Table 1. The mean square deviations of energy of power pulses

$\overline{N}$, MW	T_A/T					
	1.5	3	6	12	24	48
	σ					
0.465	0.064	0.057	0.055	0.053	0.052	0.052
0.920	0.065	0.058	0.055	0.053	0.052	0.052
1.415	0.068	0.059	0.056	0.054	0.054	0.053
1.475	0.068	0.060	0.056	0.055	0.054	0.054

As one would expect from physical reasons σ decreases at increase in the degree of inertance of the regulator T_A/T (in other words, at reduction of transfer coefficient of the regulator T/T_A). Reduction of the mean square deviation σ is sharply slowed down and practically stops at enough big inertance of the regulator. In Figures 12 − 15 the transients caused by jump of reactivity $\rho = -0.01$ are shown at various values of parameter of the regulator T_A/T (from 6 up to 48). These transients correspond to various levels of mean power of the reactor.

From figures it is visible that control time and overcorrection decrease with reduction of T_A/T. However at fast enough regulator (i.e. at small enough value of T_A/T) quality of transient worsens, transient gets oscillatory character (Figure 16).

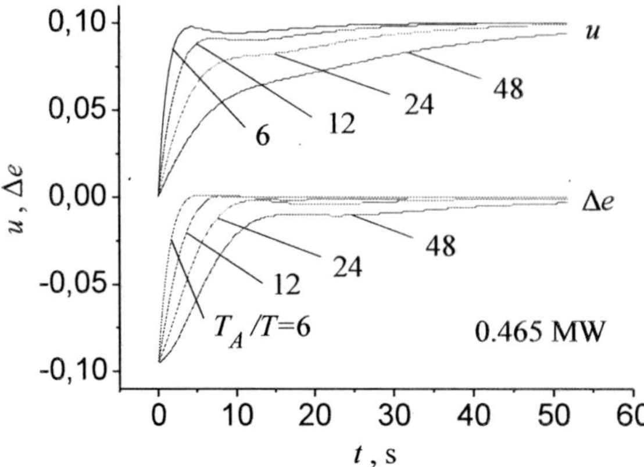

Figure 12. Transients at jump of reactivity $\rho = -0.01$ at various values of parameter of the regulator T_A/T (t – time, u – reactivity of the regulator, Δe – relative deviation of energy of power pulses, mean power of the reactor is 0.465 MW).

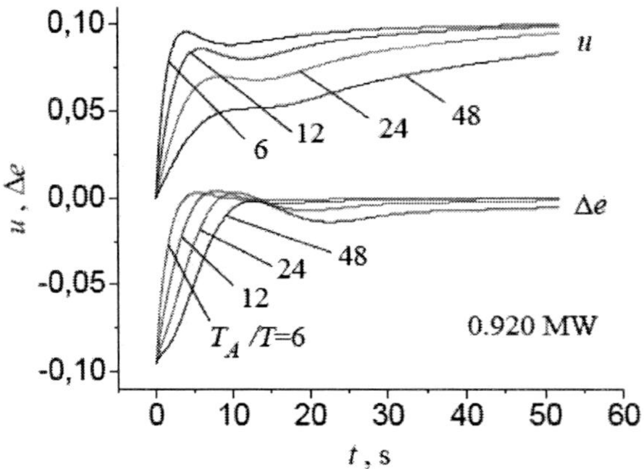

Figure 13. The same that Figure 12 but at mean power 0.920 MW.

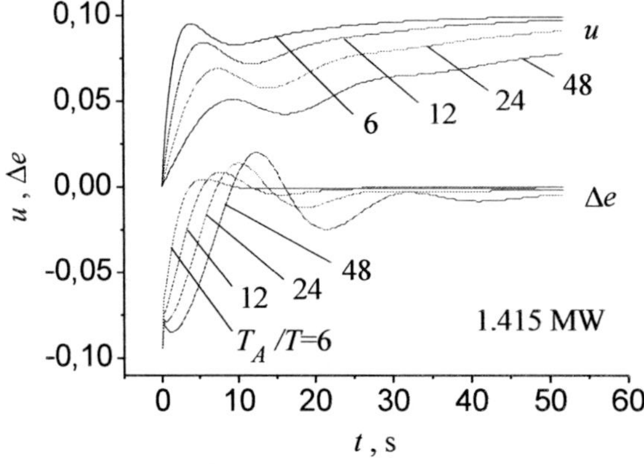

Figure 14. The same that Figure 12 but at mean power 1.415 MW.

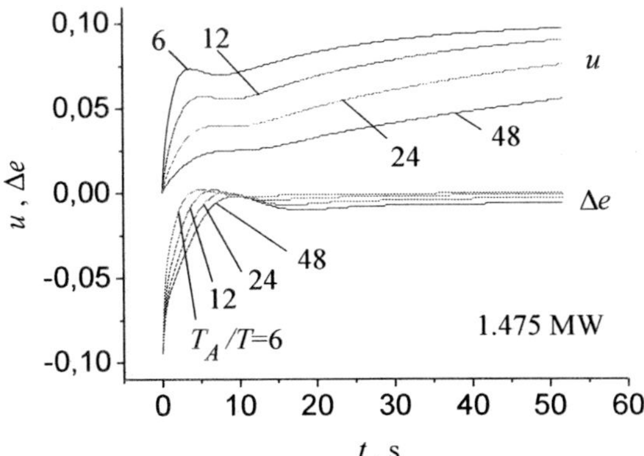

Figure 15. The same that Figure 12 but at mean power 1.475 MW.

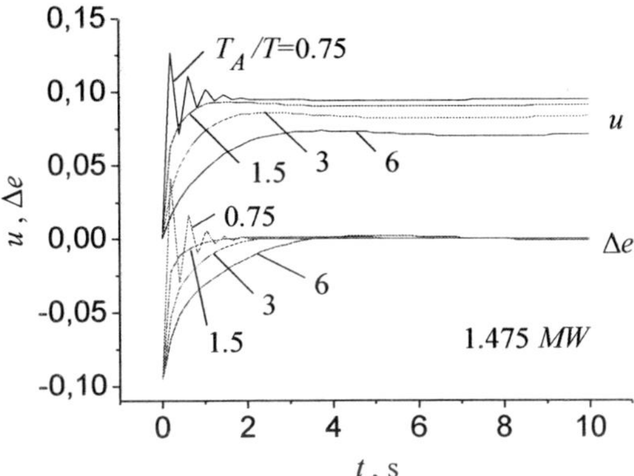

Figure 16. Transients similar to processes in Figure 15 but at faster regulator.

In Figure 17, 18 transients are shown at disturbance of reactivity in the form of the sum of the regular component (jump) and the random component. Each of these figures corresponds to the fixed value of parameter T_A/T.

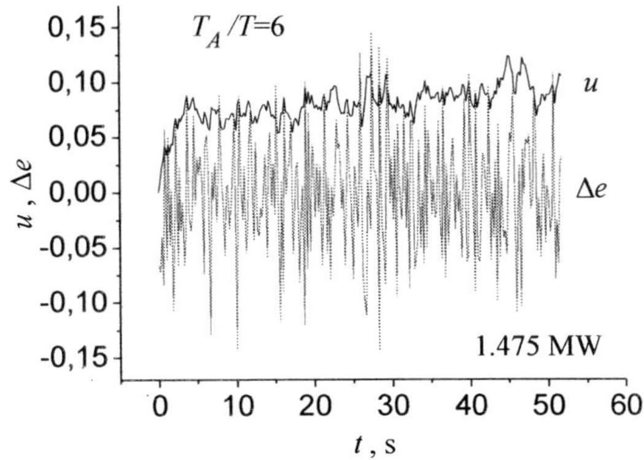

Figure 17. Transients at disturbance of reactivity in the form of the jump and the random component (u – reactivity of the regulator, Δe – relative deviation of energy of power pulses, t – time).

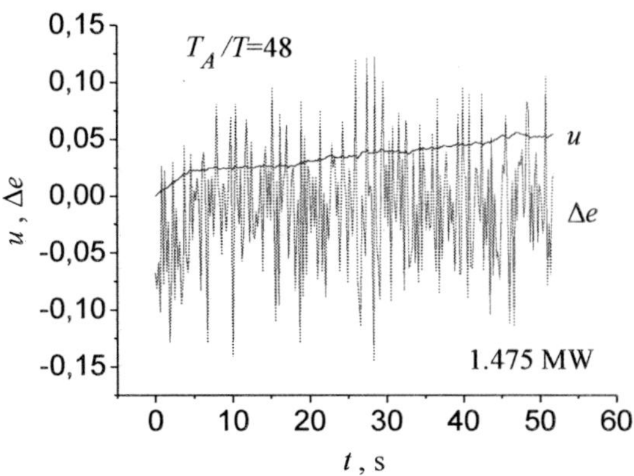

Figure 18. The same that in Figure 17 but at slower regulator.

In Figure 19 transients of reactivity of the regulator are shown at disturbance of reactivity in the form of jump and random component and various values of parameter T_A/T. Transients of power are not shown to not shade the figure.

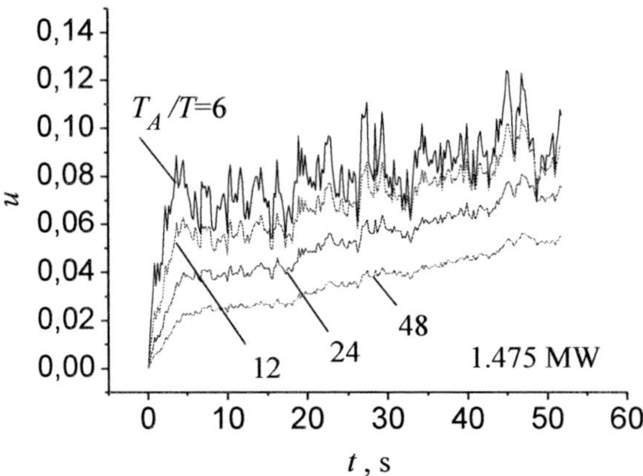

Figure 19. Reactivity of the regulator at various values of parameter T_A/T (disturbance of reactivity in the form of jump and random component).

The Figure 19 is an additional illustration of discrepancy of demands shown to a regulator when disturbance of reactivity contains simultaneously regular and random components. With increase in parameter T_A/T the regulator reacts less sharply to fluctuations of power (i.e. more slowly reacts to random disturbance of reactance) but, at the same time, compensates regular disturbance of reactivity more slowly.

At some degree of regulator inertance the further increase of inertance does not lead to appreciable reduction of mean square deviation of energy of power pulses. At the same time it noticeably worsens quality of the transients caused by sharp regular disturbances of reactivity. In this connection it is expedient to choose value of regulator parameter T_A/T from conditions of the compromise. Proceeding from results of modeling it is possible to conclude that the range of values T_A/T from 12 up to 24 quite satisfies to these compromise conditions for the IBR-2 reactor.

In Figures 20, 21 transients for the mode of reactor going up on the set period are shown at random disturbance of reactivity. For greater presentation of figures the set inverse reactor period α_R^0 differed from zero short time. During this time it was accepted $\alpha_R^0 = 0.05 \text{ s}^{-1}$ that corresponded to small enough reactor period (20 s). The Figure 20 corresponds to the regulator parameter $T_A/T = 48$, and Figure 21 – to the parameter $T_A/T = 6$.

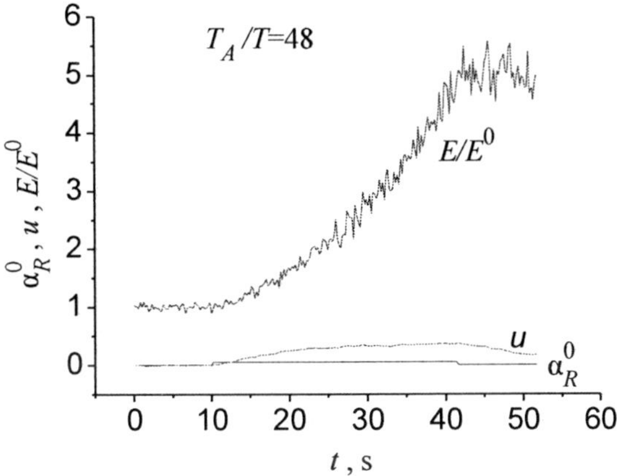

Figure 20. Reactor going up on the set inverse period α_R^0 (t – time, u – the regulator reactivity, E/E^0 – energy of power pulses in relative units).

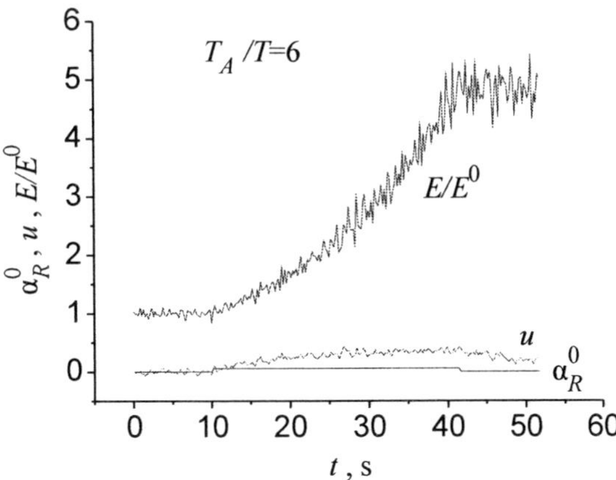

Figure 21. The same that Figure 20 but at faster regulator.

Transients of the regulator reactivity corresponding these two values of parameter T_A/T are combined (for greater presentation) in Figure 22. From Figure 22 it is visible that for the mode of reactor going up it is expedient to use more inertial regulator.

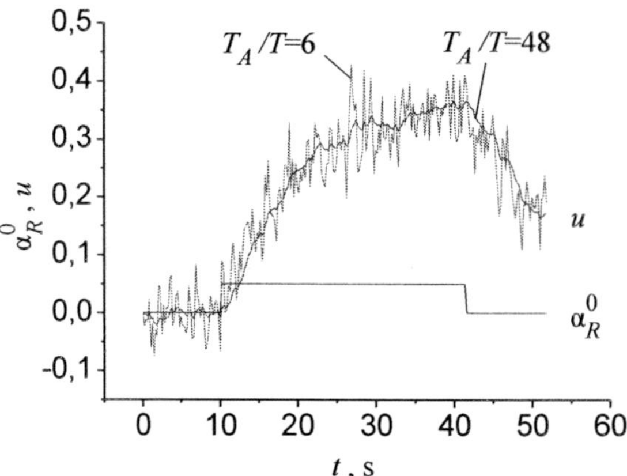

Figure 22. The regulator reactivity u at various values of parameter T_A/T for the mode of reactor going up on the set inverse period α_R^0.

CONCLUSION

The principle work and feature of the design of a pulsed reactor are the essential fluctuations of power pulse amplitudes generated by a reactor. These fluctuations reaching 40% are caused by high sensitivity of a pulsed reactor to changes of reactivity. The general problem of definition of optimal (in statistical sense) automatic control algorithms is considered for various operating modes of a pulsed reactor. A minimum of the expected mean-square deviation of a control parameter of the future power pulse from base value of this parameter is accepted as the optimality criterion. To define optimal algorithm elements of the theory of optimal systems are used. The optimal algorithm is received on the assumption, that in a regulator any information on the previous pulses can be used. Feature of definition of algorithm is the use of, in addition, an entered concept of a degree of ageing of the information. This concept has clear enough physical sense. The algorithms corresponding to the decision of a problem in the general statement are received. On the basis of the analysis of the received algorithms the proved simplification of algorithms and their reduction to the kind convenient for realization (when in formation of control action the parameters accessible to direct measurement are only used) is carried out. Structures of the regulators realizing simplified algorithms are certain. It is shown that the algorithms conceived as a result of physically proved simplifications essentially do not miss traditional algorithms. On the one hand, it specifies a correctness of the considered method of a choice of algorithm, and with another, once again confirms efficiency of the engineering intuitive approach demanded from a regulator certain inertia. On model of the reactor modeling transients is realized at random and regular disturbances of reactivity for the mode of power stabilization and for the mode of reactor going up on the set period. The estimation of influence of the parameter of the regulator on transients is given. The expediency of a compromise choice of a degree of regulator inertance is confirmed.

APPENDIX 1
ELEMENTS OF THE THEORY OF OPTIMAL CONTROL SYSTEMS

Let's state elements of the theory of the optimal control systems [7] allowing to calculate optimal (in statistical sense) algorithm of regulation on an example of the system shown in Figure A1. Figure A1 repeats Figure 6 corresponding system of stabilization of reactor power.

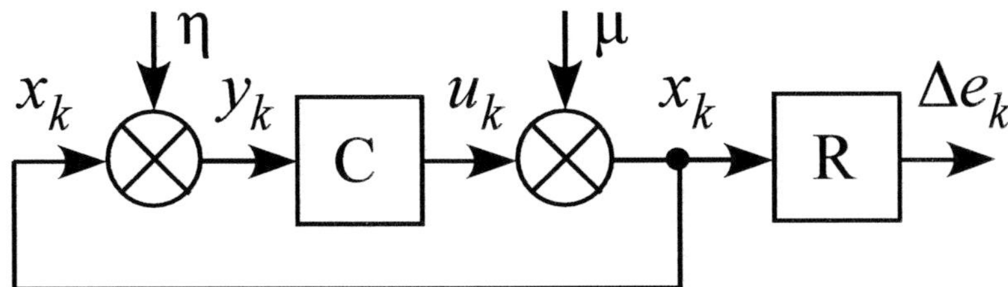

Figure A1. The block diagram of system "the regulator – the reactor" for definition of algorithm of regulation for a mode of stabilization of power (C – regulator, R – reactor).

We shall remind sense of the parameters specified in Figure A1. Δe – controlled parameter of the control object, u– control action of the regulator C on the object R, μ – extraneous influence on object (random disturbance), $x = u + \mu$ – total influence on the object, η – random error of measurement of total influence, $y = x + \eta$ – total influence x, measured with the error η, an index k – number of the pulse.

Let's enter specific loss function

$$Q_k = Q_k(k, x_k),\tag{A1.1}$$

i.e. loss function for k-th pulse. We shall name optimal system for which the specific risk being mathematic expectation of specific loss function is minimal:

$$R_k = M\{Q_k\} = \min.\tag{A1.2}$$

For a generality we believe that the regulator possesses memory and, hence, u_k represents function from values of variables y_j and u_j acted on an input of the regulator in the previous pulses ($j < k$). We shall enter time vectors

$$\begin{aligned}\vec{u}_{k-1} &= (u_0, u_1, \ldots, u_{k-1})\\ \vec{y}_{k-1} &= (y_0, y_1, \ldots, y_{k-1})\end{aligned},\tag{A1.3}$$

where the zero index concerns to a reference mark.

The problem consists in definition optimal (generally random) strategy of functioning of the regulator, in other words, in definition of optimal probability density

$$P_k(u_k) = \Gamma_k(u_k|\vec{u}_{k-1}, \vec{y}_{k-1}),$$
(A1.4)

at which the minimal value of specific risk R_k is reached. As Γ_k is a probability density then $\Gamma_k \geq 0$ and the following condition should be satisfied

$$\int_{\Omega(u_k)} \Gamma_k(u_k)d\Omega = 1.$$
(A1.5)

Hereinafter we shall designate by the letter Ω the area including set of possible values of random variables (in this case u_k) and by the symbol $d\Omega$ – infinitesimal element of this area.

For a conclusion of the formula for specific risk R_k it is necessary to find all over again expression for conditional specific risk r_k (r_k it is specific risk R_k at the fixed values of the variables acting on an input of the regulator).

$$r_k = M\{Q_k|\vec{u}_{k-1}, \vec{y}_{k-1}\} = \int_{\Omega(x_k)} Q_k(k, x_k)P(x_k|\vec{u}_{k-1}, \vec{y}_{k-1})d\Omega.$$
(A1.6)

Probability density in expression (A1.6) we shall present in the form of

$$P(x_k|\vec{u}_{k-1}, \vec{y}_{k-1}) = \int_{\Omega(\mu,u_k)} P(x_k|\mu, u_k)P(\mu, u_k|\vec{u}_{k-1}, \vec{y}_{k-1})d\Omega.$$
(A1.7)

Further according to the formula of multiplication of probabilities we shall write down

$$P(\mu, u_k|\vec{u}_{k-1}, \vec{y}_{k-1}) = P(\mu|\vec{u}_{k-1}, \vec{y}_{k-1})P(u_k|\mu, \vec{u}_{k-1}, \vec{y}_{k-1}),$$
(A1.8)

where $P(\mu|\vec{u}_{k-1}, \vec{y}_{k-1})$ – the a posteriori probability density μ for k-th pulse. At the fixed vectors $\vec{u}_{k-1}$ and $\vec{y}_{k-1}$ u_k does not depend on μ. Therefore

$$P(u_k|\mu, \vec{u}_{k-1}, \vec{y}_{k-1}) = \Gamma(u_k|\vec{u}_{k-1}, \vec{y}_{k-1}).$$
(A1.9)

Let's copy the equation (A1.7) in view of expressions (A1.8) and (A1.9):

$$P(x_k|\vec{u}_{k-1}, \vec{y}_{k-1}) = \int_{\Omega(\mu,u_k)} P(x_k|\mu, u_k)P(\mu|\vec{u}_{k-1}, \vec{y}_{k-1})\Gamma(u_k|\vec{u}_{k-1}, \vec{y}_{k-1})d\Omega$$

(A1.10)

Having substituted the equation (A1.10) in the equation (A1.6) we shall receive

$$r_k = \int\limits_{\Omega(\mu,u_k,x_k)} Q_k(k,x_k)P(x_k|\mu,u_k)P(\mu|\vec{u}_{k-1},\vec{y}_{k-1})\Gamma(u_k|\vec{u}_{k-1},\vec{y}_{k-1})d\Omega .$$

$$(A1.11)$$

To find the a posteriori probability density of μ for the k-th pulse $P_k(\mu) = P(\mu|\vec{u}_{k-1},\vec{y}_{k-1})$ we shall consider joint probability density $P(\mu,\vec{u}_{k-1},\vec{y}_{k-1})$ having presented it in the form of product:

$$P(\mu,\vec{u}_{k-1},\vec{y}_{k-1}) = P(\vec{u}_{k-1},\vec{y}_{k-1}|\mu)P_0(\mu) = P(\mu|\vec{u}_{k-1},\vec{y}_{k-1})P(\vec{u}_{k-1},\vec{y}_{k-1})$$

$$(A1.12)$$

Here $P_0(\mu)$ – the a priori probability density of μ .

From the equation (A1.12) it is found the a posteriori probability density for k-th pulse in view of the supervision made in the previous pulses:

$$P_k(\mu) = P(\mu|\vec{u}_{k-1},\vec{y}_{k-1}) = \frac{P_0(\mu)P(\vec{u}_{k-1},\vec{y}_{k-1}|\mu)}{P(\vec{u}_{k-1},\vec{y}_{k-1})} . \qquad (A1.13)$$

Let's consider more in detail probability density of the complex event $P(\vec{u}_{k-1},\vec{y}_{k-1}|\mu)$ entering in the equation (A1.13). This complex event consists in the following. At the fixed value μ it is appeared at first the pair values u_0, y_0 (the first event), then the pair values u_1, y_1 (the second event) and so on. Probability density of this complex event is expressed as product of several factors. The first factor is probability density $P(u_0,y_0|\mu)$ of the first event, the second factor is probability density $P(u_1,y_1|\mu,u_0,y_0)$ of the second event provided that there was a first event, etc. Thus,

$$P(\vec{u}_{k-1},\vec{y}_{k-1}|\mu) =$$
$$P(u_0,y_0|\mu)P(u_1,y_1|\mu,u_0,y_0)P(u_2,y_2|\mu,\vec{u}_1,\vec{y}_1) \dots P(u_{k-1},y_{k-1}|\mu,\vec{u}_{k-2},\vec{y}_{k-2}).$$

$$(A1.14)$$

In the formula (A1.14) we shall consider j-th factor $(0 < j \le k-1)$ and we shall transform it:

$$P(u_j,y_j|\mu,\vec{u}_{j-1},\vec{y}_{j-1}) = P(y_j|\mu,u_j,\vec{u}_{j-1},\vec{y}_{j-1})P(u_j|\mu,\vec{u}_{j-1},\vec{y}_{j-1}) .$$

$$(A1.15)$$

Let's consider factors in the right part of the equation (A1.15). The first factor represents the a posteriori probability density of y_j for j-th pulse provided that μ and u_j are fixed. This probability density will not change if in addition to fix time vectors $\vec{u}_{j-1}$, $\vec{y}_{j-1}$ (Figure A1), i.e. it does not depend on these vectors:

$$P(y_j|\mu,u_j,\vec{u}_{j-1},\vec{y}_{j-1}) = P(y_j|\mu,u_j). \tag{A1.16}$$

In the second factor vectors $\vec{u}_{j-1}$ and $\vec{y}_{j-1}$ are fixed. Hence, it does not depend on μ (Figure A1). Thus, the second factor represents the random strategy Γ_j of the regulator for j-th pulse:

$$P(u_j|\mu,\vec{u}_{j-1},\vec{y}_{j-1}) = \Gamma_j(u_j|\vec{u}_{j-1},\vec{y}_{j-1}). \tag{A1.17}$$

Considering formulas (A1.16) and (A1.17) we shall write down the formula (A1.15) in the form of

$$P(u_j,y_j|\mu,\vec{u}_{j-1},\vec{y}_{j-1}) = P(y_j|\mu,u_j)\Gamma_j. \tag{A1.18}$$

In view of the expression (A1.18) we shall write down the equation (A1.14) as follows:

$$P(\vec{u}_{k-1},\vec{y}_{k-1}|\mu) = \left[\prod_{j=0}^{k-1} P(y_j|\mu,u_j)\right]\prod_{j=0}^{k-1}\Gamma_j, \tag{A1.19}$$

where the designation is entered

$$\Gamma_0 = P_0(u_0). \tag{A1.20}$$

The probability density (A1.20) does not depend on supervision which during the initial moment yet was not and is a priori.

Having substituted the equation (A1.19) in the equation (A1.13) we shall receive following expression a posteriori probability density (A1.13):

$$P_k(\mu) = P(\mu|\vec{u}_{k-1},\vec{y}_{k-1}) = \frac{P_0(\mu)\left[\prod_{j=0}^{k-1} P(y_j|\mu,u_j)\right]\prod_{j=0}^{k-1}\Gamma_j}{P(\vec{u}_{k-1},\vec{y}_{k-1})} \tag{A1.21}$$

Substituting the equation (A1.21) in the equation (A1.11) we shall receive final expression for conditional specific risk:

$$r_k = \int\limits_{\Omega(\mu,u_k,x_k)} Q_k(k,x_k)P(x_k|\mu,u_k)\frac{P_0(\mu)\left[\prod\limits_{j=0}^{k-1}P(y_j|\mu,u_j)\right]}{P(\vec{u}_{k-1},\vec{y}_{k-1})}\prod\limits_{j=0}^{k}\Gamma_j d\Omega$$

$$(A1.22)$$

Let's consider values r_k at the various experiences made with system. In these experiences vectors $\vec{u}_{k-1}$ and $\vec{y}_{k-1}$ (which, generally speaking, in advance are not known) can accept various values. The probability density $P(\vec{u}_{k-1},\vec{y}_{k-1})$ represents joint probability density of these vectors. The specific risk for k-th pulse R_k is average value of conditional specific risk r_k at mass production of experiences and is defined by the formula

$$R_k = \int\limits_{\Omega(\vec{u}_{k-1},\vec{y}_{k-1})} r_k P(\vec{u}_{k-1},\vec{y}_{k-1})d\Omega .$$

$$(A1.23)$$

The formula (A1.23) after substitution in it the expression (A1.22) will become

$$R_k = \int\limits_{\Omega(\mu,x_k,\vec{u}_k,\vec{y}_{k-1})} Q_k(k,x_k)P(x_k|\mu,u_k)P_0(\mu)\left[\prod\limits_{j=0}^{k-1}P(y_j|\mu,u_j)\right]\prod\limits_{j=0}^{k}\Gamma_j d\Omega$$

$$(A1.24)$$

For definition of optimal control strategy we shall copy the equation (A1.24) in the form of

$$R_k = \int\limits_{\Omega(\vec{u}_k,\vec{y}_{k-1})} \Gamma_k(u_k|\vec{u}_{k-1},\vec{y}_{k-1})\prod\limits_{j=0}^{k-1}\Gamma_j(u_j|\vec{u}_{j-1},\vec{y}_{j-1})\times$$

$$\left\{\int\limits_{\Omega(\mu,x_k)} Q_k(k,x_k)P(x_k|\mu,u_k)P_0(\mu)\left[\prod\limits_{j=0}^{k-1}P(y_j|\mu,u_j)\right]d\Omega(\mu,x_k)\right\}d\Omega(\vec{u}_k,\vec{y}_{k-1})$$

$$(A1.25)$$

Integral inside of the braces being function $\vec{u}_k$ and $\vec{y}_{k-1}$ we shall designate κ_k:

$$\kappa_k = \kappa_k(u_k, \vec{u}_{k-1}, \vec{y}_{k-1}) =$$

$$\int\limits_{\Omega(\mu, x_k)} Q_k(k, x_k) P(x_k|\mu, u_k) P_0(\mu) \left[\prod_{j=0}^{k-1} P(y_j|\mu, u_j)\right] d\Omega(\mu, x_k). \qquad (A1.26)$$

Using this designation we shall write down the formula (A1.25) in the form of

$$R_k = \int\limits_{\Omega(\vec{u}_k, \vec{y}_{k-1})} \kappa(u_k, \vec{u}_{k-1}, \vec{y}_{k-1}) \prod_{j=0}^{k-1} \Gamma_j(u_j|\vec{u}_{j-1}, \vec{y}_{j-1}) \Gamma_k(u_k|\vec{u}_{k-1}, \vec{y}_{k-1}) d\Omega =$$

$$\int\limits_{\Omega(\vec{u}_{k-1}, \vec{y}_{k-1})} \prod_{j=0}^{k-1} \Gamma_j(u_j|\vec{u}_{j-1}, \vec{y}_{j-1}) \left\{ \int\limits_{\Omega(u_k)} \kappa_k(\vec{u}_k, \vec{y}_{k-1}) \Gamma_k(u_k|\vec{u}_{k-1}, \vec{y}_{k-1}) d\Omega(u_k) \right\} \times$$

$$d\Omega(\vec{u}_{k-1}, \vec{y}_{k-1})$$

$$(A1.27)$$

or in a following kind

$$R_k = \int\limits_{\Omega(\vec{u}_{k-1}, \vec{y}_{k-1})} \left[\prod_{j=0}^{k-1} \Gamma_j\right] \chi_k(\vec{u}_{k-1}, \vec{y}_{k-1}) d\Omega, \qquad (A1.28)$$

where

$$\chi_k(\vec{u}_{k-1}, \vec{y}_{k-1}) = \int\limits_{\Omega(u_k)} \kappa_k(u_k, \vec{u}_{k-1}, \vec{y}_{k-1}) \Gamma_k(u_k|\vec{u}_{k-1}, \vec{y}_{k-1}) d\Omega(u_k)$$

$$(A1.29)$$

So, when strategy $\Gamma_0, \ldots, \Gamma_{k-1}$ are somehow realized (i.e. are fixed) the optimal regulator should develop such strategy Γ_k for k-th pulse at which the specific risk R_k will be minimal. Apparently from the equations (A1.28) and (A1.29) it will be carried out if for any vectors $\vec{u}_{k-1}$ and $\vec{y}_{k-1}$ to pick up strategy Γ_k so that function χ_k is minimal.

Using the mean value theorem and considering the formula (A1.5) we shall write down the equation (A1.29) as follows:

$$\chi_k = \tilde{\kappa}_k \int\limits_{\Omega(u_k)} \Gamma_k d\Omega = \tilde{\kappa}_k \geq (\kappa_k)_{\min}, \qquad (A1.30)$$

where $\tilde{\kappa}_k$ – some mean value of function κ_k and $(\kappa_k)_{\min}$ – its minimal value.

Let $u_k^*(\vec{u}_{k-1}, \vec{y}_{k-1})$ is the value u_k corresponding to a minimum of the function κ_k, i.e.

$$\kappa_k(u_k^*, \vec{u}_{k-1}, \vec{y}_{k-1}) = \min_{(u_k)} \kappa_k(u_k, \vec{u}_{k-1}, \vec{y}_{k-1}).$$
(A1.31)

Let's consider delta-function

$$\Gamma_k = \delta(u_k - u_k^*).$$
(A1.32)

This function satisfies to a condition (A1.5). Function (A1.32) is the optimal strategy of the regulator at which the minimum of risk R_k is reached. To be convinced of it we shall substitute the equation (A1.32) in the equation (A1.29) and consider the expression (A1.31). As a result we shall receive

$$\chi_k\big|_{\Gamma_k = \delta(u_k - u_k^*)} = \int_{\Omega(u_k)} \kappa_k(u_k, \vec{u}_{k-1}, \vec{y}_{k-1})\delta(u_k - u_k^*)d\Omega = \kappa_k(u_k^*, \vec{u}_{k-1}, \vec{y}_{k-1}) = (\kappa_k)_{\min}$$
(A1.33)

This value is minimally possible for the function χ_k on what the expression (A1.30) points. Thus, strategy (A1.32) is optimal and, as follows from the equation (A1.32), regular.

So, the optimal algorithm of the regulator consists in a choice of value u_k^* at which the minimum of function κ_k (A1.26) is reached.

APPENDIX 2
CONCLUSION OF OPTIMAL ALGORITHM

We minimize function κ_k (A1.26) for k-th pulse on control action in this pulse. For our case (Figure 6) following expressions are fair

$$P(y_j|\mu, u_j) = P_j(\eta) = P_j[\eta = y_j - (u_j + \mu)]$$
(A2.1)

$$P(x_k|\mu, u_k) = \delta[x_k - (u_k + \mu)]$$
(A2.2)

In the equation (A2.2) δ there is a delta-function owing to what integration on x_k in the equation (A1.26) is replaced with substitution $x_k = u_k + \mu$.

At theoretical research for simplification of calculations it is possible to consider (considering the normal distribution law of μ) that μ can change from $-\infty$ up to $+\infty$. In this case using parities (19) – (22), (A2.1) and (A2.2) it is possible to result the equation (A1.26) in a following kind (to within constant factor)

$$\kappa_k = \int\limits_{-\infty}^{\infty} \left\{ \exp\left(u_k + \mu + \Delta\rho_{\theta k} + \frac{\Delta S_k}{S^0} \right) - 1 \right\}^2 \times$$

$$\exp\left(-\frac{(\mu - m_0)^2}{2\sigma_\mu^2} - \frac{\sum\limits_{j=0}^{k-1} a_{k-j}^2 [y_j - (u_j + \mu)]^2}{2\sigma_\eta^2} \right) d\mu \,. \tag{A2.3}$$

Having lead transformation of expression in braces the integral (A2.3) can be presented in the form of the sum of three integrals. Owing to the accepted infinite limits these integrals are easily taken by the formula

$$\int\limits_{-\infty}^{\infty} \exp(-p\mu^2 \pm q\mu)d\mu = \sqrt{\frac{\pi}{p}} \exp\left(\frac{q^2}{4p} \right),$$

$$(p > 0)\,.$$

As a result the equation (A2.3) will become

$$\kappa_k = \sqrt{\frac{\pi}{p}} \exp(G) \times$$

$$\left[\exp\left(2\Delta\rho_{\theta k} + 2\frac{\Delta S_k}{S^0} + 2u_k + \frac{q_1^2}{4p} \right) - 2\exp\left(\Delta\rho_{\theta k} + \frac{\Delta S_k}{S^0} + u_k + \frac{q_2^2}{4p} \right) + \exp\left(\frac{q_3^2}{4p} \right) \right]\,. \tag{A2.4}$$

Here following designations are used:

$$G = \frac{-\sigma_\eta^2 m_0^2 - \sigma_\mu^2 \sum\limits_{j=0}^{k-1} a_{k-j}^2 (y_j - u_j)^2}{2\sigma_\mu^2 \sigma_\eta^2} \,,$$

$$p = \frac{\sigma_\eta^2 + \sigma_\mu^2 \sum\limits_{j=0}^{k-1} a_{k-j}^2}{2\sigma_\mu^2 \sigma_\eta^2} \,,$$

$$q_1 = 2 + \Sigma \,,\; q_2 = 1 + \Sigma \,,\; q_3 = \Sigma \,,$$

$$\Sigma = \frac{\sigma_\eta^2 m_0 + \sigma_\mu^2 \sum_{j=0}^{k-1} a_{k-j}^2 (y_j - u_j)}{\sigma_\mu^2 \sigma_\eta^2}$$

At last taking a derivative of function (A2.4) on u_k and equating a derivative to zero we shall receive the formula for optimal control action:

$$u_k = -\Delta\rho_{\theta k} - \frac{\Delta S_k}{S^0} - \frac{\dfrac{3}{2}\sigma_\eta^2 + \left(\dfrac{\sigma_\eta}{\sigma_\mu}\right)^2 m_0 + \sum_{j=0}^{k-1} a_{k-j}^2 (y_j - u_j)}{\left(\dfrac{\sigma_\eta}{\sigma_\mu}\right)^2 + \sum_{j=0}^{k-1} a_{k-j}^2}$$

$$(A2.5)$$

REFERENCES

[1] Popov A.K. On statistically optimal regulation of pulse energy of the fast reactor. *Atomnaya energia,* 1971, v. 31, issue 3, p. 269.

[2] Popov A.K., Marachev A. A. On statistically optimal algorithms of regulation for different regimes of a pulsed reactor. *JINR Communication* P13-2002-277, Dubna, 2002.

[3] Shabalin E.P. Pulsed reactors on fast neutrons. *Atomizdat,* Moscow, 1976.

[4] Popov A.K. The impulse transfer coefficient of the IBR-2 reactor. *JINR Communication* P3-95-463, Dubna, 1995.

[5] Bondarenko I.I., Stavisski Yu.Ya. Pulsed mode of operation of the fast reactor. *Atomnaya energia,* 1959, v. 17, issue 5, p. 417-422.

[6] Pepelyshev Yu.N., Popov A.K. Investigation of reactivity effects of mobile reflectors of the IBR-2 reactor in dynamics. *JINR Preprint* P-13-2005-184, Dubna, 2005.

[7] Feldbaum A.A. Basis of the theory of optimal control systems. *Fizmatgiz,* Moscow, 1963.

[8] Popov A.K. Regulation of pulse energy in the fast pulsed reactor using an injector. *Atomnaya energia,* 1969, v. 27, issue 6, p. 554-556.

[9] Pepyolyshev Yu.N., Popov A.K. Control of reactor pulses by external neutron injection. *Annals of Nuclear Energy,* 2001, 28, p. 1613-1624.

[10] Bondarchenko E.A., Pepelyshev Yu.N., Popov A.K. Experimental and Model Study of Dynamic Features of the IBR-2 Periodic Pulsed Reactor. *Physics of Particles and Nuclei,* 2004, vol. 35, No. 4, p. 498-529.

In: Nuclear Waste Research: Siting, Technology and Treatment   ISBN 978-1-60456-184-5
Editor: Arnold P. Lattefer, pp. 143-165   © 2008 Nova Science Publishers, Inc.

Chapter 4

APPLICATION OF GAS-COOLED REACTOR TECHNOLOGY TO DESIGN OF ACCELERATOR DRIVEN SYSTEMS FOR NUCLEAR WASTE TRANSMUTATION

Alberto Abánades
Universidad Politécnica de Madrid.
Avda. Ramiro de Maeztu, 7, 28040 Madrid

ABSTRACT

Accelerator Driven Systems (ADS) are innovative nuclear concepts that are currently under study for its application to nuclear waste management. ADS are subcritical devices driven by a high power external neutron source, usually generated by spallation reactions in heavy materials induced by accelerated protons. The subcritical operation mode of the nuclear core opens the possibility to use fuel matrixes with high minor actinide contents in a safe way. In that sense, ADS are applied as nuclear waste burners, getting rid of long lived radioactive waste as Pu, Np isotopes, and some long lived fission fragments as ^{99}Tc, producing a certain amount of energy, contributing to the sustainability of nuclear energy by a better energetic profit of the nuclear materials. The development of ADS has also as an objective the reduction of the nuclear waste stockpile and, once integrated into the nuclear fuel cycle, reduce the requirements in monitoring time and volume of final disposal repositories.

The ADS development is encountering several technological challenges, as it is foreseen in the arising of an innovative machine that integrate different fields that has been traditionally separated, as the accelerator, the high energy physics and the reactor physics. The integration of those technologies implies the overcoming of some uncertainties with material damage, accelerator reliability or the fabrication and high-burn up integrity of new high actinide nuclear fuels.

Nevertheless, those technological uncertainties seem to be solved in the short-medium term. In the meantime, nuclear knowledge is reviewed to establish the best options for the practical application of ADS. One of these technologies is based on the experience of gas cooled reactors. Graphite as neutron moderator permits certain spectrum modulation in function of the moderator/fuel ratio to obtain, thermal, fast or epithermal spectrum, and therefore permits spectrum adaptation to the targeted nuclear species to increase transmutation efficiency. The high burn-up that can be obtained by coated particle ceramic fuels, the possibility of high temperature operation and the inherent safety of subcritical systems, are also positive inputs for the deployment of gas-cooled ADS. In this chapter, the current state-of-the-art of this technology will be given, including a discussion about its future development and limitations.

INTRODUCTION

Sustainable development is one of the most searched objectives of industrial, social and economic mankind evolution at the present moment. This aim is applied with special interest to the energy field, where primary resources are limited, and even exhausting, as is the case of fossil fuels in the long term. Nuclear energy is one of the main energy sources in the world and should also envisage that challenge, looking for a more efficient and less harmful exploitation of the Uranium source or any other nuclear fuel, as Thorium.

From the point of view of sustainability of nuclear energy, nuclear wastes management and its inventory minimization are the most important issues that should be addressed. Figure 1 shows the radioactivity evolution in the spent fuel after a typical nuclear reactor discharge. The radioactive activity and hazard is dominated by fission products during the first centuries and by transuranics/actinides in the long term. Radioactive waste management actions should be preserve radioactive element release in the environment during the time in which its activity could imply radiological hazard to the biosphere. This time is somehow depending on the limits that are adopted as safe, but in practice that means that radioactive materials should be confined during tens of thousands of years due to the actinide activity.

The accepted technical solution in the current state-of-the-art of nuclear technology for radioactive waste management is based on waste confinement in long-term deep geological repositories at the end of the process. Such repository should fulfil very strict requirements that should last tens of thousand years, what implies to assess the behaviour of the confinement barriers during a geological time scale. Nevertheless, mankind will be monitoring such installation and correction actions might be possible if lost of confinement barrier is expected, but uncertainties are, obviously, very high about material durability and human behaviour in a so long time period. On the other hand, those strict requirements imply a very high financial cost for deep geological disposal installations, what burden economic viability of nuclear power and nuclear waste management.

A new complementary technique to waste disposal under development is the transmutation of nuclear wastes to less offending species with lower activity in the long term. This technique will be explained more in details in next sections but basically consists on the irradiation of nuclear waste isotopes, mainly transuranics due to its longer radioactive life, to reduce its inventory and the overall activity of the resulted waste stream. Getting rid of transuranics and actinides, the confinement time in the final disposal of the nuclear wastes are reduced from geological to engineering time scale, from tens of thousands to some centuries, what is has been achieved successfully by many civilization in the past (for instance, aztec an roman constructions).

Irradiation and fission of some of the target transuranics produces energy that obviously can be used as primary power source to obtain electricity or any other energy services.

Therefore, the application nuclear waste transmutation seems to be positive for nuclear energy since might imply a reduction of the environmental impact of nuclear energy due to a lower radioactive inventory in the long term; and higher energy extraction efficiency per Uranium ore mass. This rationale might result in a reduction of the nuclear levelized energy cost. Nevertheless, detailed economic analysis is still pending to take into account the cost of transmutation facilities and the new nuclear fuel cycle implementation that is needed.

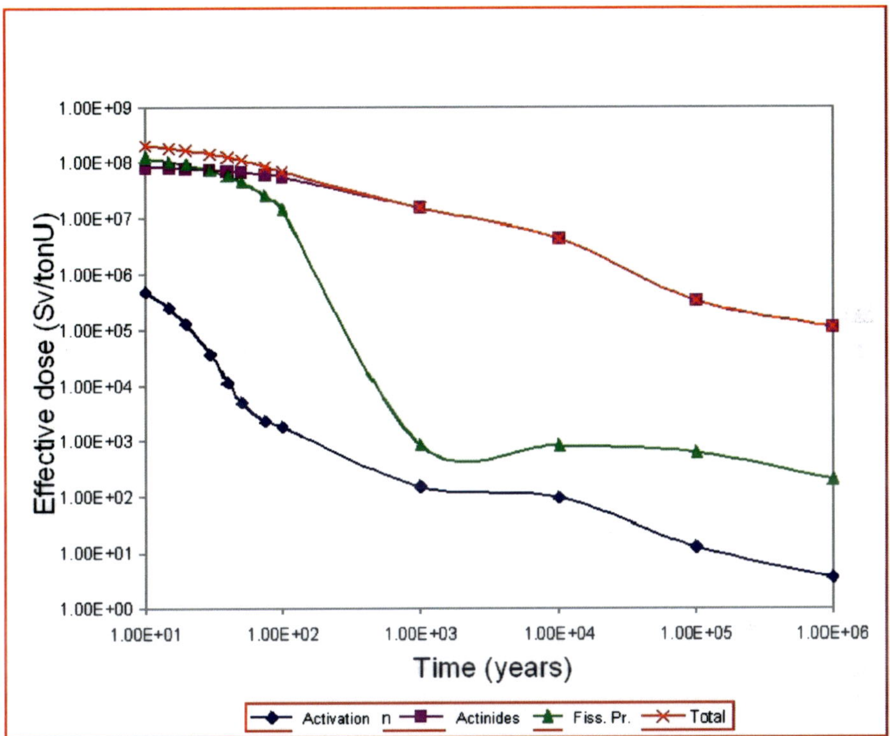

Figure 1. Radioactive waste activity after a typical PWR cycle fuel discharge by nuclide type.

TRANSMUTATION OF NUCLEAR WASTES

Any nuclear reaction in fact produces a transmutation phenomenon by the modification of the nuclear structure of the nucleus producing a transformation of a given isotope to another isotope and even to a new element; or generates two or more nuclei. From this point of view, nuclear transmutation is the basic principle of nuclear technology and its applications. ^{235}U fissions or ^{238}U neutron captures are transmutation processes.

Nuclear transmutation applied to waste management is based on the induction of two basic reaction types:

- Fission of the transuranics, transmuting these long lived elements into less harmful fission products, or neutron capture in transuranics leading to fissionable material (Figure 2). This reaction eliminates the harmful heavy element, producing fission products, reducing in fact the activity time from the typical tens of thousand years of actinides to a few centuries.

- Neutron capture in the rest of the nuclei to obtain new more stable nuclei, even with null activity. (Figure 3).

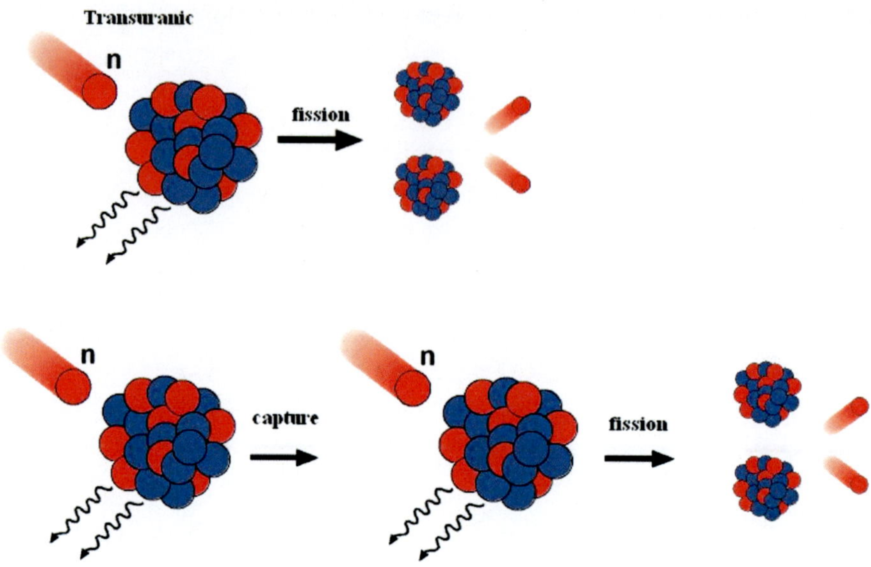

Figure 2. Transmutation process with transuranics.

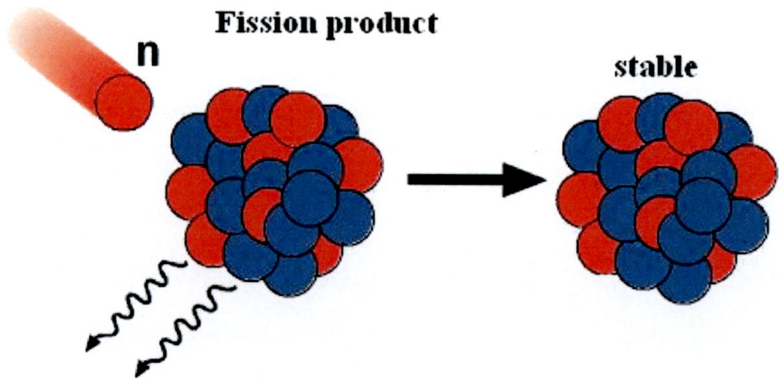

Figure 3. Transmutation process with fission products.

The practical application of this transmutation of nuclear wastes is done in nuclear systems in which transuranic fission and fission product capture will be possible under safe conditions. Fission products are neutronic poison and a large inventory in a reactor burden the neutron multiplication capability and limit the burn-up and the operating cycle, for instance, in commercial thermal power reactors. On the other hand transuranics also can reduce the neutron multiplication when they are fertile material and need a previous neutron capture to produce a fission, but its main problem is that its fraction of delay neutrons (β) is very low (for Pu, three times less than U). Therefore, safety margins before achieving core prompt criticality is much lower in nuclear cores with transuranic-based fuel than with Uranium-based fuel.

New reactor designs adapted to the neutronic requirements of transuranic-based fuels are a must for an efficient application of the transmutation of nuclear wastes. The main options under development are fast reactors, either critical or subcritical, due to the higher fission to capture ratio in most of transuranics in a fast core; or subcritical thermal cores.

ACCELERATOR DRIVEN SYSTEMS (ADS)

Subcritical reactors are the topic of this chapter and their implementation is the result of the combination of a subcritical reactor core with a high power neutron source, which is needed to keep at a certain level the neutron population and the power. This powerful neutron source is generally produced by the impinging of medium-high energy hadrons (protons or deuterons) in a heavy material block. The energetic hadrons induce spallation reactions generating a neutron rich particle cascade. For instance, in the case of 1 GeV protons, neutron yield are of the order of 30 neutrons per proton, and linearly increasing at higher energy, and of the order of 13 neutrons per proton at 600 MeV, that is the range considered in many transmutation concepts under development.

Briefly, an Accelerator Driven System is composed by:

- A medium-high energy proton accelerator: In current concepts, proton energy ranges from 380 to 1500 MeV.
- A spallation target or neutron source: Usually with Lead or Eutectic Lead/Bismuth as spallation material, where protons generated in the accelerator collide producing the neutron cascade.
- A subritical core: Neutrons produced in the high performance source enter this core, design with enough subcriticality margins (k < 0,95). The power generated in this subcritical core is driven by the neutron source and, hence, the proton beam intensity by the well known expression:

$$P_{th} = G \cdot P_{beam} = \frac{G_o}{1-k} \cdot P_{beam}$$

where k is the system criticality (including the source effect) and G_o is a parameter that depends in a first approximation on the number of neutrons produced per proton, and proton energy.

Accelerator Driven Systems has as main advantages respect to fast reactors:

- The subcritical operation mode, which gives higher fuel composition flexibility and less limiting burn-up limitation what means a higher capacity to destroy harmful isotopes, even long-lived fission fragments and Americium and Curium.
- The operation safety if design with multiplication factor low enough as to assess complete subcritical operation under any circumstances, even accidental or abnormal (k < 0.95). Redundant safety measures can be removed in that case, and even control

rods are missing in some proposed concepts, as thermal power in the core is driven by the accelerator.

The main disadvantage is that the accelerator has a non-negligible cost and operation availability in the present state-of-the-art, making fast reactor a very attractive option for a short-term deployment of transmutation of nuclear wastes, although its transmutation capability and potential seem to be lower.

NUCLEAR FUEL CYCLE WITH TRANSMUTATION OF NUCLEAR WASTES

The application of transmutation techniques into the nuclear cycle with the aim of minimising the radioactive inventory and the environmental impact of nuclear electricity requires a complete review and modification of the current nuclear fuel cycle. Nuclear waste transmutation implies the development and implementation of reprocessing facilities to make the partition of the spent nuclear fuel with the isotopic requirements for the fabrication of the fuel that should be loaded in the transmuter. That is the reason why this technology are referred in a global sense as Partitioning and Transmutation (P&T) as efficient transmutation is not possible without a detailed partitioning of the irradiated fuel, and accurate partitioning has no sense without the possibility of isotope transmutation.

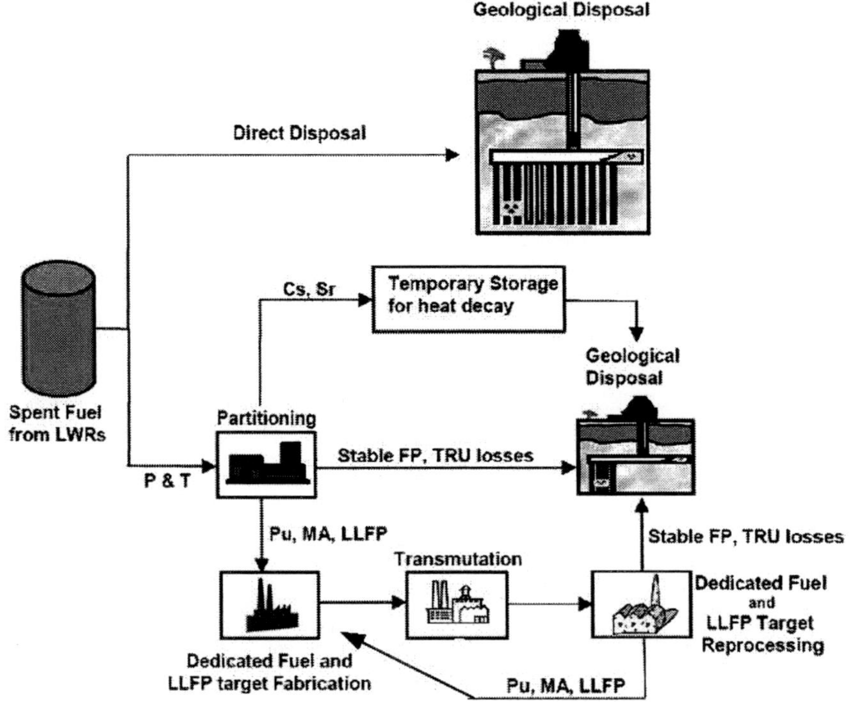

Figure 4. Second stage of the fuel cycle with P&T (Source: IP-EUROTRANS Project).

A basic scheme of the implementation of transmutation technology into a nuclear fuel cycle is depicted in Figure 4 where the alternatives to spent fuel management are shown.

In opposition to a open fuel cycle in which spent fuel is arranged in a geological disposal facility, a sort of closed cycle is introduced in which short lived isotopes are stored in a temporary storage for decay, and the stable fission products and transuranic losses from partitioning are stored in the deep geological facility. The radioactive waste inventory is minimised by Plutonium, minor actinides and long-lived fission fragments continuous reprocessing and re-irradiation in transmutation installation by means of dedicated fuel fabrication reprocessing plants. The nature of those transmutation facilities are currently under study with fast critical reactors or subcritical reactors (ADS) as main candidates.

The implementation of this fuel cycle requires a huge effort to introduced new facilities and detailed cycle integration studies to try to minimise the waste stream finally send to the geological disposal, the transition cost from current fuel cycle installations to the requirements of this new fuel cycle, and the overall cost of electricity production from nuclear source in that scenario.

RESEARCH ACTIVITIES IN THE FIELD OF TRANSMUTATION OF NUCLEAR WASTES

The research activities in the field of nuclear wastes management are focused in two technical options: deep geological disposal and transmutation of nuclear wastes. The latter has an increasing interest in the scientific community since the 90, when the ATW ('Accelerator Transmutation of Wastes') program [Bowman, 1991] and the Rubbia´s Energy Amplifier project [Rubbia, 1995] were initiated.

Transmutation itself is not a new concept and experimental work for that purpose were initiated in the very beginning of nuclear science and technology when Rutherford transmuted ^{14}N to ^{17}O using α-particles [Rutherford, 1920]. During the development of the nuclear technology, the combined use of particle accelerators with nuclei were the practical procedure to create artificially isotopes to be studied to establish the valuable nuclear data that have contributed to the knowledge of decay chains of nuclei. Even today particle accelerators are used for the creation of artificial nuclei and the behaviour of existing ones under particle interaction.

The application of particle accelerator in power subcritical reactors was discarded in the past due to the high performance requirements in terms of availability and particle production. Accelerator technology has evolved in the last decades with the development of powerful accelerators, as it is the case, for instance of the 380-MeV cyclotron at PSI (Switzerland), high excellent facilities as the CERN accelerator complex or high power LINACs (Linear Accelerator). Those developments have given the possibility of integration of accelerator and nuclear technology for the development of high power neutron sources and subcritical systems for power generation.

In Figure 5 we see a general overview of the transmutation processes that can be accomplish in nuclear isotopes. Such processes can be induced by high energy charged particles by direct spallation phenomena, or even by photons in some cases. The most effective particle, in any case, has been found to be the neutron. Neutrons are the main driven particle in existing nuclear reactors and also in new reactors, with Accelerator Driven Systems (ADS) among them. ADS devices can be designed to operate with fast, thermal or epithermal neutron spectra.

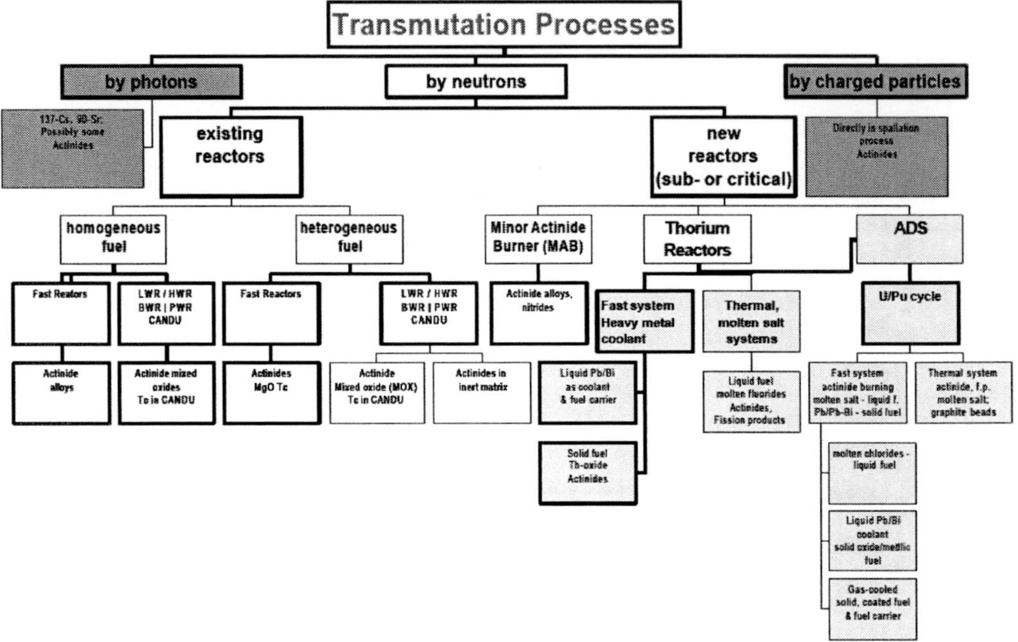

Figure 5. Nuclear transmutation technological overview [Gudowski, 1999].

In the current state-of-the-art of the ADS technology the main material core is still open, being investigated from molten salts to liquid metals as coolants with different fuel matrixes. Among the ADS designs, the reference designs are highlighted and are based on fast cores with oxide fuel cooled and moderated with liquid eutectic Lead-Bismuth.

Research activities about transmutation of nuclear waste are in progress in many countries. In Europe, most R&D activities in the area of P&T are performed within the EURATOM framework: the 5th (FP5, 1998-2002), the current 6th framework programme (FP6, 2002-2006) which activities has been extended to 2007, and support is also foreseen in FP7.

Outside the European Union, P&T activities are also pursued in China, India, Japan, the Republic of Korea, Russia and some other countries of the Commonwealth of Independent States (mostly within the framework of the International Science and Technology Centre ISTC, details given farther down), and in the USA. The rationale for these activities varies from country to country, and so do the transmutation strategies envisaged. However, no matter whether energy production or long-lived waste transmutation is considered to be the main driving force, and independent from the particular short, medium, and long-term strategies, there are many common generic R&D activities that are pursued by most groups, and for which international collaboration is sought.

One of the alternative options is gas cooled accelerator driven systems as a result of the possibility of establish synergies with critical reactor development currently in progress and its proven basic reactor technology in comparison with liquid metal technology, although the neutron source is based on liquid metal, what implies the understanding of nuclear and thermo-mechanical phenomena that are involved in the design of this component.

GAS TECHNOLOGY IN THE NUCLEAR FIELD

The first nuclear power plant (Calder Hall, UK, 1953, Figure 6) in the world was based on gas technology with an electric power of 50 MW. The reasons for that are the cooling simplicity when based on a gas as coolant, and the more advanced characterization of graphite as moderator at the beginning of the nuclear technology (In fact, the Chicago Pile 1, historically considered the first nuclear reactor has graphite as main material). Their limitation in terms of neutronic economy, and, mainly, on thermal power transfer capabilities to design high power units (more than 1000 MW) were overcome in the 70's by water reactors and most of the nuclear plants in the world are now based on PWR and BWR concepts.

The application of gas technology to the design of nuclear reactor with graphite as moderator opens the possibility to different neutron spectra: fast, thermal, and even epithermal depending on the average neutron energy sought in the nuclear core, trough the graphite/fuel ratio and the negligible effect in the core neutronics of gases.

Nevertheless, the thermo-physical properties of gases, and helium in particular, limits the core power density due to its cooling features to a few cents W/cm^3, what implies huge cores for high power nuclear devices, that are limited in volume by vessel pressure.

The generation IV initiative is an example of this design flexibility. Gas cooled thermal reactors (VHTGR) and gas fast reactor (GFR) are proposed. Other gas cooled concepts are the Pebble-bed Modular Reactor (PBMR) in South Africa [Koster, 2003], or the High Temperature Reactor (HTR) in China [Zhang, 2006] . Such renewed interest is a combined effect of the need to design medium sized reactors, and its potential for Hydrogen production and non-nuclear applications as desalinization.

The operation at high temperatures are another of the advantages of gas cooled reactors as conversion efficiency from thermal to electric energy is higher than in water reactors, contributing to a higher economic competitiveness.

The application of gas technology to transmutation of nuclear wastes is considered among the technical options for its future development because of the neutron spectra design flexibility, the synergies with the development that are in progress for critical reactors and its safety advantages.

GAS COOLED DESIGNS OF NUCLEAR WASTES TRANSMUTERS

The transmutation of nuclear wastes should be integrated in a nuclear fuel cycle in which each family of nuclei (fission products and transuranics) should have a suitable neutron spectrum for its elimination. Gas technology can be applied to obtain very fast neutrons in the core by gas cooled reactors without moderators, or epithermal or thermal cores in function of the moderator content, usually graphite, and core design.

Figure 6. Calder Hall Power Plant.

The PDS-XADS project, supported by the 5th Framework Program of the European Union, considers a gas-cooled option for an installation devoted to ADS technology demonstration, among other options based on Lead cooled cores. This project has as one of its technical requirement a fast spectrum core. The design of the gas cooled PDS-XADS was proposed by CEA and core design is extensively based on the experience from Sodium cooled reactors, its European Fast Reactor and the general context of its future reactor program.

The basic characteristics of the ADS Demonstrator proposed as gas cooled option for nuclear waste transmutation by CEA (Figure 7) include a thermal power around 100 MWt for the sub-critical core, and a proton beam of 1 GeV with an intensity of up to 10 mA.

Taking into account the large number of new techniques to investigate, and time constraints, the design of the ADS demonstrator is based on the use, as far as possible, of well-known and proven techniques. It shall also have sufficient flexibility so that future innovative features can be tested. Solid fuel is therefore preferred to liquid fuel based on molten salts techniques. Similarly, it is chosen to physically separate the spallation target and the reactor.

The spallation target produces the neutrons required to maintain the fission reaction in the surrounding subcritical core. It consists of a long cylindrical shell filled up with a liquid metal (lead-bismuth eutectic), enclosed in the helium pressure resistant reactor thimble. The fluid heated by the beam impact is circulated under forced convection, and is evacuated outside the module towards external pump and heat exchanger systems. The eutectic Lead-Bismuth is separated from the vacuum of the accelerator by a wall called window.

Fuel sub-assemblies will be based on proven technologies extrapolated from the fast neutron reactor techniques (fuel within pins, same fuel pellets, and same cladding material). Later, new fuels based on the high temperature reactor technology with particles coated with ceramic or included in ceramic matrices have the potential for allowing to strongly increasing the temperature limits. This would allow both to increase the capability for passive core cooling in accidental conditions, and high thermal cycle efficiency.

In the preliminary ADS Demonstrator, the fuel sub-assemblies are arranged in an annular array of three rows surrounding a 60-cm diameter target cavity. The subcritical level is $k_{eff} = 0.95$, and the neutron flux is about 10^{15} n/s-cm2. The active core is surrounded by an outer region of about seven rows of subassemblies dedicated to transmutation targets, lateral shielding and in vessel fuel subassemblies storage.

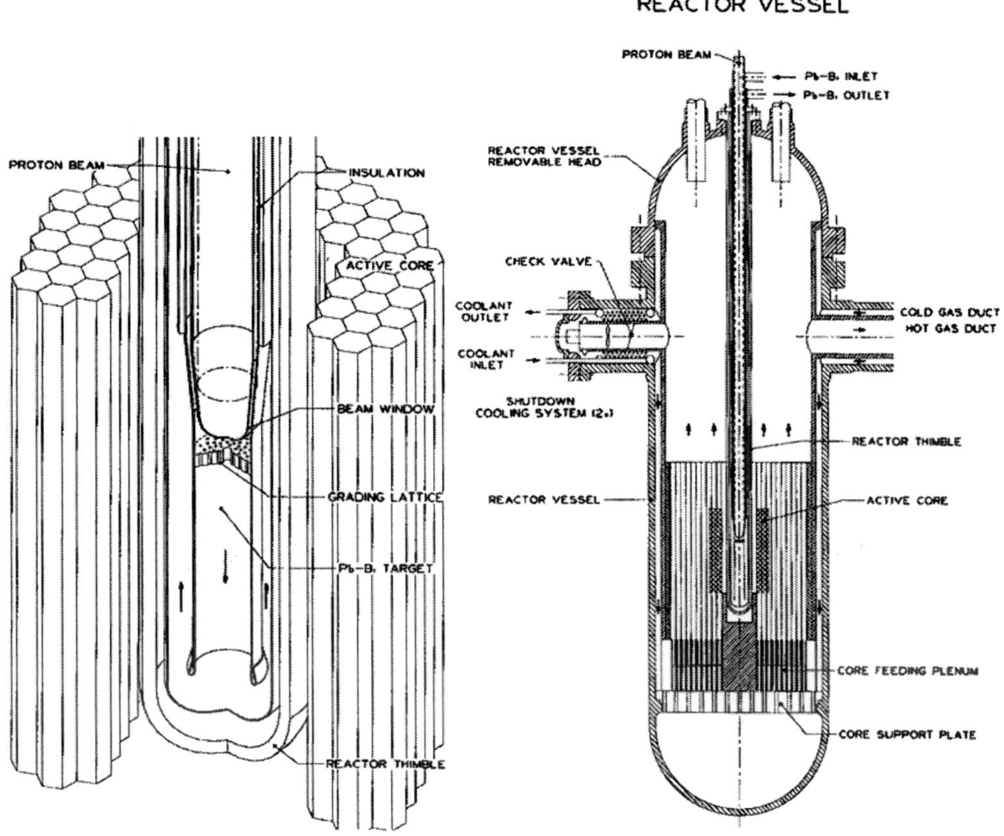

Figure 7. Description of the CEA gas cooled ADS concept [Carluec].

Evaluations of the modular thermal gas cooled reactor (GT-MHR) and of gas-cooled fast reactors show the potential of the direct cycle in terms of cost reduction and safety. The cooling medium is helium pressurised at 60 bar. The core inlet temperature is 200 °C, to avoid any freezing hazard of the spallation target. Core outlet temperature is limited to 450 °C to avoid creep effect on structural materials.

During power operation, helium is cooled outside the reactor vessel by an intermediate helium/water heat exchanger. The turbocompressor achieving the forced helium circulation is housed together with heat exchanger in a Power Conversion Vessel. During handling operation, the primary circuit is depressurised and the core is cooled by a shutdown cooling system with blowers and intermediate helium/water heat exchangers located inside the reactor vessel.

In case of loss of the forced circulation of helium, the decay heat can be removed by natural circulation using the heat exchangers of the shutdown cooling system. In case of loss of pressure, the blowers and the intermediate heat exchangers of the shutdown cooling system remove the decay heat. The arrangement of the internal structures results from the choice of a standing core, a cold vessel and an upward core flow selected to enhance natural circulation.

The two redundant shutdown cooling systems (blower and heat exchanger) are located within nozzles facing the primary helium nozzle. Distance between these nozzles and the top of the active core provides sufficient natural circulation capabilities. Reactor internals (core support plate/core feeding plenum/inner vessel), shutdown cooling equipment and target unit sub-systems are potentially accessible and removable. Refuelling will be undertaken with the reactor shutdown. An in-reactor fuel handling machine (pantograph system) can be located in dedicated penetrations located within the reactor vessel closure head. The height of the plenum above the core is determined by fuel handling system requirements.

In compliance with the range of helium temperatures, modified 9%Cr ferritic steel (T91) is selected for pressure containing boundary vessels and type 316 LN austenitic stainless steel for reactor internals. This design is clearly preliminary and several alternative options such as a solid target and/or accelerator beam entering at the bottom of the reactor vessel will be considered for further optimisation.

Of great interest is the proposal given by General Atomics [Rodriguez, 2000; Baxter, 2001]. The concept under study is based on the use of thermal neutrons and fast neutrons in the same transmuter to achieve high burnup of plutonium and minor actinides in the same assembly. The main features and performance characteristics of the transmuter under study are summarised in the following paragraphs.

The general layout of the transmuter concept being studied is illustrated in Figure 8. As shown in this figure, the transmuter consists of a steel vessel housing, inside of which there is an annular nuclear transmutation region that operates in the thermal neutron energy regime. In this annular region, plutonium and minor actinides from PWR waste are fission together in ceramic-coated (TRISO) particles. Most of the mixture, about 90%, is plutonium. The remaining 10% are minor actinides.

Fission neutrons in this annular region are thermalised in graphite blocks where the TRISO particles are arranged. There is an inner and an outer graphite neutron reflector surrounding the annular thermal region. This thermal region operates in the critical mode for approximately three years, or 75% of its cycle time, after which it becomes subcritical.

In the centre of the inner reflector, there is a cylindrical region that operates in the fast energy neutron energy regime. This region is approximately 15% the size of the active thermal region, and consists of an assembly of tungsten tubes that house TRISO particles already transmuted in the thermal region. Therefore, they contain mainly minor actinides, and the fission products from the transmuted plutonium.

A major motivation for including this fast region assembly inside of the thermal assembly is to take advantage in the fast fission process of the large heat storage and conduction heat removal capabilities of the thermal assembly, which provides conduction cooldown paths from the fast assembly to the outside of the vessel.

The fast region assembly is designed so that, by itself, it is subcritical. However, neutrons produced in the thermal region operating in the critical mode can travel through the inner reflector, cause fission in the fast region assembly, and get amplified, thus creating subcritical transmutation. Neutrons generated by fission are born fast, and remain fast in the metallic components of the fast assembly. In this mode, the fast fission process is driven by the critical reaction in the thermal region. After the thermal region becomes subcritical, the entire transmuter (fast and thermal) is then driven for a fourth year to cause deep transmutation by neutrons generated in the spallation target located in the centre of the fast region assembly. In this mode, the target is driven by the proton beam illustrated in Figure 8. Deep transmutation can be achieved with no reprocessing thanks to the encapsulation in ceramic-coated microspheres of the materials to be transmuted, which accommodate the production of fission gas products within internal expansion volumes. Furthermore, the fact that the transmuter needs the proton beam for only 25% of its operating time allows four transmuters to time-share one accelerator. Three out of the four transmuters operate in the critical mode while the fourth one is driven in the subcritical mode to deep burnup (transmutation) by the spallation source and the proton beam.

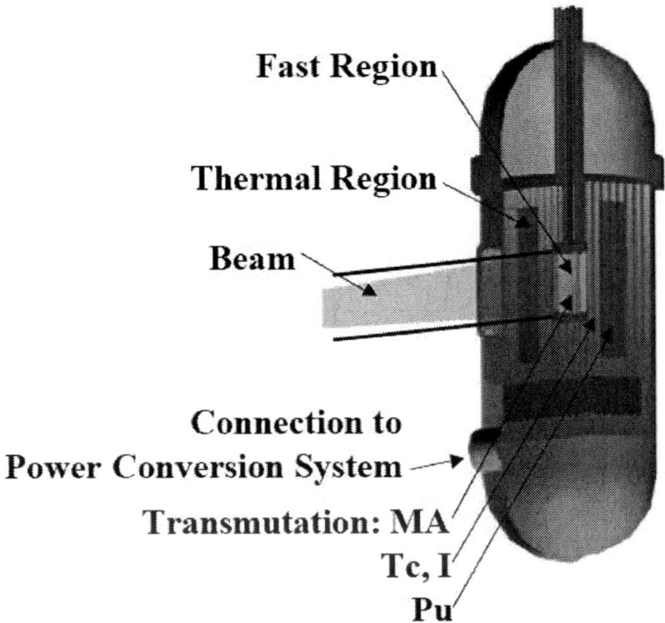

Figure 8. Integrated fast-thermal transmutation core [Baxter, 2001].

Still referring to Figure 8, there is shown a coaxial duct in the lower part of the transmuter. The outer part of the duct brings in cold cooling helium to remove fission heat from the transmuter. This helium flows upward in an annular space between the inside of the vessel and the outside of a steel barrel that contains the thermal-fast assembly.

Helium then flows downward through cooling channels in the fission regions of the transmuter, and carries the heat at a temperature of 850 °C through the central part of the coaxial duct to a direct-cycle gas-turbine-generator system that generates electricity. The high operating temperatures and the characteristics of this direct-cycle power conversion system allow electric generation with a high net thermal efficiency of approximately 47%.

Each transmuter is rated at 600 MWt, and the proton accelerator produces an 800 MeV proton beam with a current in the 15 to 20 mA range. The high power conversion efficiency, the fact that 75% of the transmutation is done in a critical operating mode, and the fact that the proton accelerator is time-shared by four transmuters lead to a favourable revenue-cost balance and the potential to attract private investment for the deployment of these units. Burnup calculation results for this fuel cycle are shown in Figure 9 for an initial 1,000-kg charge of weapons-grade plutonium. As the figure indicates, most of the Pu transmutation is accomplished in the thermal-critical regime.

When this is followed by a one-year irradiation step in a thermal subcritical regime (accelerator - driven), essentially all fissionable Pu is gone. At this time, a three-year step of transmutation in a fast subcritical regime leads to Point C in the chart, where most of the initial charge is gone.

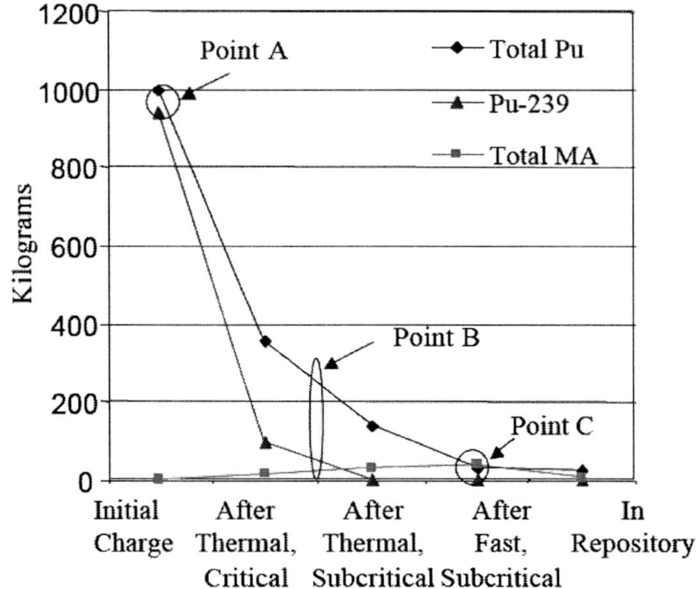

Figure 9. Transmutation performance in the GA reference design.

If this remaining material is placed in a repository for 200 years, only 60 out of 1000 kg of the initial charge are left. The significance of Point B in the above figure is that it shows the level of transmutation demonstrated in a test at the Peach Bottom 1 reactor. In that test, plutonium fuel particles coated in Silicon Carbide were irradiated to levels exceeding 700,000 MW-days per metric ton of Pu fuel, which transmuted over 95% of Pu-239, and corresponds to the Point A-to-Point B trajectory in Figure 3.2. The above performance corresponds to 99.9% destruction of Pu-239, over 90% destruction of total Plutonium, and toxicity levels of the transmuted material encapsulated in ceramic coatings that are below natural uranium toxicity levels in a few hundred years.

Similar calculations are now being performed for an initial charge of plutonium and minor actinides separated from spent reactor fuel. As part of the accelerator transmutation process evaluations in the US, a thermal-fast test facility not very different from but smaller than the one described here is being considered. It would be rated at approximately 100 MWt, and driven by a 400 MeV, 20 mA proton beam. Most importantly, it would have similar geometrical and engineering features to demonstrate safety and performance, and allow the design to be extrapolated to the 600 MWt size without crossing singularities or discontinuities, thus simplifying safety analyses and reviews, and the licensing process.

The complete system is illustrated in Figure 10. As shown in this figure, the system includes a front-end separation stage in which uranium in the waste stream is separated and recycled to the commercial fuel cycle, or disposed of as low-level (Class C) waste. This leaves approximately 1% of plutonium and minor actinides as the main elements to be transmuted. It is estimated that by the year 2015, approximately 70,000 metric ton of civilian power reactor nuclear waste will be available in the US. This will yield a nominal amount of approximately 675 metric ton of plutonium, and 75 metric ton of minor actinides after steps 3a, 3b and 3c in Figure 10.

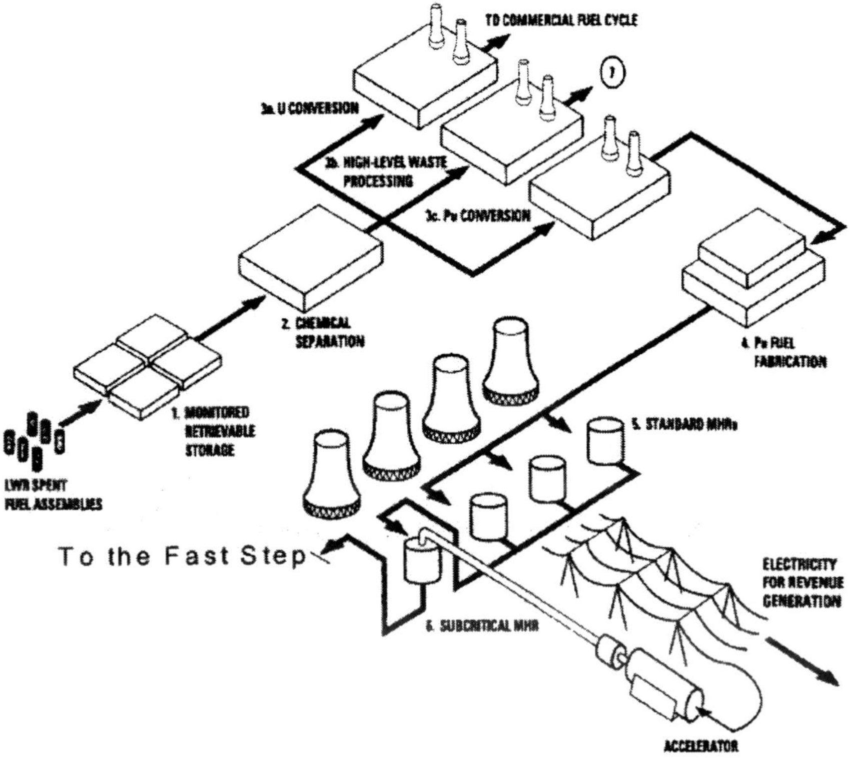

Figure 10. Thermal transmutation scheme proposed by GA ([Baxter, 2001]).

Another initiative concerning gas-cooled reactor technology was proposed by LAESA, a Spanish consortium that was established in Saragossa (Spain) at the end of the 90's. The LAESA team proposed a few-MW gas-cooled pebble-bed subcritical device as experimental facility to test the technical viability of this technology to transmutation of nuclear wastes in a once-trough cycle scenario, as it is the case of Spain and many other countries.

The experimental prototype proposed by LAESA [Abánades, 2007] has a thermal power production of 10 MW. In Figure 11 we can see a sketch of the system. The main components of the PBT are: (a) the accelerator system, generator of the proton beam, with a beam power of 3,8 MW, able to produce 10 mA of 380 MeV protons; (b) a conical shaped spallation target, which couples the accelerator proton current to the nuclear system, providing neutrons that keep the neutron population in the whole system; and (c) the subcritical nuclear core, designed to work far away from criticality (k_{eff} of the order of 0.8).

The nuclear core is a cylinder containing the fuel pebbles. The central part is occupied by the spallation target, which acts as a neutron source. The core is filled with graphite pebbles containing the nuclear fuel, similar to the ones that have been developed for the HTGR reactors and used in the previously presented gas cooled ADS concepts. The fuel is confined in 3-cm-radius pebbles. The external layer of the pebble is made of pyrolytic graphite with a thickness of 5 mm, while the inner 2.5-cm-radius volume is filled with 1-mm-diameter TRISO micro-spheres containing the fuel material. Intensive neutronic simulations were performed to optimise the fuel structure in order to obtain a convenient epithermal neutron spectrum.

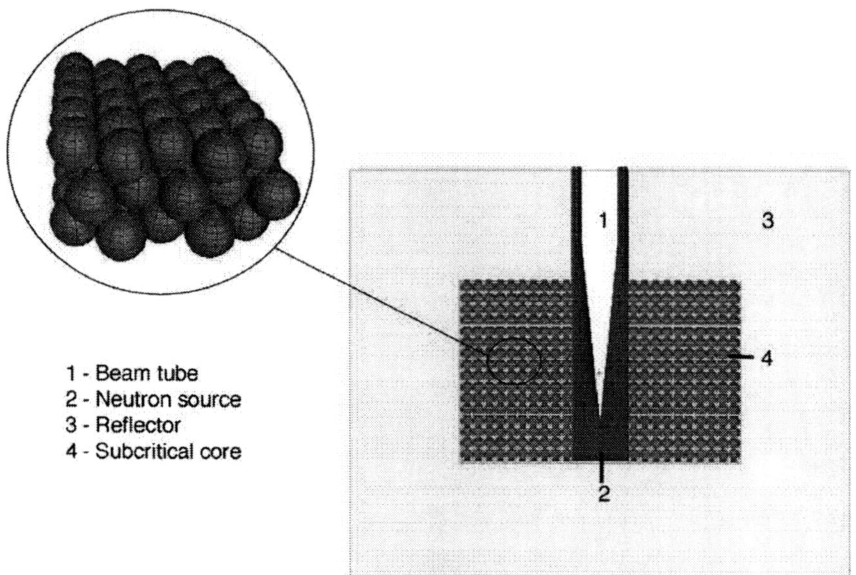

Figure 11. Schemetic description of the LAESA Pebble Bed Transmutator.

Regarding transmutation performance of this subcritical gas-cooled system, plutonium isotopes suffer a considerable mass reduction (Figure 12). In particular, the most abundant isotopes (^{239}Pu, ^{240}Pu and ^{241}Pu) are practically eliminated. ^{242}Pu and ^{238}Pu increase slightly their masses, but the radio-toxicity of ^{242}Pu is very small. The mass of ^{237}Np is also considerably reduced. ^{241}Am is eliminated while ^{243}Am grows (Figure 13). The quantity of curium isotopes increases due to their small cross sections in comparison to the rest of core materials.

FUEL DEVELOPMENT FOR GAS-COOLED REACTORS

One of the key issues in the development of gas cooled reactors for the future, including ADS, is the fuel development to get a fuel matrix that is intended to reach very high burn-up and high temperature.

The ceramic materials are stable at high temperatures and have very high melting points. This provides large thermal margins to ensure reactor integrity during loss of cooling events. Moreover, the coated particles are nearly spherical and include gas expansion volumes, which will accommodate fission gas products and result in lower internal pressures. The spherical shape easily withstands mechanical stresses due to these pressures. The composite effect is that the particles can tolerate high levels of irradiation and deeper bum-up, which leads to a higher degree of transmutation without reprocessing. This capability has been demonstrated in multiple reactor irradiations. In addition, the ceramic coatings are far more durable than metallic coatings, better assuring waste integrity in the repository.

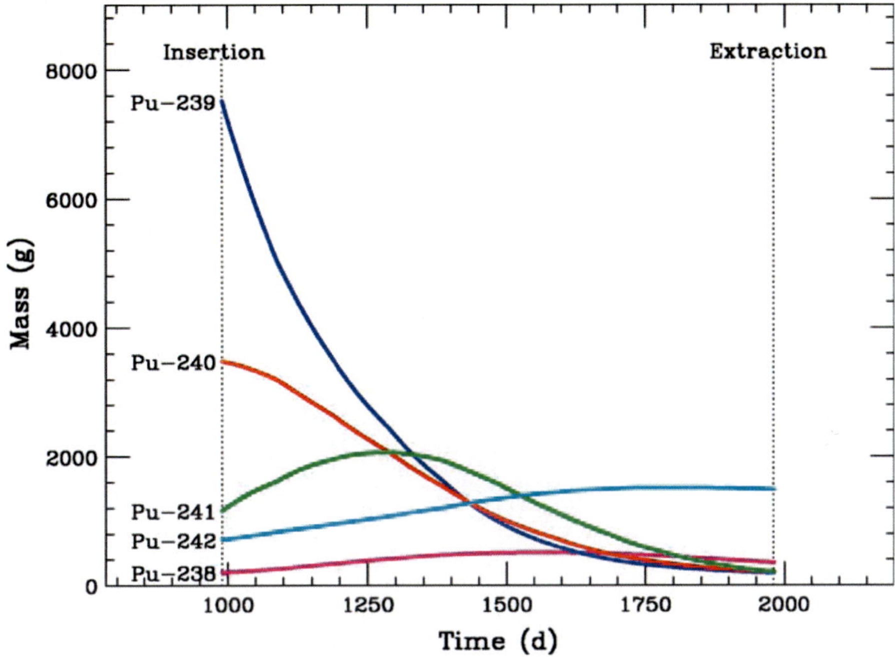

Figure 12. Plutonium elimination vs. irradiation time in the LAESA PBT.

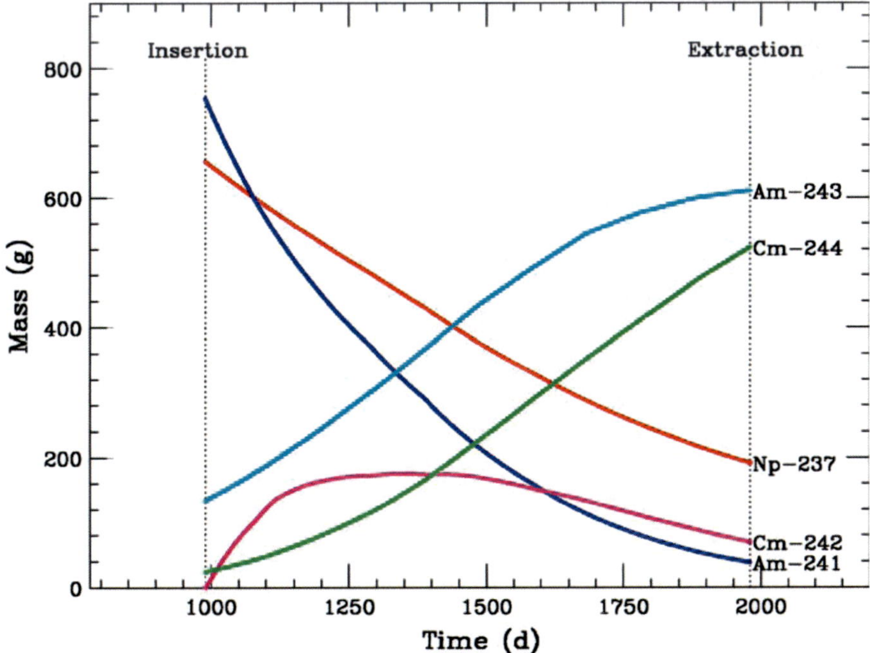

Figure 13. Minor actinides elimination vs. irradiation time in the PBT.

Basic coated particle structure is (Figure 14) composed of:

- Oxide kernel, containing the fissile isotopes. Current state-of-the-art TRISO particle technology is based in PuO_2 heavy metal content. In the case of it application to transmutation of nuclear waste different isotope content is envisaged, with a non-negligible proportion of other minor actinides as americium and curium.
- A porous pyrolitic carbon buffer, which porosity should be carefully determined to serve as gaseous fission product trap. The design of this layer is one of the most important issues to achieve high burn-up.
- A Silicon Carbide barrier layer, that assures pebble mechanical and thermal integrity.
- An Outer isotropic pyrolitic carbon, that acts as moderator and thermal carrier. Its thickness is one of the free parameters to tune neutron spectra.

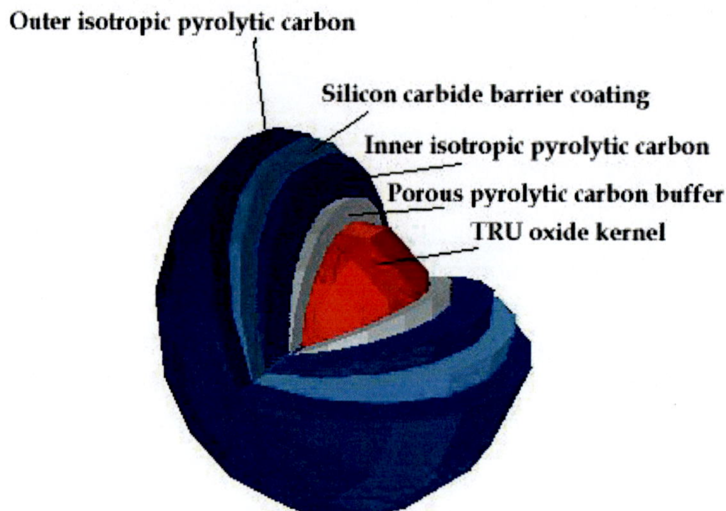

Figure 14. Schematic view of a TRISO coated particle.

Most of the gas-cooled reactor concepts used this TRISO coated particle either arranged in prismatic or pebble-bed packed cores. Both arrangements have different operation modes. In the case of prismatic cores, TRISO particles are fixed in the core during irradiation and its burn-up strategy is similar to PWR or BWR reactors, in which refuelling is programmed at given times. In the case of pebble-bed cores, continuous refuelling is possible, as it is the case of the german critical HTR or the LAESA proposal.

The ceramic TRISO particles are a very robust fuel matrix, able to held more than 700 MWd/kg Heavy Metal, as it was tested in Peach Bottom 1 reactor (Figure 15), and reported phenomena under severe irradiation include failure due to pressure vessel failure, irradiation-induced cracking in the SiC layer. In particular, failures in the pyrolitic carbon are strongly related to anisotropy and coating porosity with a strong influence on the shrinkage and swelling behaviour. The porosity of the layer has an impact on the strength of the interfacial bond between the SiC and the pyrolitic carbon.

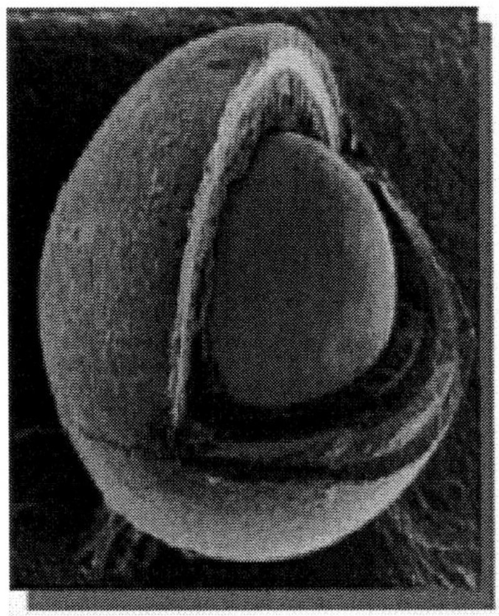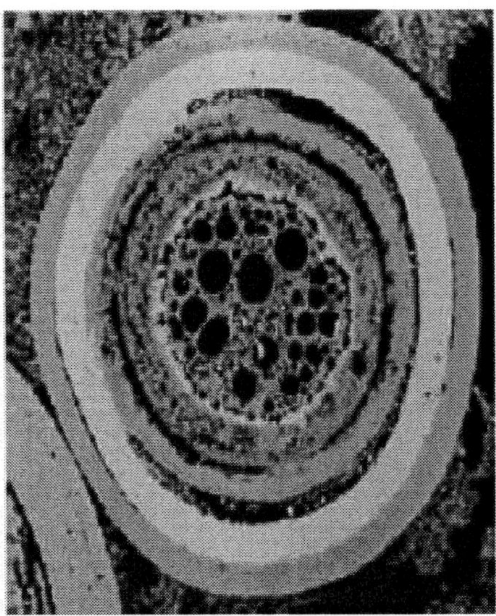

Figure 15. TRISO coated particle after 747 MWd/kg burn-up.

Fission product and impurity attack of the SiC and kernel migration are thermally driven phenomena that are strongly influenced by fuel burnup, temperature, and temperature gradient across the pebble. The temperature gradient is a strong function of the power density. Another key issue of this kind of ceramic fuel is reprocessing technology. As it was point out, partitioning is one of the main subjects to solve for the development of transmutation. The partitioning and reprocessing of TRISO coated particle is a wanted activity that should be done to increase the global efficiency of the nuclear fuel cycle with this kind of fuel, as it will opens the possibility to a sort of closed cycle, and technical options will not be limited to once-through.

Figure 16 shows the reprocessing scheme for TRISO coated particles under development in Oak Ridge National Laboratory. The first step is a mechanical treatment in which the fuel elements in this case are crushed, drying out fission product gases that are retained in the spent fuel particles, and extracting much of the carbon.

A nitric acid treatment separates the actinides and fission products of this mechanical chopped stream for a later chemical separation. After this chemical step, the final stream is compacted and treated as waste.

NUCLEAR FUEL CYCLE WITH GAS COOLED ADS INTEGRATION

The availability of nuclear systems in which nuclear fuel in form of ceramic fuel, the development of adequate fuel fabrication techniques, and the development of separation techniques for this kind of fuel can be integrated in a new fuel cycle that could envisage a more efficient profit of nuclear primary sources.

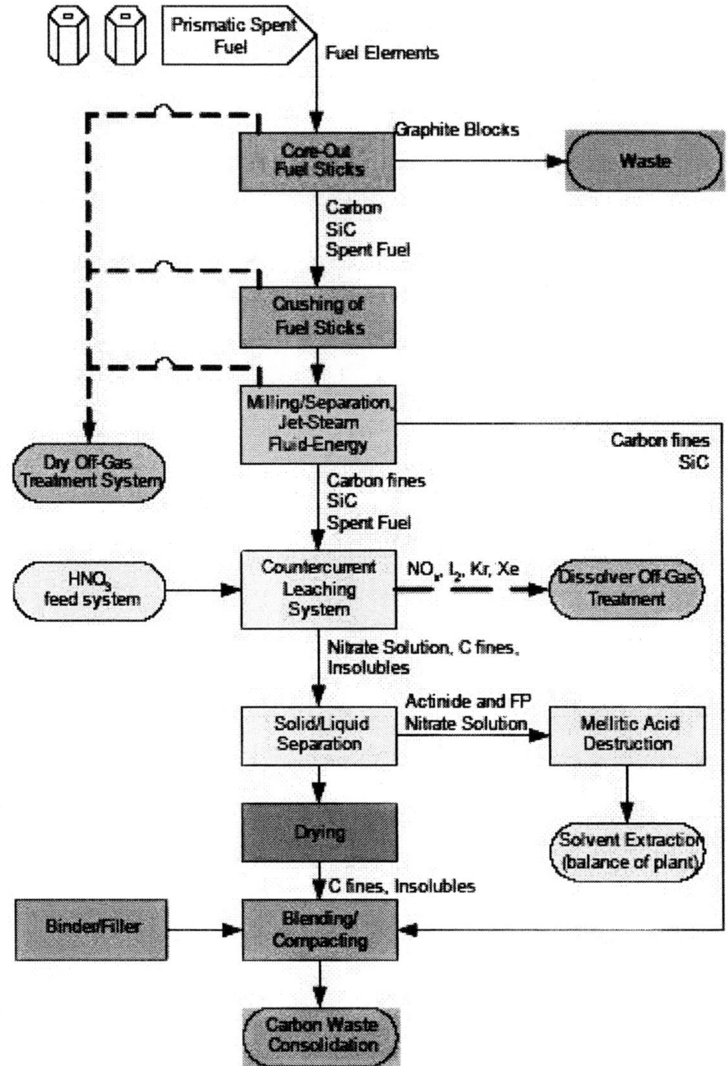

Figure 16. Reprocessing scheme for TRISO coated particle fuel.

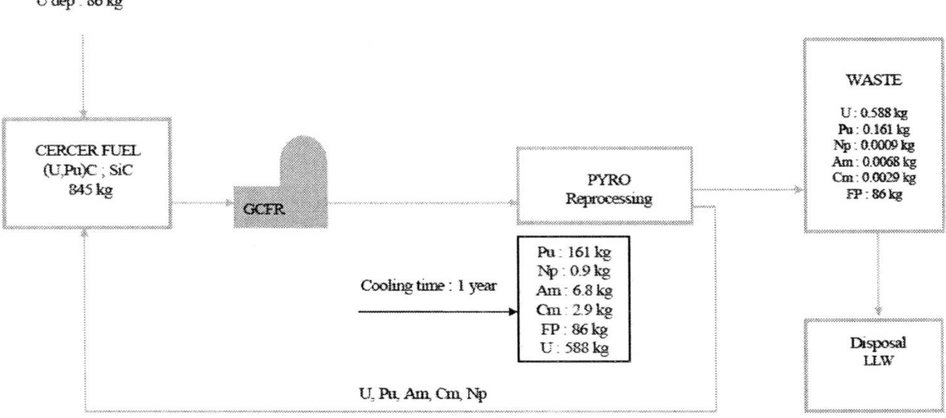

Figure 17. Close fuel cycle with gas cooled ADS. [Cavedon, 2004].

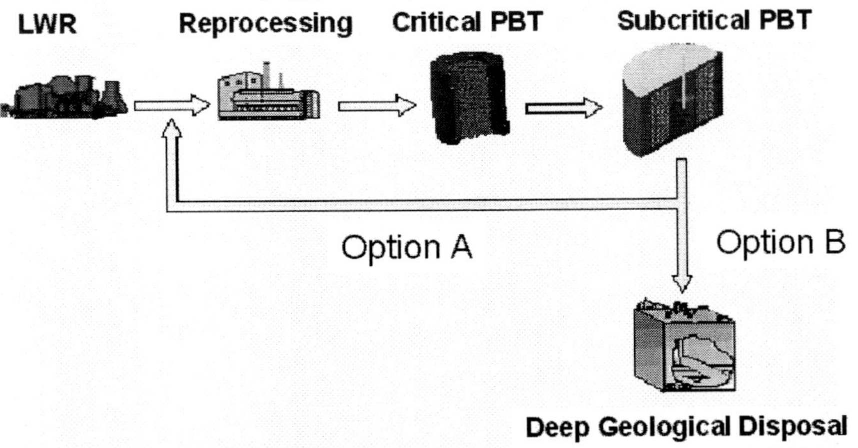

Figure 18. Once-trough fuel cycle with gas cooled ADS.

A scenario in which all these components could be integrated is presented in Figure 17. The final waste stream is reduced some orders of magnitude respect to the waste to manage in the final disposal facility. The application of this fuel cycle is very much dependent on the development of high efficiency reprocessing and fuel fabrication technology, some of them available in the medium term.

Another approach is the once-trough cycle as realistic TRISO particles reprocessing are still pending of an efficient development (Figure 18) and is not practical. In this case, the high burn-up capability of the fuel implies that a significant waste reduction can be achieved specially in plutonium inventory. In this kind of cycle, spent fuel made of ceramic particles are directly sent to the final disposal. Nevertheless, techno-economic viability studies should be carried on to establish the volumen and mass reduction that could be achieved in this case.

SAFETY ASSESSMENT OF GAS-COOLED ADS

Gas-Cooled transmuters share many of the benign safety characteristics of the modular gas-cooled reactors. Furthermore, combining thermal and fast regions in one assembly, they can be configured to:

- Obtain a negative temperature reactivity coefficient under any operating mode.
- Allow passive coolant circulation and conduction cooldown,
- Ensure transmuted wastes confinement in strong ceramic coatings that provide highly stable barriers against release in a deep geological repository.

The similarities between critical gas-cooled reactors develop in generation 3+ or IV initiatives and Gas Cooled Accelerator Driven Systems makes foreseen a similar licensing procedure for an industrial prototype.

SUMMARY

The present state-of-the art of the transmutation of nuclear wastes techniques with gas cooled Accelerator Driven Systems has been presented. It has been shown the most important concepts that has been proposed and reported in the scientific literature. Transmutation rates of such kind of systems are very promising, eliminating practically all Pu isotopes, except the less harmful ^{242}Pu. Safety advantages of refractory materials, moderation tuning and subcriticality are on of the strong points of this technology. Main technological uncertainties are concerning isotopic separation applied to TRISO coated particles. A successful partitioting of such ceramic fuel seems to be the show-stopper for the development of closed nuclear fuel cycle based on gas cooled technology.

The development of gas cooled reactors in the future will be connected to the progress that could be done among its relatives concepts in the Generation IV initiative, and the technological problems that can be found in the application of liquid metal technology in subcritical reactors.

REFERENCES

Abánades A., Pérez-Navarro A..'Engineering design studies for the transmutation of nuclear wastes with a gas-cooled pebble-bed ADS'. *Nuclear Engineering and Design,* 237 (2007) 325-333

Ball S. 'Sensitivity studies of modular high-temperature gas-cooled reactor postulated accidents'. *Nuclear Engineering and Design* 236 (2006) 454-462.

Baxter A., Rodriguez C. 'The application of gas-cooled reactor technologies to the transmutation of nuclear waste', *Progress in Nuclear Energy*, 38, No 1-2. pp 81-105 (2001).

Bowman E.D., Arthur E.D. et al., *'Nuclear energy generation and waste transmutation using Accelerator-Driven Intense Thermal Neutron Source'*, LA-UR-2601 (1991).

Carluec B., Anzieu P.*'Proposal for a gas-cooled ADS Demostrator'*

Cavedon, J.M. *'Impact on advanced nuclear fuel cycle options on waste management policies'* 8th PT IEM Las Vegas 9-11 November 2004.

Forsberg C. 'The advanced high-temperature reactor: High- temperature fuel, liquid salt coolant, liquid-metal-reactor plant'. *Progress in Nuclear Energy*, 47, 1-4. pp 32-43 (2005).

Gudowski W. 'Accelerator-driven Transmutation Projects. The importance of Nuclear Physics Research for Waste Transmutation'. *Nuclear Physics* A 654 (1999) 436-457.

Hejzlar P. Pope M.J, Williams W.C. and Driscoll M.J.'Gas cooled fast reactor for generation IV service' *Progress in Nuclear Energy*, 47, 1-4. pp 271-282 (2005).

Koster A., Matzner H.D., Nicholsi D.R. 'PBMR design for the future'. *Nuclear Engineering and Design* 222 (2003) 231-245.

Landeyro P. et al.'Review of the European Project - Impact of accelerator-based technologies on nuclear fission safety (IABAT)'. *Progress in Nuclear Energy*, 38, 1-2, pp. 135-151 (2001).

Maki J.T., Petti D.A., Knudson D.L., Miller G.K.'The challenges associated with high burnup, high temperature and accelerated irradiation for TRISO-coated particle fuel'. *Journal of Nuclear Materials* 371 (2007) 270-280.

Rodriguez C, and Baxter A: "Transmutation of Nuclear Waste Using Gas-Cooled Reactor Technologies". Invited Paper, ICONE 8, *Thermal Hydraulics*-Part B, April 2-6, 2000, Baltimore, MD USA.

Rubbia, C., 1995. *'Conceptual design of a fast neutron operated high power energy amplifier'*. CERN/AT/95-44(ET).

Rutherford E. 'Bakerian lecture: Nuclear Constitution of the Atoms'. Lecture delivered June 3rd, 1920

Taiwo. T, Gohar Y., Finck P.J.*'Core physics performance of recycled LWR discharge TRU oxide fuel in a GT/MHR'*.IAEA-TECDOC-1356.

Zhang Z, Wu Z, Sun Y, Li F.'Design aspects of the Chinese modular high-temperature gas-cooled reactor HTR-PM'. *Nuclear Engineering and Design* 236 (2006) 485-490

In: Nuclear Waste Research: Siting, Technology and Treatment   ISBN 978-1-60456-184-5
Editor: Arnold P. Lattefer, pp. 167-187   © 2008 Nova Science Publishers, Inc.

Chapter 5

ARGILLITE/CONCRETE AND ARGILLITE/STEEL INTERACTIONS: EXPERIENCE GAINED FROM THE TOURNEMIRE UNDERGROUND RESEARCH LABORATORY

I. Devol-Brown, E. Tinseau, F. Marsal, D. Pellegrini,
A. Mifsud, S. Lemius, D. Stammose and J. Cabrera
Institut de Radioprotection et de Sûreté Nucléaire (IRSN),
B.P. 17, F-92262 Fontenay-aux-Roses Cedex, France

ABSTRACT

Deep argillaceous formations are potential host rocks for high-level radioactive waste repositories due to favourable properties, particularly their low permeabilities and high sorption capacities for radionuclides. Disposal concepts considered in several countries involve steel and concrete components, which could react with such clayey material and thus induce changes in the containment properties of clays.

This safety issue is addressed by the French Institute for Radiological Protection and Nuclear Safety (IRSN) through an experimental and modelling program aiming at assessing the intensity and expansion of such geochemical perturbations. For this purpose, IRSN Underground Research Laboratory (URL) in Tournemire (Aveyron, France) which includes several man-made facilities (century-old railway tunnel, 1996 and 2003 drifts, many boreholes) crossing the Toarcian argillite offers various opportunities. The present paper goes over the main outlines of the program developed in that context since several years on indurated argillite sampled in Tournemire URL.

The research program devoted to argillite/concrete interactions is twofold: laboratory experiments (batch and diffusion) with alkaline fluids and field investigations dealing with so-called "engineered analogues" of pre-existing argillite/concrete interfaces for times up to 125 years. This experimental platform offers two kinds of hydraulic conditions: (1) zones with water circulating from the upper aquifer ("wet" context) and (2) zones excluding any influence of this aquifer (so-called "dry" context though argillite is partly of fully water saturated).

Regarding the perturbations undergone by argillite in contact with steels, the program is focused on other "engineered analogues" sampled in the Tournemire tunnel. Three types of steels were introduced in boreholes drilled in 1998 in a zone located far from any mechanical disturbance in "dry" conditions as well as in two different areas of the Excavated Disturbed Zone (EDZ) around the tunnel. After two or six years of contact, both steels and clay were recovered by overcoring.

This paper illustrates the added value provided by the Tournemire URL, allowing to combine various experimental approaches as well as different interaction time scales, completed with modelling which help us understanding what processes are responsible for the propagation of disturbances induced in the argillite by concretes or steels. Though some features worth being further investigated, the data acquired in the framework of this program combined to others reported in the literature provide with valuable information regarding the extrapolation to the geochemical evolution of clay materials due to the presence of concrete or steel for periods lasting several tens of thousands of years.

INTRODUCTION

Deep argillaceous formations are presently considered in several European countries (e.g. Belgium, Switzerland, Germany, France) as potential host rocks for the disposal of high-level radioactive waste. Their very low permeability and diffusivity and a high sorption capacity make these claystones very favourable for hosting a repository. However, the construction of such facilities may lead to geochemical perturbations within claystone due to massive uses of cementitious and iron-based materials (Andra, 2005). The concrete matrix is in disequilibrium with the geochemical conditions prevailing in an argillaceous host rock and therefore could induce substantial changes in clay mineralogy and pore water chemistry. The release of iron from metalic components could also modify the properties of clays. Though many experiments have been conducted in laboratory on bentonite materials, there are still pending issues regarding the nature, intensity and expansion of those perturbations. Furthermore, very few *in situ* experiments on argillite under natural conditions are reported in the literature.

The French Institute for Radiological Protection and Nuclear Safety (IRSN), which is in charge of the technical review of the safety assessment files related to nuclear facilities and of the research supporting its expertise, has developed an experimental program since 5 years so as to address the above mentioned safety issue and thus to reinforce its appraisal on the potential consequences of clay/concrete and clay/iron interactions. This program combines investigations performed in IRSN's surface level laboratory and in its Underground Research Laboratory (URL) in Tournemire (Aveyron, France). The Tournemire URL is based on a easy access of a one century-old tunnel and adjacent galleries excavated in 1996 and in 2003 crossing a Toarcian argillaceous formation. The entire program was conducted on this Toarcian argillite, which presents geochemical characteristics close to those of the Callovo-Oxfordian argillite studied by the French Agency for Nuclear Waste Management (Andra) in its URL in Meuse/Haute-Marne.

The aim of this paper is to illustrate the added value provided by the combination of performed in situ in the Tournemire URL and in surface laboratory various experimental approaches as well as different interaction time scales completed with modelling, for assessing the perturbations undergone by claystones due to the presence of concrete and steel. In a first section, the geological and structural background of the Tournemire URL is presented. Then, in the second and third sections, mineralogical modifications of argillite after interactions with concrete or steel respectively are presented along with the experimental work and numerical modelling. Finally, the concluding section summarizes the contribution of the Tournemire URL to the knowledge acquired on argillite/concrete and argillite/steel interactions.

IRSN UNDERGROUND RESEARCH LABORATORY IN TOURNEMIRE

The Tournemire URL is located in a Mesozoic marine basin on the southern border of the French *Massif Central* and at the western limit of the *Causse du Larzac*. The studied argillaceous formations, 250 m-thick, are composed of sub-horizontal argillites and marls of Toarcian age (200 m thick) and marls of Domerian age (50 m thick) (Figure 1).

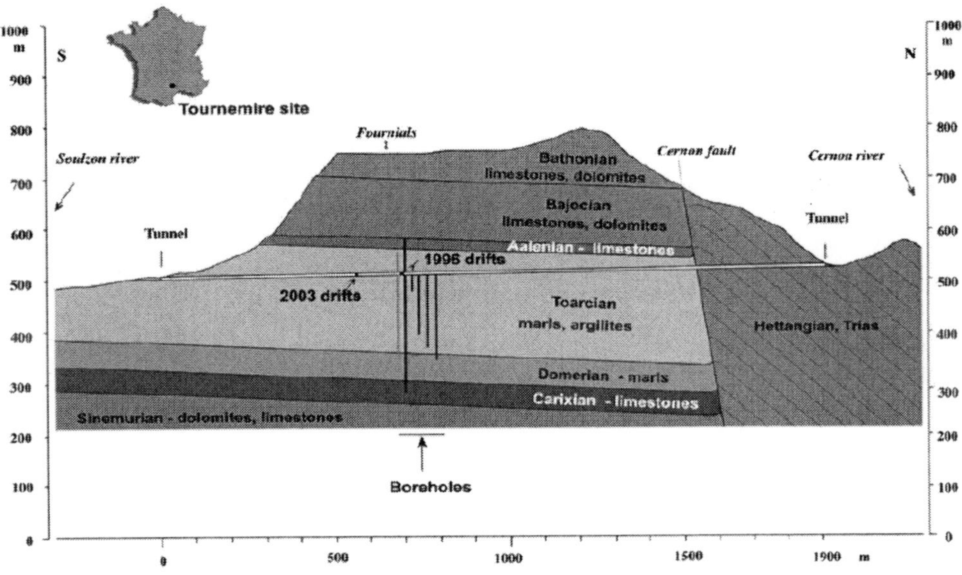

Figure 1. Geological cross section of the Tournemire platform (Cabrera et al., 2001; Matray et al., 2007).

The upper Toarcian serie is crossed by a 1885-m-long and one century-old railway tunnel excavated between 1882 and 1886. This tunnel and more recent cavities drilled by IRSN (1996 and 2003 drifts with many boreholes) provide an excellent opportunity to easily access to the argillaceous formation and to develop an independent research program for training experts and addressing safety key questions related to the behaviour of compacted claystones.

The Tournemire massif is a monocline structure with a mean dip angle of about 4° to the North. The lower (Hettangian to Carixian series) and upper (Aalenian to Bathonian series) karstified aquifers are 300 m and 250 m thick respectively, and essentially composed of limestone and dolomite. The argillaceous formations consist of well-compacted and thinly bedded argillites and marls. The clay fraction, ranging between 20 and 50 wt. % of the total rock, is mainly composed of illite (5 to 20 %), illite/smectite mixed-layer minerals (5 to 15

%), kaolinite (10 to 20 %) and chlorite (1 to 5 %). The claystone also contains 10 to 30 % of quartz grains, 10 to 40 % of carbonates (mainly composed of calcite, dolomite and siderite being present in smaller proportions) and 1 to 7 % of pyrite (Cabrera et al., 2001).

The Tournemire massif is affected by a regional reverse fault namely the Cernon fault (80 km long, Figure 1). This fault is oriented West-East and enables the communication between the two surrounding karstic aquifers. The massif is also owned by a main fault and secondary subvertical faults of hectometric extension and oriented NNW - SSE (Figure 2). The tectonic fractures are generally filled with calcite. Minor and decametric faults sometimes present geodic cavities which enable the vertical transfer of fluids. The Cernon fault and these fractures are the only way of collecting free water in contact with the clay formation.

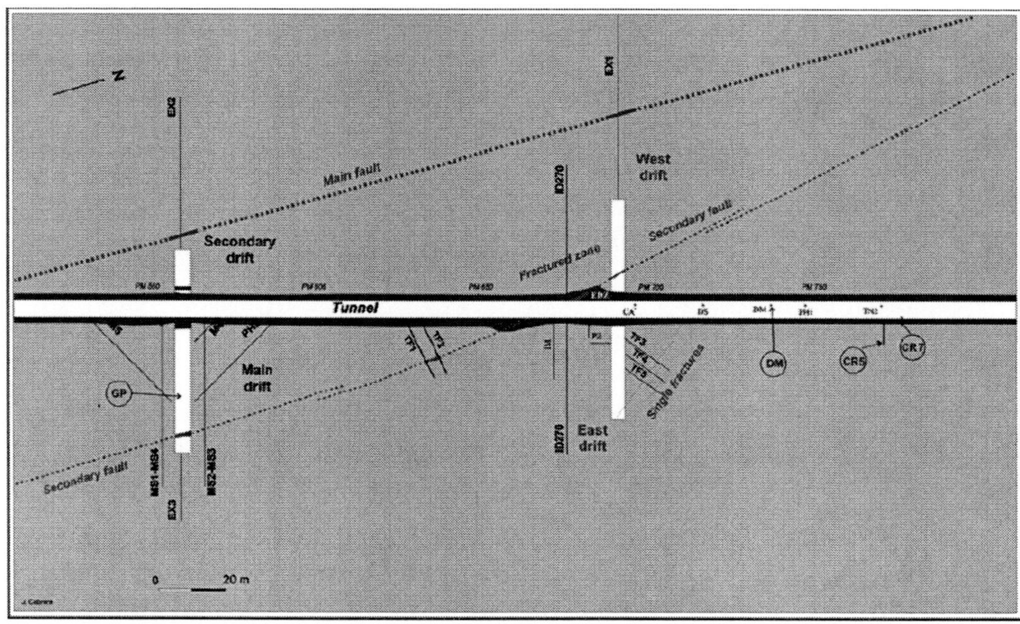

Figure 2. Plane view of drifts and boreholes in the Tournemire URL.

Two other types of fracture networks (no tectonic fractures) are observed at the Tournemire URL that may play a role on water flow and transport of dissolved species in the nearfield of excavations. These fissures are located around the tunnel and drifts. The first type, due to the stress redistribution during excavation and subsequent rock convergence, is mainly observed in the tunnel. It consists in a combination of unloading joints and fractures in the Excavation Disturbed Zone (EDZ) which may be interconnected. The second type of network, observed only on drifts, is characterized by subhorizontal fractures at the drift wall and developed parallel to the bedding (several ten centimetres deep each with a millimetric aperture). The degree of aperture of this subhorizontal fissures network is attributed to variations of the chemical potential of the interstitial solutions responsible of the swelling/shrinking cycles (Valès et al., 2004) and depends on the seasonal variations of the drift atmosphere (hygrometry and temperature).

ARGILLITE/CONCRETE INTERACTIONS

The experimental program devoted to argillite/concrete interactions is composed of two parts addressing phenomena at different space and time scales: experiments performed in surface laboratory using alkaline fluids (batch and diffusion experiments (Devol-Brown et al., 2007; Motellier et al., 2007)) and *in situ* engineered analogues providing with argillite/concrete interfaces recovered in the Tournemire URL (Tinseau et al., 2006). This program is being performed with several partners cited below.

LABORATORY EXPERIMENTS

Experimental Approach

Alkaline solutions representative of concrete leaching fluids were used in laboratory experiments on clay suspensions in order to accelerate the clay transformations with respect to solid concrete. Cylindrical cores from different boreholes were used for this experimental work (see Figure 2 for the borehole location).

In batch experiments, a given mass of argillite, provided from the EX1-borehole (horizontally air-drilled from the western drift of the tunnel – Ø(core sample) = 63 mm), was stirred with a known volume of alkaline fluid for one, three or six month periods in a closed vessel at fixed temperatures (25 and 70 °C). Two synthetic alkaline solutions were prepared to mimic the leaching of a young concrete (type I fluid) and an evolved concrete (type II fluid):

– type I: pH = 13.25; $[K] = 10^{-1}$M; $[Na] = 8 \times 10^{-2}$M; $[Ca] = 10^{-3}$M
– type II: pH = 12.5; $[Ca] = 3 \times 10^{-2}$M, $[K] = 10^{-2}$M; $[Na] = 8 \times 10^{-3}$M

The argillite was introduced into the solution as a powder or in compact lump. Powdered argillite offers a significant reactive surface which induces pronounced dissolution and precipitation phenomena whereas compacted samples allow the observation of these phenomena at the solid/solution interface. At the end of the experiment, the solid and liquid phases were separated. The solutions were filtered and analysed by (i) pH-metry, (ii) atomic absorption spectrophotometry (AAS: cations) and (iii) ion chromatography (IC: anions). The chemical and mineralogical composition of the crushed solids was determined by (i) inductively coupled plasma mass spectrometry (ICP-MS) at the Centre de Recherche de Pétrographie et Géologie (Vandoeuvre-lès-Nancy, France) and (ii) X-ray diffractometry (XRD) by Danièle Bartier at the Muséum National d'Histoire Naturelle (Département Histoire de la Terre, Paris, France). The compacted samples were only analysed by scanning electron microscopy coupled to energy dispersion spectrometry (SEM/EDS) because the quantities of altered argillite (located at the samples surfaces) were insufficient to apply techniques such as XRD (below the limit of detection).

The study concerning diffusion of species from an alkaline fluid (Na, K, Ca, Mg, OH) was conducted with Sylvie Motellier and Dominique Thoby (Commissariat à l'Energie

Atomique, DRT/LITEN/DTNM/L2T, Grenoble, France) using through-diffusion experiments on 5 mm thick samples (Figure 3).

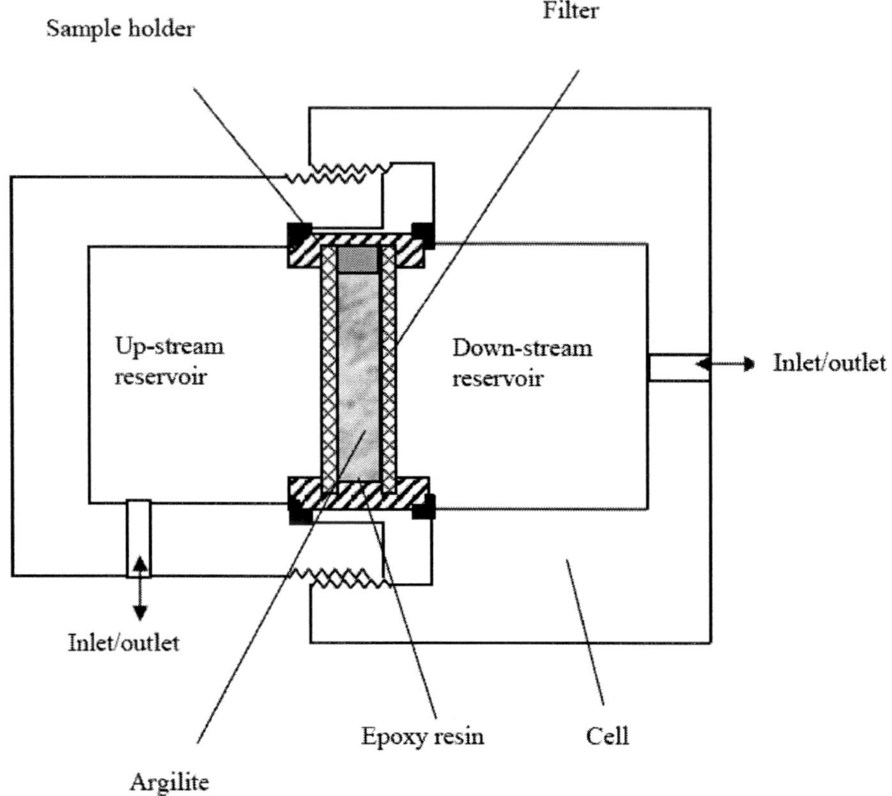

Figure 3. Schematic drawing of a through-diffusion cell (Melkior et al., 2005; Motellier et al., 2007).

The Tournemire URL provides with three kinds of samples that allows us investigating the possible influence of fractures on the diffusion process (Figure 4).

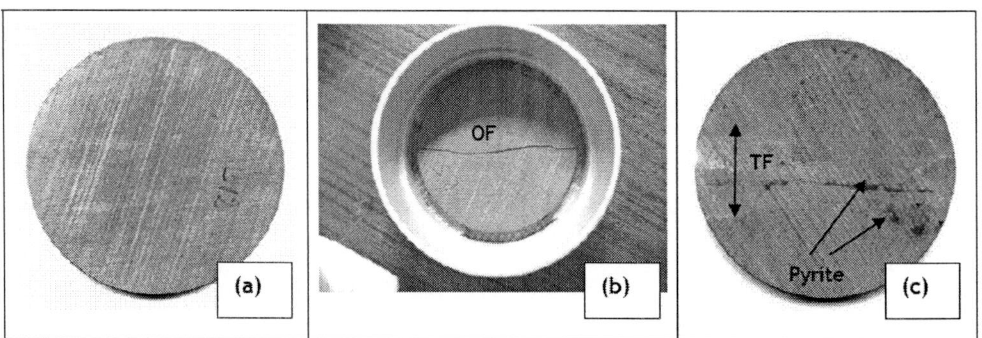

Figure 4. Photographs of sliced samples (Ø = 6.3 cm) placed in a through-diffusion cell and displaying: (a) no fracture (EX1 borehole); (b) an opened fracture (EX1 borehole); (c) a large tectonic fracture (GP borehole) (Motellier et al., 2007).

EX1-borehole gave core samples without any fracture (NF - Figure 4a). A fracture was artificially created in one core sample (so-called opened fracture, OF) with the aim of estimating its influence on diffusion (Figure 4b). GP-borehole ($\emptyset$(core sample) = 59 mm) was drilled in the 2003 main drift during its excavation. This borehore had an inclination angle of 60 °; it crossed both an unfractured area and a zone containing a tectonic fracture (TF – Figure 4c). The specific inclination angle of this borehole was chosen for intercepting the tectonic fracture in a orthogonal angle. Samples from EX1 were taken at 170 m from the borehole head and were located some 300 m away from the GP borehole.

A pre-equilibration procedure was required i) to (re-)saturate the clay with water and ii) to force its equilibration with a fresh synthetic pore water ([Na] 12.5 mM, [K] 0.14 mM, [Ca] 0.37 mM, [Mg] 0.29 mM, [Cl] 9.3 mM, [SO_4] 0.21 mM, [F] 0.23 mM) prior to diffusion experiments. It was performed by renewed contacts of fresh synthetic pore water (every two weeks) with the clayey samples. The renewal was stopped when the composition of the withdrawn solution was similar to that imposed by the synthetic solution, i.e. when the re-equilibration between the rock and the solution was complete. Once the pre-equilibration steps were achieved, tritiated water (HTO) considered as a reference tracer was introduced in the upstream reservoir to determine the HTO effective diffusion coefficient through each sample (Motellier et al., 2007). The procedure followed for alkaline water diffusion was then somewhat different from the so-called "through-diffusion" process designed to determine the transport parameters of a diffusing species in steady chemical conditions throughout the system (reservoirs + sample). In these experiments, the initial solutions in the upstream and downstream reservoirs were different. Steady chemical conditions were maintained in the upstream reservoir filled with an alkaline (type I) solution while the downstream reservoir, initially filled with synthetic pore water, was left free of any particular constraint. Hence, the composition of the downstream solution was expected to evolve until it reached that of the input alkaline solution, eventually forcing the whole system to equilibrate with it. The diffusion of the chemical species present in the alkaline solution (Na, K, Ca, Mg, OH) was monitored via pH-metry and ion liquid chromatography analyses using samples taken regularly from the downstream reservoir. Finally, tritiated water was introduced in the upstream reservoir to determine the change of diffusion coefficient for HTO after contact with an alkaline fluid. Microscopic investigations of the solids before and after alteration by the diffusion experiments were also performed for assessing the dissolution-precipitation phenomena induced by the alkaline plume. Each experiment was repeated 3 times.

Results and Interpretation

The batch experiments (Devol-Brown et al., 2007) show that whatever the temperature (25 and 70 °C) and the initial composition of the alkaline solution (type I or II), the solution composition is modified after being in contact with the argillite: decrease in pH which is more significant when the contact duration is longer, more or less marked decrease in potassium, sodium, calcium and magnesium concentrations, release of silica. However, no significant influence of the composition of the alkaline solution and temperature is detected with respect to the reaction processes involving the solids: indeed, the observed dissolved and precipitated phases are almost the same. On the other hand, the analysis of the solutions indicates that the reactions are intensified under the most aggressive conditions (70 °C and/or type I solution).

Nevertheless, while a period of interaction shorter than one month produces changes in the solution composition, it is not sufficient to generate detectable neoformed minerals neither on XRD patterns nor through SEM observations. Longer interaction periods used in the study (6 months) helped us to identify mineral transformations by means of XRD and microscopy approaches. However, some of the modifications are still too weak to be quantified by XRD.

Some differences are observed between reacted solutions from powdered and compact solids, in coherence with a weaker evolution in altered compact solids than in powdered ones which can be attributed to the difference of reactive surface between the two solids in contact with solutions. Dissolution/precipitation phenomena are less pronounced with compact solids than with dispersed clay: phenomena occurring on altered crushed and compact solids do not correspond to the same advancement stage and are not identified by the same techniques (XRD could be only used for crushed solids).

Major cation concentrations and pH monitoring during diffusion experiments show a rapid evolution in the downstream reservoirs. The pH value and alkaline cation concentrations increase until reaching the values imposed by the composition of the alkaline solution. Alkaline-earth cations evolve quite differently: their concentrations drastically drop, probably due to precipitation of carbonate phases. No significant differences can be observed between the zones displaying no fracture or an opened fracture. This result may indicate that the discontinuities induced by unloading and stress redistribution in the samples may be readily sealed by the swelling property of the clayey rock when fully hydrated. The tectonic fractured-zone slice shows larger heterogeneity in the diffusion measurements than the two others samples: it is assumed that the occurrence of calcite and pyrite veins in this zone plays a role in the diffusion processes inasmuch as these minerals reduce both porosity and cation exchange capacity of the clay rock. Effective diffusion coefficients for HTO decrease during the alkaline diffusion process (Table 1).

Table 1. Effective diffusion coefficients for tritiated water (HTO) before and after alkaline fluid interaction

Effective diffusion coefficient (De) for HTO $\times 10^{-11}$ ($m^2.s^{-1}$)	Non fractured samples (NF)	Samples with opened fracture (OF)	Samples with tectonic fracture (TF)
De(HTO) before alkaline interaction (Motellier et al., 2007)	2.62 ± 0.12	2.80 ± 0.14	2.17 ± 0.09
De(HTO) after alkaline interaction	1.92 ± 0.08	2.04 ± 0.10	1.42 ± 0.06

The decrease of the effective diffusion coefficient for HTO is around 26 – 28 % for samples containing no fracture or an opened fracture but it is more pronounced in presence of a calcite-filled fracture (35 %). This tendency may be induced by mineralogical modifications during the contact with the alkaline fluid, in particular the presence of some neoformed phases (Table 2). Nakayama et al. (2004) observed the same kind of phenomenon on compacted bentonite-sand samples, by measuring porosity during an alkaline fluid diffusion. Modelling of a clayey rock in contact with a fresh cement also explains this progressive porosity "clogging" at the interface with the hydrated cement by calcite precipitation (De Windt et al., 2004; Gaucher et al., 2004).

The solid samples from the batch and diffusion experiments generally present similar results (Table 2).

Table 2. Synopsis of the results from observation of dissolved and precipitated phases

Process	Mineral	Batch (Devol-Brown et al., 2007)	Diffusion
Dissolution	Dolomite	x	-
	Quartz	-	-
Precipitation	Calcite	x	x
	Quartz	-	rare
	Interlayered clay	x	-
	Zeolites	Non systematic	Non systematic
	Feldspar	Non systematic	Non systematic

Most of the solid samples suffer of dolomite dissolution, to greater or lesser extents. The XRD and SEM observations do not reveal any significant dissolution of aluminosilicate phases. The decrease of silica and aluminum in the solid phases, concomittant with the increased concentration of these elements in solution, is however consistent with the dissolution of one or several aluminosilicate phases for the set of studied samples. The XRD analyses are unable to reveal the nature of this (these) phase(s), as the sensitivity of this technique is too weak. Calcite systematically precipitates. The neoformation of interlayered clay minerals is also detected for all the samples of batch tests but not in these from diffusion experiments. However, it is not excluded that this type of neoformation occurs inside the thin argillite slices, whereas only their external surface were examined due to technical constraints. Neoformation of zeolites or feldspars (especially potassic) is discrete and only observed in some samples. No clear correlation was found between the experimental conditions and the non-systematic precipitation of one of these phases. Furthermore, several authors (Adler, 2001; European Commission, 2005) report calcium silicate hydrates precipitation (CSH or CASH), which goes undetected in this study. This could be explained by the selected alkaline solutions, the systematic absence of silica in fresh fluids used in batch and diffusion experiments and/or by the limits of the analytical techniques used. In fact, these poorly or not crystallised phases are not easily detected by XRD and moreover are not stable under the SEM electron beam.

ENGINEERED ANALOGUES RECOVERED IN THE TOURNEMIRE URL

Sampling

The Tournemire URL presents several opening excavations (one century-old railway tunnel, 1996 and 2003 drifts, boreholes - Figures 1 and 2) in the Toarcian argillite. Some of these excavations were lined with concrete walls and offer therefore the possibility of studying transformations undergone by argillite in presence of cement in an *in situ* context for longer times (from few to 120 years) than those usually accessible by surface laboratory experiments.

The reference argillite sample was taken from the horizontal borehole EX1 at a distance of 17 m from the 1996 drift wall to minimize any mechanical and chemical disturbances.

Different contact samples were taken in the EDZ, for a first characterisation campaign (Tinseau et al., 2006):

(1) dry interfaces were sampled at the tunnel masonry/argillite contact (contact time ~ 120 years),
(2) wet interfaces were taken close to the drained areas i) just below the tunnel roadbed in contact with the canal draining the Cernon fault water (contact time ~ 15 years - overcoring DM borehole), ii) at the tunnel masonry/argillite interface at a distance of 70 m from the Cernon fault, a high flow rate and transmissive regional fault (contact time ~ 120 years).

The upper part of the DM borehole (sampled between 1 m and 1.30 m in depth) is slowly drained by water seeping from a small canal located under the tunnel roadbed, whereas the lower part of the borehole (below 1.5 m) is assumed to be more representative of argillite/concrete interactions in confined conditions). A second characterisation campaign was therefore focused on this lower part of the DM borehole.

All interface samples were vacuum sealed under inert atmosphere in Al-coated plastic bags just after drilling.

The nature of the cementitious materials is different according to the location: the tunnel masonry (limestone blocks) is bedded with lime, actually siliceous (its initial chemical composition is missing), whereas the different cores are of CEM II 32.5 concrete (ISO 9002), which is constituted by clinker, gypsum and calcium carbonate.

Waters were recovered, filtered through a 0.2 µm membrane (Millex®) and preserved on-site. Major cations (Na, K, Ca, Fe and Mg) were analysed by atomic absorption spectrometry, Cl and SO_4 by ion chromatography, silica by colorimetry and total alkalinity by acidimetry. The chemical composition of crushed solids was determined by spectrometric analysis (ICP-AES and ICP-MS) for major and trace elements by the Centre de Géologie de la Surface in Strasbourg (France). Semi-quantitative XRD measurements and SEM observations combined with EDS were also performed on crushed solids, to compare the mineralogy of initial and reacted solids.

Characterisation

First Characterisation Campaign

Tinseau et al. (2006) have shown that the composition of cementitious material (siliceous lime or CEM II 32.5) had likely an impact on the production yield of neoformed mineral phases. For example, in contact with CEM II concrete, clay neoformation is clearly more marked than with siliceous lime.

In dry conditions (contact time ~ 120 years), no significant alteration of argillite is observed in the presence of "old cement" (siliceous lime). Precipitation of gypsum as acicular crystals at the surface of argillite is due to atmospheric pyrite oxidation. Brown colour areas enriched in iron oxy-hydroxides around pyrite crystals indicate limited iron diffusion in the nearby environment of oxidised pyrite.

In wet conditions, the water renewal appears to be a key parameter. The argillite samples close to the Cernon fault (high flow rate - contact time ~ 120 years) present important dolomite precipitation due to the fault water composition and the presence of adjacent dolomitic formations (Cabrera et al., 2001). Decrease in quartz and kaolinite contents and slight increase in K-feldspars are also noticed. K-feldspar overgrowths should be linked to the kaolinite dissolution. Chlorite has totally disappeared. Oxidation of pyrite, induced by the air contact, also occurs leading to precipitation of Fe-oxy-hydroxides. The lack of relatively soluble gypsum in this context is explained by very high flow rates. It is possible that the important dolomite neoformation prevents the percolation of water in argillite and therefore limits mineralogical modifications in the claystone.

The water flow rate is assumed to be reduced for argillite samples located far from the Cernon fault and in contact with old cement. But equilibrium has not yet been reached, as well as for the samples submitted to rapid flow rate where the reaction path is influenced by leaching. No dolomite precipitation is observed in these samples. Moreover, dolomite initially present in argillite is partly dissolved, while important neoformation of calcite is evidenced both by SEM and XRD. Furthermore, no variation in quartz and chlorite concentration is observed in these samples. An increase of mixed layer illite/smectite is suspected through the deconvolution of the XRD patterns. Finally, oxidation of pyrite and precipitation of iron oxy-hydroxides and gypsum are observed.

In the DM overcoring zone (contact time 15 ~ years), changes are observed amongst both accessory minerals and the clay fraction. At first, pyrite oxidation, accompanied by gypsum and iron oxy-hydroxides precipitation, occurs and is confirmed by the presence of sulfate in water: this is interpreted as an oxidative perturbation due to interaction with air. A cross section shows a zonation on the 2.5 centimetres thick of argillite in contact with concrete. Calcite precipitation appears in the very first millimetres; the next centimetre shows neoformed clays associated with calcite crystalisation while the last centimetre is characterized by calcite precipitation and feldspar overgrowths. The neoformation of few Na-zeolites (recognizable thanks to the EDS spectrum and the mineral morphology) is also noticed through SEM observations in samples from the upper part of the DM overcoring (at around 1.10 m under the tunnel level). Simultaneously the quartz content significantly decreases, with a more pronounced depletion in the first millimetres in contact with concrete. These SEM observations are supported by XRD semi–quantitative analysis. Moreover, XRD deconvolution and semi-quantitative measurements indicate an increase of illite/smectite (I/S) content with the distance to concrete. All these observations suggest the recrystallisation of mixed-layer I/S.

Additional DM Sample Characterisation

The alkaline perturbation propagation is analysed in DM samples selected below 1.55 m at 3 locations as illustrated in Figure 5: a first zone around the borehole wall (P1 profile) allows to study the propagation of the disturbance within the argillite matrix perpendicular to the concrete; a second zone (P2 profile) on both sides of macroscopic fractures initially filled by concrete perpendicular to the fracture planes; and a third zone along the decompression fractures showing a visible blue front along the fractures planes, sub-perpendicular to the concrete (P3 profile).

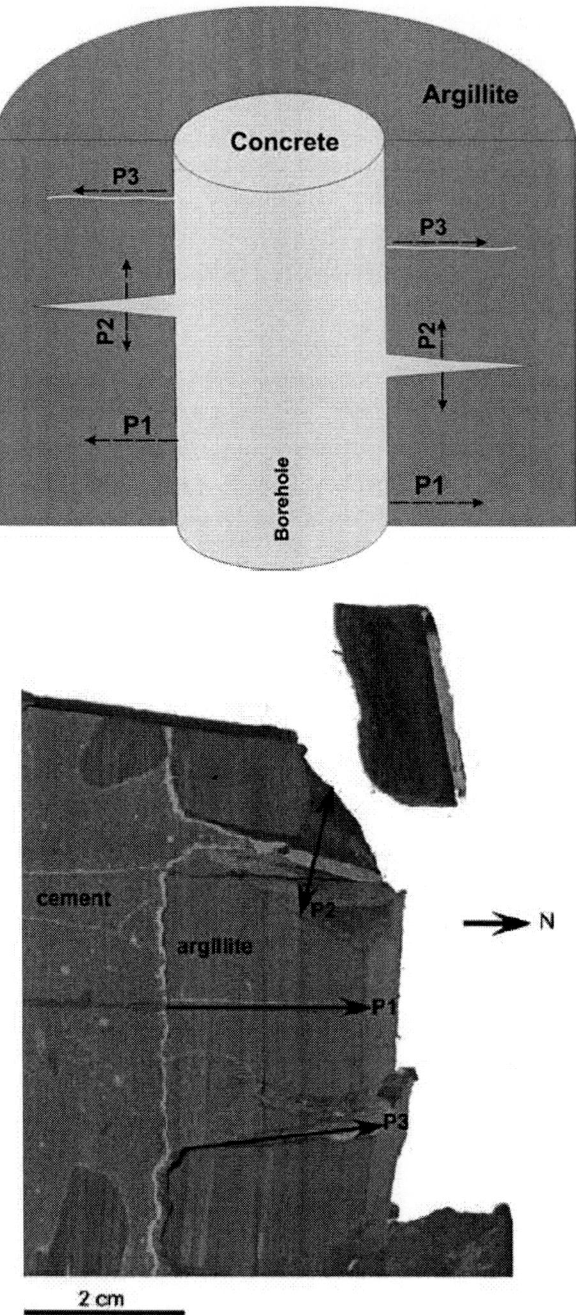

Figure 5. The three profiles of interest for the alkaline disturbance into argillite at 1.55 m deep. P1 profile in the argillite matrix, P2 profile on both sides of a fracture filled by concrete and P3 profile along decompression fractures.

In order to characterise these different profiles of the alkaline propagation, microstructural, petrography and mineralogical analyses by XRD, SEM, TEM (CRMCN CNRS, UPR 7251, Marseille, France), ICP-AES (Laboratoire Hydr'ASA, UMR CNRS 6532, Poitiers, France) techniques coupled to isotopic measurements (Sr, C, O) (Isabelle Techer - GIS/CEREGE, UMR CNRS 6635, Nîmes, France) were performed.

(a) Petrography Study

The macroscopic observations of a sample located at 1.55 m and focused on the P1 and P3 perturbations show a change in the argillite texture at a distance of 1 cm and 1.5 cm from the concrete contact for P1 and P3, respectively. Beyond theses distances, the argillite is again well bedded and pyrite is observed. Thin sections observed by optical microscope indicate that the mineralogical transformations are difficult to determine except when microcrystals of calcium carbonate are neoformed. Dark fronts are mainly composed of fine microscopic particles (micrite), whereas grey zones are mostly composed of coarse grained (10 mm) calcite rich (microsparite). SEM observations show that the optical fronts correspond to textural and mineralogical changes of the argillite (Figure 6). More fronts are observed along P3 profile than into the matrix P1.

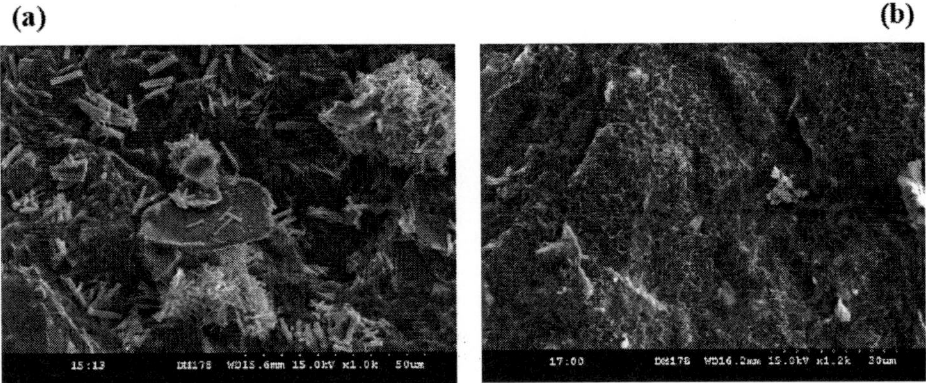

Figure 6. SEM picture showing: (a) ettringite and gypsum minerals observed from 0 to 2 mm from the concrete and (b) honeycomb clays observed from 2 to 5 mm, in P1 profile.

(b) XRD Study

X-ray whole rock diffractograms of P1 and P3 profiles are characterized by predominant phyllosilicates (illite, micas, kaolinite, chlorite), carbonates (vaterite, calcite, dolomite), K and Na-Ca feldspars, pyrite and gypsum (oxidation product). No phase of calcium silicate hydrates (CSH, CASH) can be observed, the main peaks of diffraction being explained by common minerals. If such phases exist, they are present in insufficient quantity to be detected, or they are amorphous.

X-ray diffractograms of the fraction with a grain size $< 2\mu m$ obtained under air-dried condition and treated by ethylene glycol for the two profiles P1 and P3 display the same clay mineralogy assemblage. They present a reflection at ~14.2 Å assigned to chlorite. A reflection situated at ~10 Å and its harmonic at ~5 Å are attributed to non-swelling dioctahedral layers (illite or mica). The reflections situated at ~7.2 and ~3.5 Å indicate the presence of kaolinite. A broad diffraction domain, observed between 11 and 12 Å on air-dried samples, is assigned to illite/smectite mixed layer (I/S) minerals. The ethylene glycol saturated samples show minimal modification indicating that I/S minerals are mainly composed of illite layers.

Nearly the same mineralogical transformations are observed for P1 and P3 profiles except for the presence of the visible blue front in P3 corresponding to the precipitation of Si, Ca phases and of Ca-rich phases.

(c) Isotopic Study

The coupling of $\delta^{13}C$ and $^{87}Sr/^{86}Sr$ ratios was chosen as effective isotopic tool for tracing paleocirculation of alkaline fluids. $^{87}Sr/^{86}Sr$ ratios measured in the argillite at 20 - 22 mm from the concrete are close to those determined for the Toarcian argillite formation. Close to the concrete interface, Sr isotopic values increase towards the concrete Sr isotopic signature (0.7083 - 0.7084) (Figure 7).

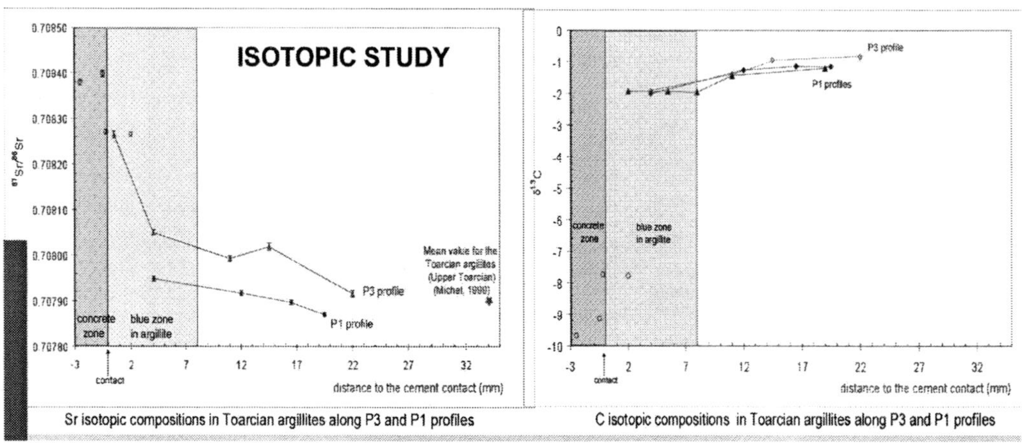

Figure 7. Sr and C isotopic compositions in Toarcian argillites along P3 and P1 profiles (errors bars are reported in both cases).

Both variations along the P1 and P3 profiles show a correlation between $\delta^{13}C$ values and the distance to the concrete contact. Isotopic (Sr and C) disturbances of the Toarcian argillite are more important for P3 (along a fracture). For both studied profiles, these results suggest the interaction between the argillite and a fluid in equilibrium with the concrete. This interaction is isotopically detectable over a distance of about 10 mm.

(d) Conclusion

Alkaline perturbations into the argillite are put into evidence in the matrix and along fractures. Mineralogical and textural transformations, as well as their expansion, are less developed within the argillite matrix (P1, zone of major mineralogical and textural transformations is 1 cm thick) than along fractures (P3, zone of major mineralogical and textural transformations is 1.3 - 1.5 cm thick). For both types of profile, mineralogical transformations are identified beyond the zone of major transformation without any apparent change of the texture.

Nearly the same mineralogical transformations are observed for P1 and P3 profiles except for the presence of the visible blue front in P3 corresponding to the precipitation of Si, Ca phases and of Ca-rich phases. However, for P3, fronts are better defined and a gradient of crystallization of calcite indicating the propagation of an alkaline fluid (confirmed by isotopy) is observed. Moreover very angular K-feldpars (possible overgrowths?) have been often encountered in the last front for P1 and P3; no zeolites were observed. Ongoing analyses are performed in order to complete the mineralogical sequence and to access the microporosity: they consist in µ-XRD, isotopic analyses, cathodoluminescence and X-ray tomography.

DM Sample Modelling

In addition to the characterisation of the 15 years old DM samples described in the previous sections, simulations of the main experimental outcomes were performed (De Windt et al., 2007) so as to help understanding the processes involved in the observed alkaline perturbation. These calculations carried out with the reactive transport code HYTEC (van der Lee et al., 2003) in collaboration with Laurent de Windt (Ecole des Mines de Paris (Centre de Géosciences, Fontainebleau, France)) mainly focused on P1 and P3 profiles, which were simulated using respectively a 1D - radial and a 2D - cylindrical grid. Though not observed in these profiles, zeolites were allowed to precipitate since this kind of mineral was identified from the first characterisation campaign. Solute transport was considered driven by pure diffusion in both profile cases. Either local thermodynamic equilibrium or full kinetic-control was assumed.

The CEM II concrete filling the borehole was modelled by a mixture of Portland cement phases (CSH, sulfo-aluminates; 35 wt. %) and calcareous aggregates (calcite; 65 wt. %). The interstitial water was a K–Na–OH fluid of pH 13.2 (at 15 °C, the average temperature within the tunnel). The undisturbed argillite was represented by: muscovite (for illite), kaolinite, Ca-montmorillonite (smectitic part of the argillite), quartz, K-feldspar, calcite and dolomite with contents according to mineralogical composition given in the first section of the present paper. Its pore water was of Na-Cl-(HCO$_3$) type with pH close to 8. The tritiated water effective diffusion coefficient and porosity were set respectively to 2×10^{-11} m.s^{-1} and 9.5 % according to Motellier et al. (2007), whilst values of 3×10^{-12} m.s^{-1} and 13 % were assigned to these parameters for concrete assuming non porous aggregates and typical Portland cement characteristics.

The kinetic simulations show a better agreement with the experimental results than the thermodynamic equilibrium ones, in particular with respect to the expansion of the perturbation as shown in Figure 8. The global trend of kinetics is to smooth the sharpness and intensity of the mineralogical transformation fronts, to delocalize them deeper inside the argillite matrix (< 1 cm) and the fracture (≤ 2 cm) as well as to lessen clogging within argillite porosity.

The calculated mineralogical evolution is globally similar in the argillite matrix and in the fracture filling, in accordance with the experimental observation. The secondary mineral sequence simulated from concrete to clay is: CSH/ettringite/carbonates at the concrete/argillite interface, followed by a zone of neoformation of clay-like phases and calcite, then a last zone with precipitation of calcite and (scarce) zeolite. Calculations show that the carbonated precipitate (with predominance of vaterite over calcite) observed at the concrete/argillite interface follows the Ostwald's empirical rule and can be more specifically attributed to the alteration of concrete rather than to that of argillite.

These simulations have thus pointed out that kinetics is likely to govern the propagation of the perturbation for such interaction duration and that main mineralogical modifications highlighted by the characterisation campaigns could be explained. However, there are still discrepancies between modelling and experimental results, e.g. the precipitation of muscovite (illite) instead of K-feldspar. Indeed, such modelling work also raises issues that require complementary studies, so as to reinforce the interpretation in terms of interaction processes and consequences on argillite evolution, notably with respect to the nature of the secondary clay-like phases, the relevance of the selected kinetic parameters and the so-called "clogging" process.

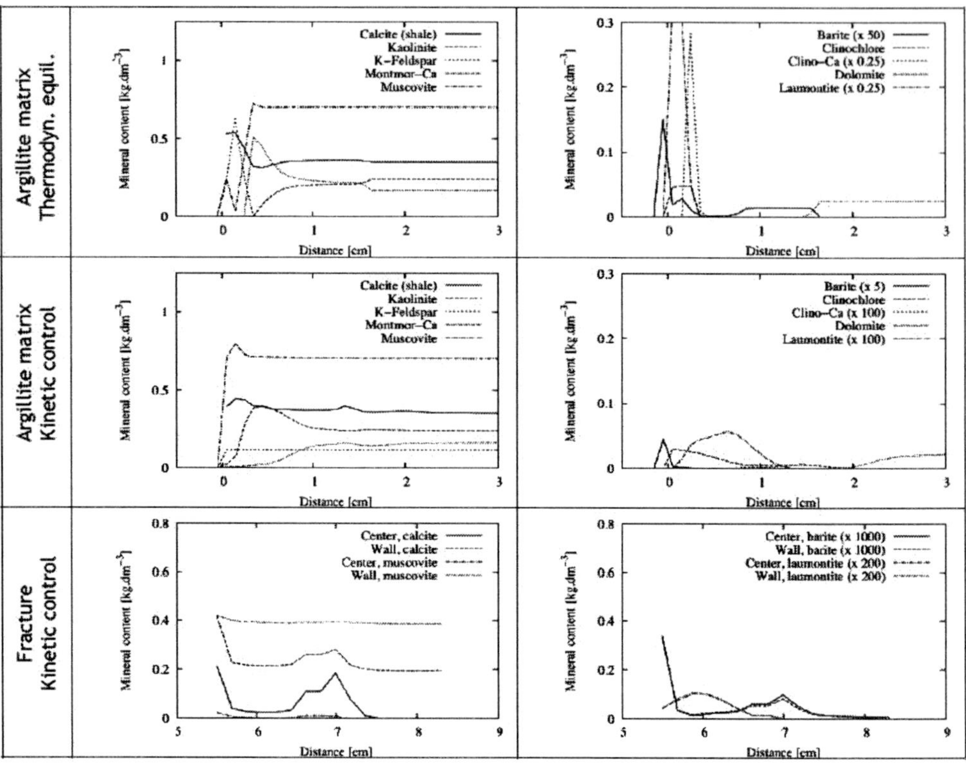

Figure 8. Profile of mineral contents calculated at 15 years in the argillite matrix assuming local thermodynamic equilibrium (top) or kinetic control on mineral dissolution/precipitation (middle) and in the fracture assuming kinetic control (bottom). Note that for the matrix case, concrete/argillite interface is at x = 0 cm, whereas for the fracture case, concrete/argillite interface is at x = 5.6 cm.

ARGILLITE/STEEL INTERACTIONS

Experimental Concept

Steel samples were introduced in 1999 within the argillaceous Tournemire massif through vertical (CR7) and horizontal (CR5) drillings performed from the railway tunnel (Figure 2) with the aim of assessing argillite/steel interactions. The samples were emplaced far away from the tunnel EDZ (at a distance of 9 m) under dry conditions in the CR5 case, or in the EDZ slowly drained by water seeping in the CR7 case. Two austenitic stainless steels 309S and 316L (samples BI5, BR2) and one carbon steel A42 (samples BC3, BC4) were selected for the experiment because they are potential materials for the High-Level Waste (HLW) containers. Re-compacted crushed Toarcian argillite from the Tournemire URL with a nearly isotropic texture was mixed in contact with the steel samples and was used to backfill the CR5 borehole. In parallel, original Toarcian argillite with an undisturbed anisotropic texture (pieces of cores) was used for the vertical borehole located in the EDZ zone (CR7).

After two years (Foct et al., 2004) and six years of interaction respectively, both steels and clay interfaces were recovered by overcoring. One of these samples is shown on Figure 9.

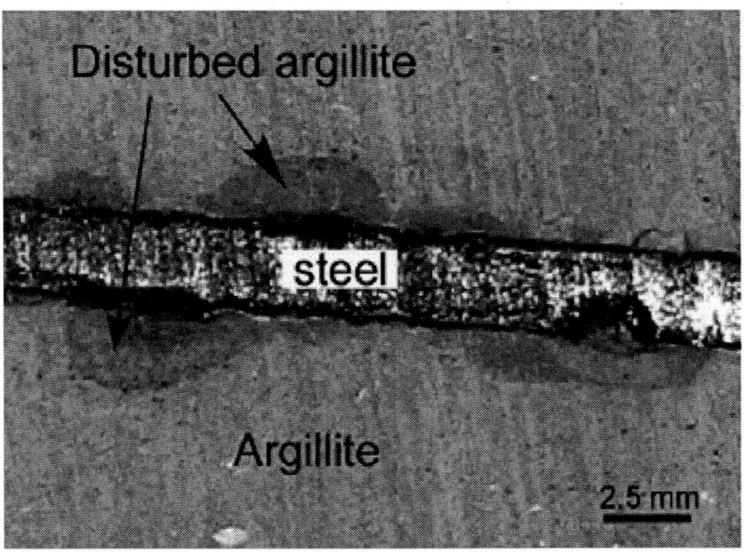

Figure 9. Argillite – carbon steel sample.

In order to determine the main interaction processes, the study focused both on petrographic and mineralogical characterisation (XRD, SEM, TEM and ICP-AES) and on geochemical modelling of modifications in the argillite in contact with steel samples. It was mainly performed by Anne Gaudin (Laboratoire de Planétologie et Géodynamique, UMR-CNRS 6112, Nantes, France) and Stéphane Gaboreau (Laboratoire Hydr'ASA, UMR CNRS 6532, Poitiers, France).

Characterisation

No significant changes are observed for argillite in contact with the stainless steels. For argillite in contact with carbon steel, the development of a reddish zone is observed within the argillite which appears more extended for the re-compacted crushed Toarcian argillite (CR5, up to 2 mm thick) than for the pieces of cores (CR7, < 1 mm). The extension of this zone is irregular and fissures are observed in some samples. Fractures enriched in Fe are also detected for the EDZ samples in the CR7 borehole. Whatever the extension of this zone, the composition is quite similar.

According to petrological SEM observations combined with EDS element mapping, this altered zone shows a clear Fe-enrichment, a strong decrease in Ca-content and a slight decrease in Si, Al and K-contents (Figure 10). Some partly dissolved framboïdal pyrites are also observed in the Fe-rich zone.

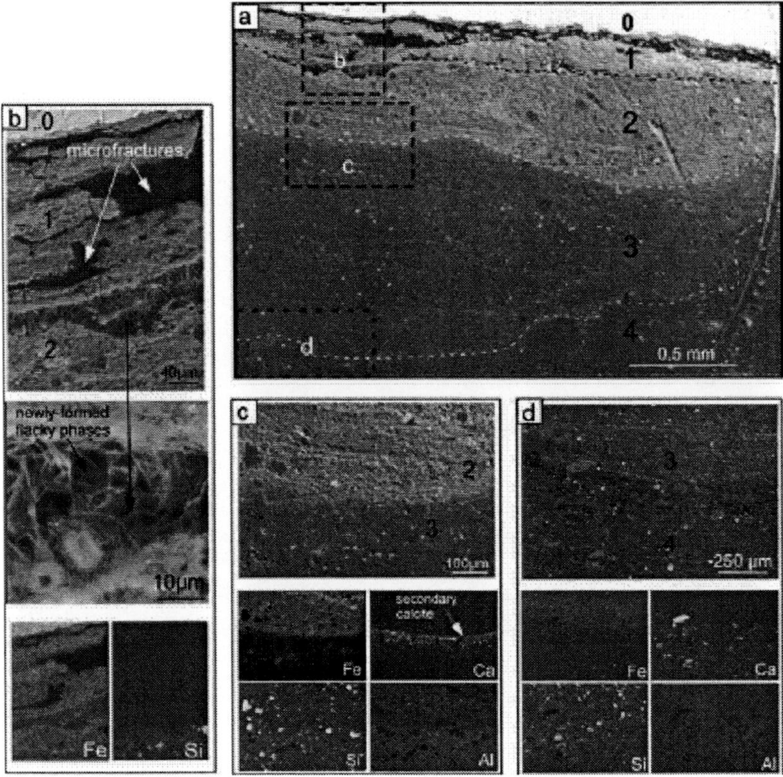

Figure 10. SEM images and EDS element maps of the steel/argillite contact. (a):(0) the unaltered steel, (1) the corroded steel layer, (2 and 3) the altered argillite, and (4) the unaltered argillite. The corroded steel layer (1) is composed of iron and oxygen-rich phases. The development of newly-formed Fe, O-rich phases is locally observed (b). The argillite is characterized by the diffuse occurrence of silicon, calcium, magnesium, aluminum and potassium elements, and displays successive Fe-enriched alteration zones.

The Fe-rich zone analysis by semi-quantitative XRD indicates the formation of goethite, lepidocrocite and gypsum and the dissolution of a part of calcite and clay minerals. Moreover, a relative increase of the quartz and feldspar contents is noted. At last, melanterite and magnetite are also detected in trace. Regarding the clay fraction, the deconvolution of the "illite + I-S" XRD reflection spectrum of the oriented clay fraction required three elementary components: a first one is assigned to a well-crystallized illite, a second one to a poor-crystallized illite and a third one is attributed to a mixed layer illite-smectite with a high proportion of illite layers (more than 70 %). A small decrease is observed in relative intensity of the I-S component from 30 % for the initial argillite down to 20 % for the disturbed argillite. Moreover, a slight shift of the position of this I-S component towards the illite end-member observed for the altered argillite indicates a relative enrichment of the illite proportion within the I-S mixed layers.

The formation of Fe^{3+}-rich minerals (lepidocrocite and goethite) and gypsum suggests oxidizing conditions during the steel-argillite interaction. Moreover, the calcite dissolution suggests a pH decrease which is probably due to a proton release related to the oxidation of the carbon steel and pyrite. This pH decrease could be responsible of the dissolution of a part of the smectite-illite mixed layers of the disturbed argillite, and the enrichment of the proportion of the illite layers in the mixed layers would be due to a preferential dissolution of the smectite layers.

Modelling

Two sets of geochemical modelling were performed in collaboration with P. Vieillard (Laboratoire Hydr'ASA, UMR CNRS 6532, Poitiers, France), using a thermodynamic approach (KINDIS code) as regard to anoxic and oxic conditions to understand the formation of the different alteration oxides (goethite, magnetite). In the first one, a circulation of free water with a constant oxygen fugacity (atmospheric condition) was considered. In the second one, water progressively equilibrated with the argillite without contact with air was taken into account.

Modelling of the interaction of the fluid with argillite and iron at constant oxygen fugacity leads to the oxidation of iron and pyrite combined with the precipitation of goethite. Then, the fluid reaches equilibrium with trioctahedral smectite. This precipitation of Fe-smectite is accompanied by the dissolution of montmorillonite and chlorite. At the end of the run, precipitation of gypsum corroborates the pH decrease.

Modelling the interaction of the fluid with argillite and iron at decreasing oxygen fugacity (fO_2) leads also to the oxidation of iron. This simulation differs from the precedent one by the neoformed phases. First, the fluid reaches progressively saturation with kaolinite, pyrite and quartz, respectively. But continued reaction is observed, resulting in precipitation of magnetite and trioctahedral Fe-phyllosilicates (chamosite, saponite). Precipitation of Fe-phyllosilicate and equilibrium between calcite and fluid induce an increase in pH and thus the destabilization of chlorite.

The two sets of assumptions give some indications on the conditions responsible of the formation and the dissolution of phases present in the argillite. Fe-oxides are sensitive to fO_2. The formation of newly formed Fe^{3+} - rich mineral (goethite) suggests oxidizing conditions during all the time of experiment in CR7 borehole. The presence of magnetite in contact with steel, in the CR5 borehole, could be now explained by the reductive conditions, when all the oxygen occluded has been consumed. The formation of magnetite could be explained either by the partial oxidation of steel or by the reduction of goethite. Precipitation of Fe-phyllosilicates is related to the presence of others elements in the presence of free water (advection-dispersion transport).

CONCLUSION

The Tournemire URL, a 2 km long and 125 years old railway tunnel built through the Toarcian formation, completed with more recent drifts, is used to carry out a research program developed to address key safety questions related to the modifications of the Toarcian claystone due to the presence of concrete or steel. This URL provides with complementary studies at different space and time scales, whether through experiments in surface laboratories or from in situ engineered analogues.

Regarding the influence of concrete, the characterisation of engineered analogues (15 years old) allow confirmation of data obtained in laboratory for short durations (max. 18 months) and, in most cases, at higher temperature than those measured in the tunnel (mean annual value of 15°C). Experimental and modelling results indicate textural, petrographic, mineralogical and chemical changes of the initial argillite resulting from precipitations (mainly calcite and interlayered clays, zeolites and feldpars on some samples) and dissolutions (dolomite and quartz on some samples). These results are in good agreement with those reported in the literature (e.g. Adler, 2001; European Commission, 2005; Gaucher and Blanc, 2006), except for the hydrated calcium silicate precipitation (CSH or CASH), which goes undetected in this study. The present study also highlights the control of the perturbation process by kinetics on the investigated durations. However, the non systematic and presently misunderstood observation of slow kinetics neoformations, such as argillaceous minerals, zeolites and feldspars, has not revealed a unique reaction scheme. A further characterisation of the neoformed minerals would give a better understanding of the interaction mechanisms involved and their dependency on different experimental or predominant *in situ* conditions.

The experimental study of the 6 years argillite/steel interaction indicates clear textural, petrographic, mineralogical and chemical modifications of the initial nature of the argillite. The mineralogical, chemical and petrographic studies of the Fe°/argillite interfaces indicate the development of a Fe-rich front within the argillite. It is due to the diffusion of iron from the corroded steel associated with mineralogical changes of the surrounding argillite. This Fe-enrichment, close to the steel surface, is accompanied by crystallization of significant amounts of goethite/lepidocrocite (FeOOH) and some traces of magnetite close to the steel surface. The formation of Fe-rich phases within the argillite is accompanied by dissolution-crystallization processes. Thus, after XRD and SEM observations of bulk sample, these processes consist in dissolution of calcite, phyllosilicate (I/S) and pyrite and crystallization of some traces of gypsum and melanterite ($FeSO_4.7H_2O$). These observations are corroborated with thermodynamic modelling, giving indications on the conditions of dissolution and precipitation of mineral phases. Kinetic simulations are in progress to assess the evolution of this system on longer interaction periods. The outcomes will be compared to steel samples still emplaced in other Tournemire boreholes.

This research program therefore illustrates the benefit of combining different approaches (laboratory and *in situ* conditions, solid and liquid characterisation techniques, bulk and local analyses, experiments and modelling...) in order to obtain complementary information and thus to reduce uncertainties, due e.g. to experimental and analytical problems, specific to each approach. Such a pool of various indications, furthermore in general agreement with the results described in literature, increases the level of confidence in the existing tools. At last, though some features worth being further investigated, the data acquired in the framework of this program combined to others reported in the literature provide with valuable information regarding the extrapolation issue to geochemical evolution of clay materials due to the presence of concrete or steel for periods lasting several tens of thousands of years.

The Tournemire URL provides further possibilities to investigate *in situ* argillite/concrete and argillite/steel interactions. This program will be continued by the characterisation of others *in situ* engineered analogues, not recovered yet. In parallel, *in situ* experiments could be planned, e.g.: (i) experiments based on the circulation of an alkaline fluid in a packed-off section of a vertical borehole; (ii) "passive experiments" making use of vertical or horizontal boreholes filled with concrete/steel/bentonite. Assessment of others geochemical issues such as the redox buffering capacities of clays is also planned within the Tournemire programmes.

ACKNOWLEDGEMENT

Reviewed by Pierre De Cannière, SCK•CEN, Mol (Belgium) who is fully acknowledged by the authors for his comments and suggested revisions, which significantly improved this manuscript.

REFERENCES

Andra. Dossier 2005 Argile - Evolution phénoménologique du stockage géologique. Collection « Les rapports », *Andra* (France); 2005; 253 pp.

Adler, M. *Interaction of claystone and hyperalkaline solutions at 30 °C: a combined experimental and modelling study*; Ph.D. thesis, Bern University, CH, 2001; 120 pp.

Cabrera, J.; Beaucaire, C.; Bruno, G.; De Windt, L.; Genty, A.; Ramanbasoa, N.; Rejeb, A.; Savoye, S.; Volant, P. Projet Tournemire – Synthèse des programmes de recherche 1995-1999; *Rapport IPSN DPRE/SERGD 01-19*, Paris, France, 2001; 202 pp.

Devol-Brown, I.; Tinseau, E.; Bartier, D.; Mifsud, A.; Stammose, D. *Phys. Chem. Earth.* 2007, 32, 320-333.

De Windt, L.; Pellegrini, D.; van der Lee, J. *J. Contam. Hydrol.* 2004, 68, 165-182.

De Windt, L.; Marsal, F., Tinseau, E.; Pellegrini, D. *Phys. Chem. Earth.* 2007, submitted

European Commission. Ecoclay II: Effects of cement on clay barrier performance – Phase II. Final report. Contract FIKW-CT-2000-00028; 2005; 364 pp.

Foct, F.; Cabera, J.; Dridi, J.; Savoye, S. *Proceeding of the 2ⁿᵈ International Workshop Eurocorr*, Nice, France. 2004; p.68.

Gaucher, E.; Blanc, P.; Matray, J.M.; Michau, N. *Appl. Geochem.* 2004, 19, 1505-1515.

Gaucher, E.; Blanc P. *Waste Manage.* 2006, 26, 776-788.

Matray, J.M.; Savoye, S.; Cabrera, *J. Eng. Geol.* 2007, 90, 1-16.

Melkior, T.; Yahiaoui, S.; Motellier, S.; Thoby, D.; Tevissen, E. *Appl. Clay Sci.* 2005, 29, 172-186.

Michel, O. *Caractérisation isotopique Rb/Sr et Pb/Pb des roches totales, des minéraux de remplissage de fracture et des eaux des formations sédimentaires jurassiques de Tournemire (Aveyron, France). Implications sur les interactions Eau/Roche passées et actuelles.* Ph.D. thesis, Montpellier II University, 1999, 289 pp.

Motellier, S.; Devol-Brown, I.; Savoye, S.; Thoby, D.; Alberto, J.C. *J. Contam. Hydrol.* 2007, 94, 99-108.

Nakayama, S.; Sakamoto, Y.; Yamaguchi, T.; Akai, M.; Tanaka, T.; Sato, T.; Iida, Y. *Appl. Clay. Sci.* 2004, 27, 53-65.

Tinseau, E.; Bartier, D.; Hassouta, L.; Devol-Brown, I.; Stammose, D. *Waste Manage.* 2006, 26, 789-800.

Valès, F.; Nguyen Minh, D.; Gharbi, H.; Rejeb, A. *Appl. Clay Sci.* 2004, 26, 197-207.

van der Lee, J.; De Windt, L.; Lagneau, V.; Goblet, P. *Comput. Geosci.* 2003, 29, 265-275.

In: Nuclear Waste Research: Siting, Technology and Treatment   ISBN 978-1-60456-184-5
Editor: Arnold P. Lattefer, pp. 189-205   © 2008 Nova Science Publishers, Inc.

Chapter 6

NUCLEAR TRANSMUTATION THROUGH LASER COMPTON SCATTERING GAMMA-RAY

Dazhi Li and Kazuo Imasaki

Institute for Laser Technology,
2-6 yamadaoka suita, Osaka, Japan

Shuji Miyamoto, Sho Amano and Takayasu Mochizuki

Laboratory of Advanced Science and Technology for Industry, University of Hyogo,
3-1-2 Kouto, Kamigori-cho, Ako-gun, Hyogo, Japan

ABSTRACT

In order to effectively eliminate or reduce the hazards of long-lived activity from nuclear waste, an approach of laser Compton scattering gamma-ray based nuclear transmutation is proposed. Comparing with the conventional bremsstrahlung gamma-ray, laser Compton scattering gamma-ray features a peak in energy spectrum that can merit the coupling of gamma-ray and nuclear giant resonance to induce the photonuclear reaction. The advantage of laser Compton scattering gamma-ray in the application of transmutation is discussed, and the reaction rate of transmutation is theoretically analyzed. According to the proposal, a laser Compton scattering gamma-ray facility has been developed on NewSUBARU storage ring; fundamental experiments concerning nuclear transmutation is carried out; the measurement of reaction rate based on $^{23}Na^{127}I$ target is described and the experimental data is close to the simulation result. To improve the transmutation efficiency, the second target is considered by using the neutrons generated from the first target, therefore, a complex targets set is proposed. The related investigation is introduced.

1. INTRODUCTION

Some fission products (FP) and transuranic (TRU) mainly from the nuclear power have a long-lived (more than millions of years) activity, even longer than the life of the barrier to contain them for geological disposal, consequently, such nuclear wastes dissolve readily in ground water and move easily throughout the ecosystem posing radiological hazards. So, it is necessary to effectively eliminate the radioactivity of those nuclear wastes or convert them to less hazardous forms.

Transmutation is considered as an optional approach by reducing the radioactive life of the nuclear waste through converting the nuclei of long-lived activity to its corresponding isotope of short-lived activity. Several ways have been proposed to realize the nuclear transmutation, such as through (n, 2n) and (p, n) reactions, based on nuclear reactors and accelerator setup, respectively. High-energy photons, like gamma-ray, also can induce nuclear reaction, which is regarded as an alternative to make transmutation. And generating the strong gamma-ray beam becomes possible by today's technology. This method was originally proposed to use the bremsstrahlung gamma-ray inducing the photonuclear reaction (γ,n), however, the poor coupling between the photon beam and the giant resonance is inevitable, since the bremsstrahlung photon spectrum is very wide.

In order to improve the coupling, a gamma-ray with an appropriate energy spectrum matching the characteristic of nuclear giant resonance is required. Fortunately, laser Compton scattering (LCS) gamma-ray looks like a good choice [1,2]. The LCS gamma-ray is generated from the collision of a laser light to a high-energy electron beam, and it holds a peak in the energy spectrum, which can overlap the peak of nuclear giant resonance and hence realize a good coupling. Furthermore, this peak can be shifted easily by altering the parameters of electron beam, laser light or collision conditions, to match the peak of the nuclear giant resonance of various nuclei. So, the LCS gamma-ray seems promising in nuclear transmutation.

2. FUNDAMENTAL THEORY

The Compton scattering or Compton effect describes the interaction of light with matter, usually refers to the interaction involving only the electrons of an atom. The injected photon can gain energy when the inverse Compton scattering occurs. With the development of the accelerator technology, high-energy electron beam is available. And a scheme to generate strong gamma-ray from the collision of laser light and high-energy electron beam was proposed. Such a gamma-ray is called the laser Compton scattering gamma-ray.

2.1. Laser Compton Scattering

The laser Compton scattering, as an effective method to produce high-energy photons, has been developed and researched in many laboratories [3,4]. The kinematics of the laser Compton scattering is shown in Figure 1. When a laser photon with energy E_L is scattered on an electron with energy E_e at a collision angle α, it is scattered with the energy E_γ at angle ϑ relative to the motion of the incident electron. The energy of the scattered photon is given as [5]

$$E_\gamma = \frac{E_{max}}{1+(\vartheta/\vartheta_0)^2},$$

(1)

where

$$E_{max} = \frac{x}{x+1}E_e \; ; \; \vartheta_0 = \frac{mc^2}{E_e}\sqrt{x+1} \; ; \; x = \frac{4E_e E_L \cos^2(\alpha/2)}{m^2 c^4}$$

where E_{max} is the maximum scattered photon energy.

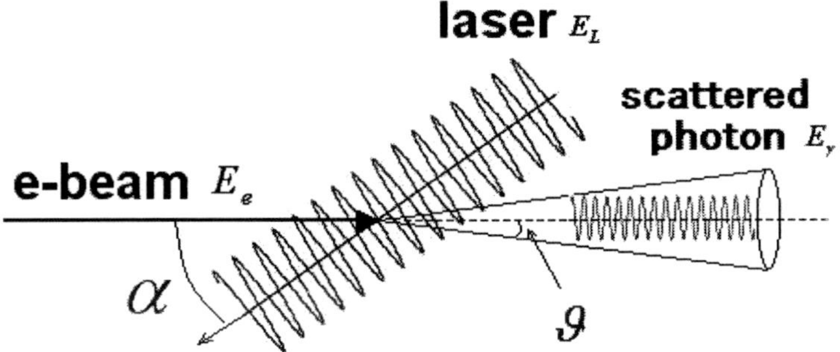

Figure 1. Kinematics of the laser Compton scattering.

It is evident that the maximum gamma-ray energy E_{max} can be tuned by the incident electron energy, the laser wavelength and the collision angle; when the electron beam energy and laser wavelength are fixed, the head-on collision ($\alpha = 0$) gives the extreme photon energy. From Eq. 1, we know that the photon's energy is dependent on the scattered angle ϑ; the maximum-energy photons are on axis $\vartheta = 0$, and the photon's energy decreases with the increase of this angle.

The energy spectrum of the scattered photons is defined by the cross section [5]

$$\frac{d\sigma}{dy} = \frac{2\sigma}{x}(\frac{1}{1-y}+1-y-4r(1-r))$$

(2)

where

$$y = \frac{E_\gamma}{E_e} \; ; \; r = \frac{y}{x(1-y)} \; ; \; \sigma_0 = 2.5 \times 10^{-25}\, cm^2.$$

Here we give an example to show the characteristics of LCS gamma-ray. Providing $E_e = 1.0\,\mathrm{GeV}$, $E_L = 1.17\,\mathrm{eV}$ and a head-on collision, the energy distribution and spectrum of the scattered photons are as shown in Figure 2 and Figure 3. From Figure 2 we understand that the dependence of the gamma-ray energy on the scattered angle gives attractive possibility to produce quasi-mono energetic gamma-ray beam by using an axial collimator. In Figure 3, it is shown that the peak appears at the maximum energy, 17.6 MeV in this case.

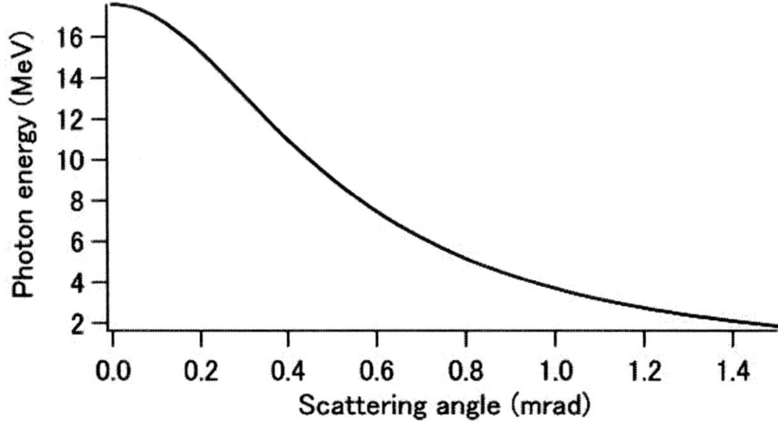

Figure 2. Energy of scattered photon vs. scattering angle.

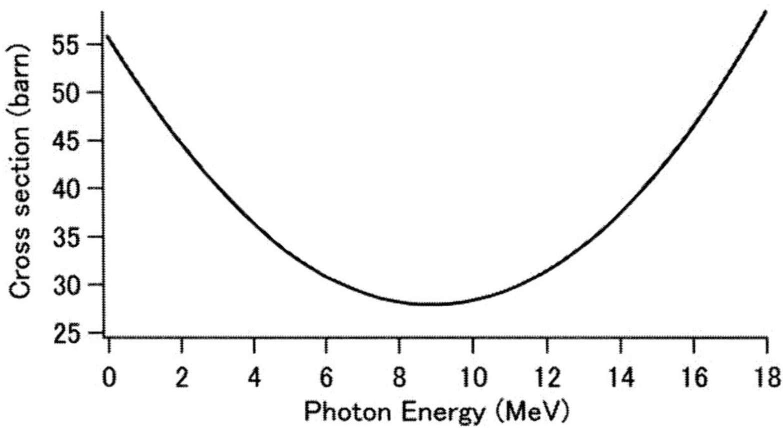

Figure 3. Energy spectrum of the laser Compton scattering photon.

2.2. Estimation of Reaction Rate

Unlike the bremsstrahlung gamma-ray, the LCS gamma-ray has a better energy spectrum for coupling between the nuclear giant resonance and the gamma photons. As an example, we plot the nuclear giant resonance spectrum, i.e., photonuclear cross-section, of ^{129}I together with LCS energy spectrum in Figure 4. The LCS gamma-ray spectrum in this plot is same to Figure 3. As mentioned above, we can shift the LCS gamma-ray peak by changing the electron beam energy or colliding angle to optimize the overlap of the two peaks in Figure 4. Next, let us estimate the reaction rate through the analytical method.

When a beam of gamma rays strikes on a target, the photonuclear reaction rate R ($\text{s}^{-1} \cdot \text{m}^{-3}$) could be written as

$$R = N_t \sigma_{pn}(E)I(E,z) \tag{3}$$

where N_t is the number of atom per volume, σ_{pn} is the photonuclear cross section, $I(E,z)$ is the LCS gamma ray intensity with respect to the photon's energy E and z, the distance of gamma ray penetrating in the target. The photon's energy E is not constant; it varies with emitted angle. In section 2.1 we understand the LCS gamma-ray features a circular transverse profile, naturally, we should consider a cylindrical target, which is noted with radius a and length b, and then the integral of volume results in the total reaction rate

$$\widetilde{R} = N_t \int \sigma_{pn}(E)I(E,z)dV \tag{4}$$

The gamma-ray intensity could be expressed as

$$I(E,z) = \varsigma \sigma_{LCS}(E)e^{-\mu z} \tag{5}$$

where ς relates to the properties of incident electron beam and laser light, $\sigma_{LCS}(E)$ is the laser Compton scattering cross section, and μ is the total linear attenuation coefficient given by

$$\mu = N_t[\sigma_{pp}(E) + Z\sigma_{cm}(E) + \sigma_{pn}(E)] \tag{6}$$

where Z is the atomic number, $\sigma_{pp}(E)$, $\sigma_{cm}(E)$ and $\sigma_{pn}(E)$ denote cross sections of pair production, Compton and photonuclear reactions. To the energy of our interest, the contribution from the photoelectron reaction is so small that it can be neglected. The three cross sections are plotted in Figure 5. The pair production effect is predominant; most of the gamma photons injected in the target are converted to electron-positron pairs. The Compton effect also absorbs many photons.Compared to the pair production and Compton effect, the photonuclear cross section is small.

Substituting Eq. (5) into Eq.(4) and considering the cylindrical target we rewrite Eq.(4) as

$$\widetilde{R} = N_t \varsigma \int_0^b \int_0^a \sigma_{pn}(E)\sigma_{LCS}(E)e^{-\mu z} \cdot 2\pi r dr dz \tag{7}$$

The total gamma photons injected on the target can be written as

$$F = \varsigma \int_0^a \sigma_{LCS}(E) \cdot 2\pi r dr \tag{8}$$

As mentioned above, the LCS gamma-ray energy is related to the emission angle, which is described by the radial position r and the distance between the target location and the interaction point of LCS. With the help of Eq.(7) and (8), the reaction rate is finally expressed as

$$\eta = \frac{N_t \int_0^b \int_0^a \sigma_{pn}(E)\sigma_{LCS}(E)e^{-\mu z} \cdot 2\pi r dr dz}{\int_0^a \sigma_{LCS}(E) \cdot 2\pi r dr} \qquad (9)$$

For a given target, the Eq.(9) is calculable, with using the analytical expressions of $\sigma_{pp}(E)$, $\sigma_{cm}(E)$ and $\sigma_{pn}(E)$ [1].

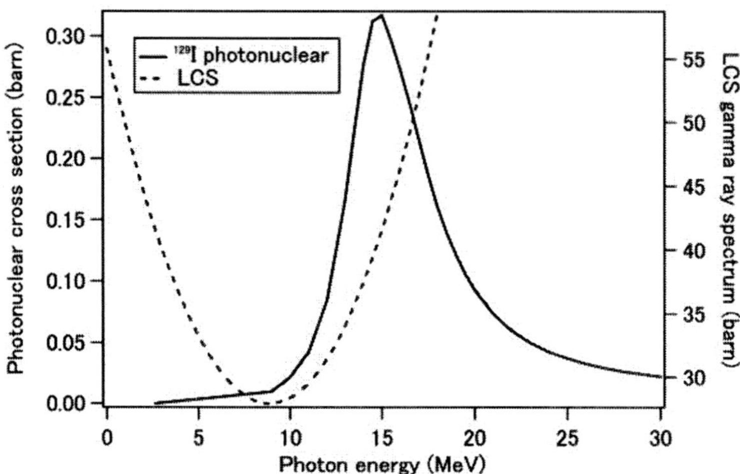

Figure 4. Coupling of laser Compton scattering gamma-ray to nuclear giant resonance.

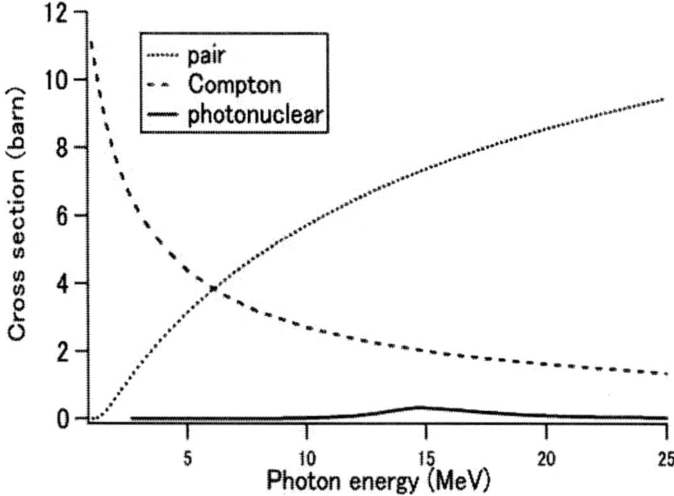

Figure 5. Cross sections of photon induced reactions for [129]I.

Actually, the injected photons induce secondary shower particles, i.e., neutrons, electrons and photons, and these particles also can make transmutation. The whole shower process is complicated and is not involved in the analytical analysis. So, Eq. (9) is just a rough estimation. A more precise estimation including the nuclear shower effect can be expected through the Monte Carlo simulation, which will be addressed later.

3. STORAGE RING BASED LCS GAMMA-RAY

We have developed laser Compton scattering setup on NewSUBARU storage ring [6]. NewSUBARU is a racetrack shape electron storage ring synchrotron radiation facility. The circumference of the ring is about 118 m, and that have two 12 m long straight sections. Electron beam is injected from 1 GeV linac at Spring-8 facility. One of the straight sections of the ring was chosen to build interaction vacuum chamber, where the electron beam collides with the incoming laser light in a head-to-head manner. Thus, high energy photons are produced due to the Compton backscattering effect, going along the incident electron moving direction in a forward cone of angle $1/\gamma$, where γ is the relativistic factor of electron, namely, 0.5 mrad for 1 GeV electron beam. The laser light with wavelength of 1064 nm comes from a Nd:YVO laser, consequently, the produced high-energy photons are gamma-ray photons, and the maximum gamma-ray energy on axis are 17.6 MeV for 1 GeV electron energy. Quasi-monochromatic gamma-ray beam is obtained by using a collimator at the axis.

3.1. Newsubaru LCS Setup

The interaction point was designed at the center of the straight section, where both the electron beam and the laser light transverse profiles were focused to the minimum. A reflected mirror is located at the downstream end to guide the laser light traveling along the beam line through the interaction point, and the light is reflected out of the chamber by another upstream mirror [7,8]. The produced gamma-ray photons would go through the downstream mirror and reach the detector or irradiate the target. A schematic setup graph is as shown in Figure 6.

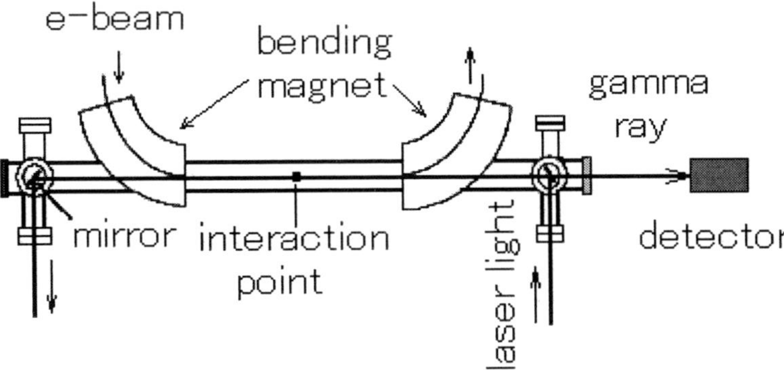

Figure 6. Schematic graph of experimental setup.

The laser is installed at the outside of the shielding wall and is injected into the vacuum duct using six mirrors deliberately arranged and a convex lens with focal length of 5 m in a well-designed position, 7.5 m away from the laser and 15 m away from the center point of the straight section. This results in a focused spot of light with radius of 0.82 mm. The electron beam size is determined by the β function and emittance. For the NewSUBARU storage ring, at the center point of the straight section, these parameters are characterized as

$\beta_x = 2.3\,\text{m}$, $\beta_y = 9.3\,\text{m}$, $\varepsilon_x = 40\,\text{nm}$, and $\varepsilon_y = 4\,\text{nm}$, resulting in the electron beam size of 0.30 mm for the horizontal direction and 0.19 mm for the vertical direction. Consequently, the size of electron beam is smaller than that of the laser beam at the interaction point.

3.2. Newsubaru LCS Gamma-Ray Properties

The gamma-ray detector is located about 16 m from the electron– photon collision point. A high-purity Germanium coaxial detector is used with the detection efficiency of 45%. The measurements of gamma-ray photons are carried out at a lower current of several mA, to avoid saturations at the detector. An example of measured spectrum is shown in Figure 7, with a lead collimator of 24 mm in diameter in front of the detector. The clear difference between the two signals of laser Compton scattering gamma-ray (the laser ON) and the background (the laser OFF) shows a good signal-to-noise ratio. The background signal is due to the bremsstrahlung by the residual gases in the vacuum duct of the straight section. The maximum energy appears around 17MeV, which is in agreement with the theoretical prediction.

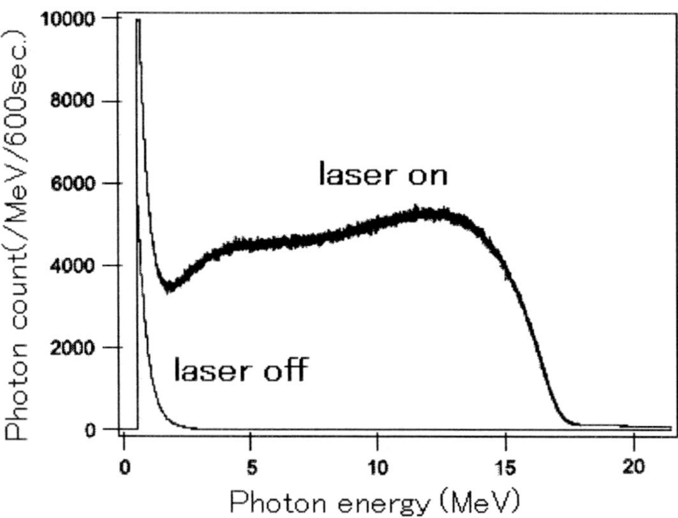

Figure 7. Experimental result of laser Compton scattering gamma-ray.

We simulated the process of generated gamma-ray transport from the source to the detector including photons passing through the reflected mirror, output window, collimator, and being detected by the detector, by employing the EGS4 code. After processing the experimental data with 10 mm in diameter collimator, we achieved the actual gamma-ray generation rate of 5×10^3 photons/mA/W/s. This yield is obtained by the gamma-ray energy of 12.4– 17.6MeV. Then, the maximum gamma-ray photons' yield of more than 10^7 photon/s is expected by NewSUBARU under the condition of storage current 500 mA and laser power 5 W.

4. NUCLEAR TRANSMUTATION THROUGH NEWSUBARU LCS GAMMA-RAY

We proposed the concept of shortening the long radioactive life of nuclear waste by transmuting their nuclei to an unstable isotope through irradiation by laser Compton scattering gamma-ray based on a storage ring. The concept points out the reaction rate of nuclear transmutation, induced by the gamma-ray photons coupling to nuclear giant resonance of a certain material. In accordance with the proposed scheme, experiments are developed to investigate the fundamental issues concerning nuclear transmutation on the NewSUBARU LCS gamma-ray facility.

The final goal of this research is to understand the feasibility of transmuting the long-lived fission products in nuclear wastes such as ^{129}I and ^{135}Cs, through (γ,n) reaction. Figure 8 shows the process of gamma-ray photonuclear reaction of ^{129}I. As is shown, ^{129}I is transmuted into ^{128}I by ^{129}I (γ,n) ^{128}I reaction, and the generated ^{128}I is unstable and is transmuted into stable nuclei ^{128}Xe with a short half-life of about 25 min by β^- and ε decays. In such a way, ^{129}I can quickly loss its radioactive poison.

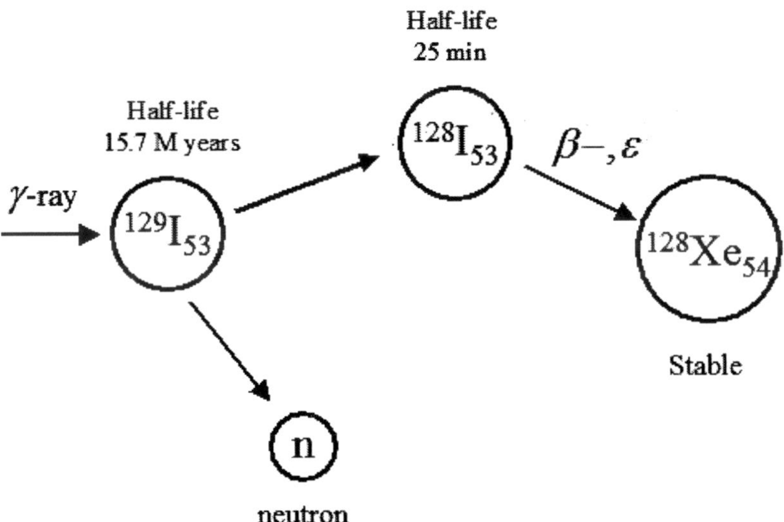

Figure 8. Transmutation process of ^{129}I.

As the first step of this study, we aim at exploring the reaction rate of gamma-ray to the nuclear giant resonance, which is defined as the number of transmutation nuclei by per gamma-ray photon.

4.1. Experimental Target

Actually, it is hard to use the target of ^{129}I directly because of its hazard. Instead, we consider ^{127}I for the transmutation experiment since its photonuclear cross section is very near to that of ^{129}I. Usually, the pure ^{127}I is unavailable, and its compound, sodium iodide ^{23}Na^{127}I, is available to make a practical target. The transmutation processes for ^{23}Na and ^{127}I are given in Figure 9.

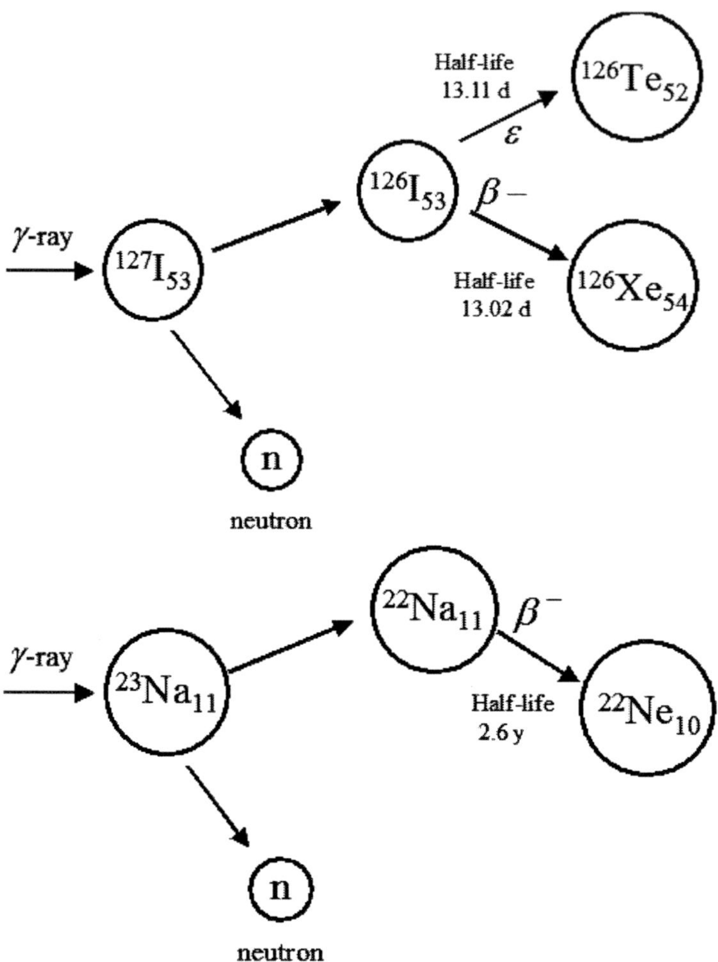

Figure 9. Transmutation process of ^{127}I and ^{23}Na.

With absorbing a gamma photon, the ^{127}I nucleus possibly releases a neutron to undergo (γ,n) reaction, and is transmuted to ^{126}I. Furthermore, ^{126}I is unstable, and eventually is transmuted to ^{126}Xe and ^{126}Te in the way of radioactive decay β^- and ε, respectively. The β^- in this process radiates 388.63 KeV photons with a half-life of 13.02 day, while ε decay radiates 666.33 KeV photons with a half-life of 13.11 day. The unstable ^{22}Na transmuted from ^{23}Na also radiates photons with energy of 1274.53 KeV through β^- decay, and its half-life is 2.6 year. In measurement, the radiations from ^{126}I and ^{126}Na can be separated by the difference of photon energy and half-life. The photonuclear cross section of ^{127}I is larger than that of ^{23}Na, which is as shown in Figure 10.

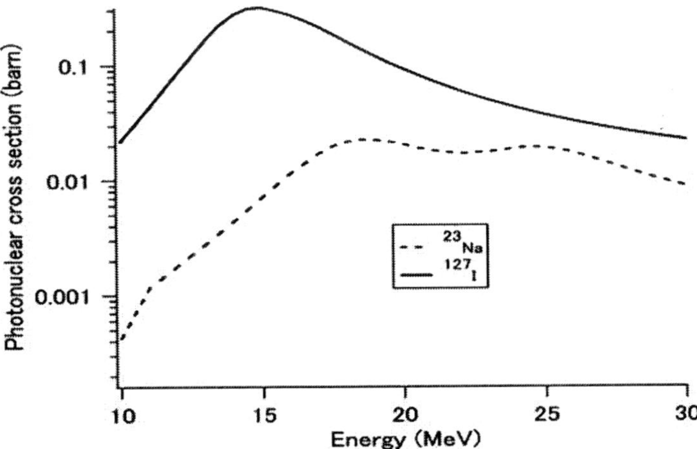

Figure 10. Photonuclear cross sections of ^{127}I and ^{23}Na.

4.2. Reaction Rate Simulation

The estimation of reaction rate is a complicated problem, since it is related to the gamma-mamma-ray energy, the geometry of target and the distance between the target and the origin point of gamma-ray generation. A Monte Carlo code, MCNP5, is used to simulate the whole process. MCNP5 is a general-purpose Monte Carlo N-particle code that can be used for neutron, photon, electron, or coupled neutron/photon/electron transport. The code treats an arbitrary three-dimensional configuration of materials in geometric cells.

Based on the setup of NewSUBARU LCS gamma-ray, the simulation model is described as shown in Figure 11. A cylindrical target is located at the center of the gamma-ray beam and 18.49 m away from the interaction point of the electron beam and laser light.

The target is taken 5 cm in length, so that most of the gamma-ray photons are absorbed in the target. According to the definition mentioned above, the reaction rate depends on the target radius. The simulation result is given in Figure 12. The maximum reaction rate 1.32% appears at 0.5 cm radius. Also plotted in Figure 12 is the result for pure ^{127}I target, which shows higher reaction rate than ^{23}Na^{127}I target.

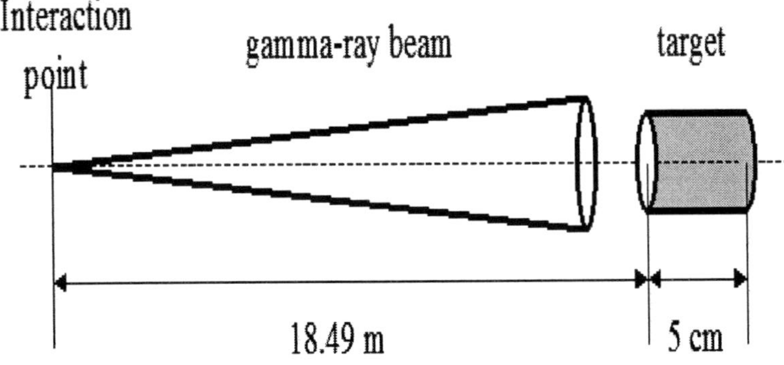

Figure 11. Schematic graph of target radiation.

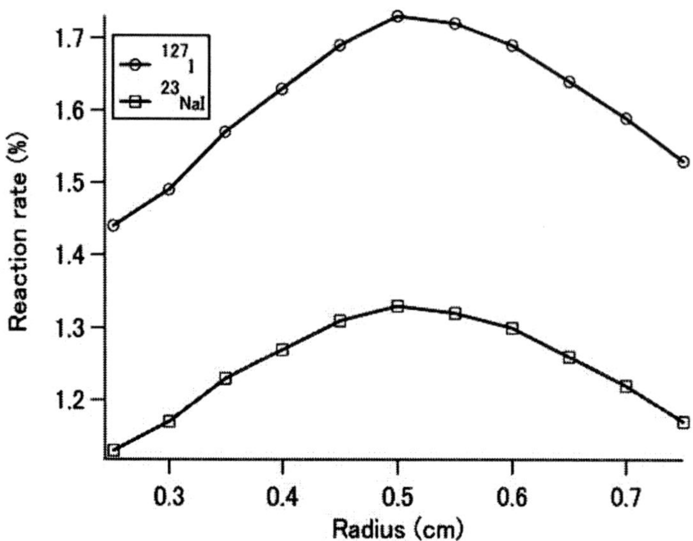

Figure 12. Simulation result of reaction rate.

4.3. Irradiation Experiment

The gamma-ray is introduced to the hatch from the tunnel to irradiate the target. An image plate is placed before the target. The image plate can record the transited photons, therefore, the total photons enter the target can be deduced.

Through the measurement of radioactivity of irradiated target, the number of transmuted nucleus at the terminus of irradiation can be deduced according to the decay law

$$N_i = \frac{\Delta N e^{\lambda t}}{1 - e^{-\lambda \Delta t}} \tag{10}$$

where N_i is the number undecayed nuclei, ΔN is the number of decays in the duration of Δt, λ is the decay constant, and t is the time interval from the end of irradiation to the beginning of activity measurement. The way to get the reaction rate is sketched in Figure 13.

The cylindrical ^{23}Na^{127}I target, 5 cm long and 0.5 cm in radius, was irradiated for 8 hours, and the emission spectrum is given in Figure 14. According to Figure 14, the decay curve can be worked out as shown in Figure 15. Comparing to the acknowledged value, the half-life deduced from our experiment is a little bit short. The loss of the radioactivity inside the target was estimated by the MCNP5 code, and after data processing, the reaction rate is achieved as shown in Figure 15.

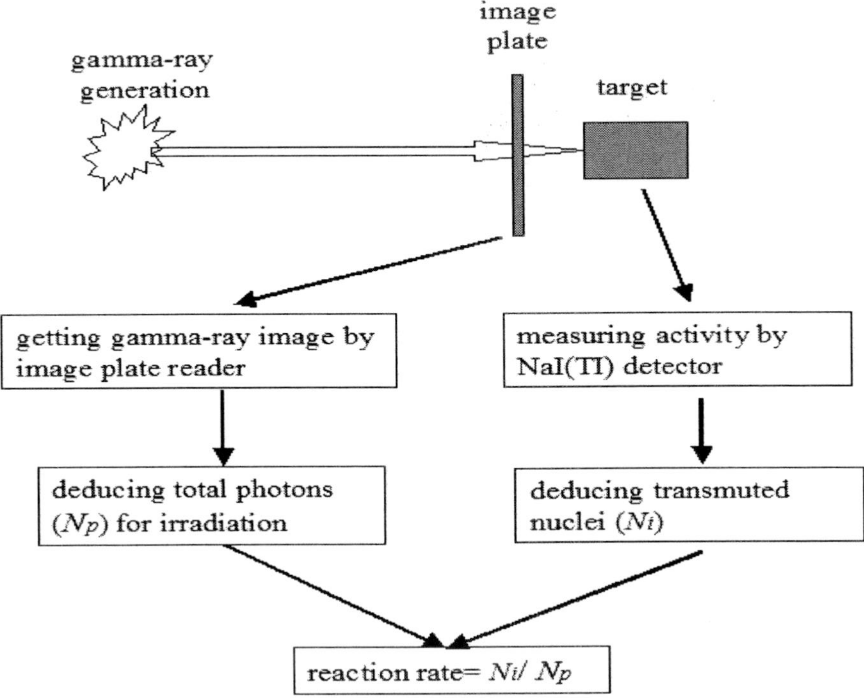

Figure 13. Schematic graph of determination of reaction rate.

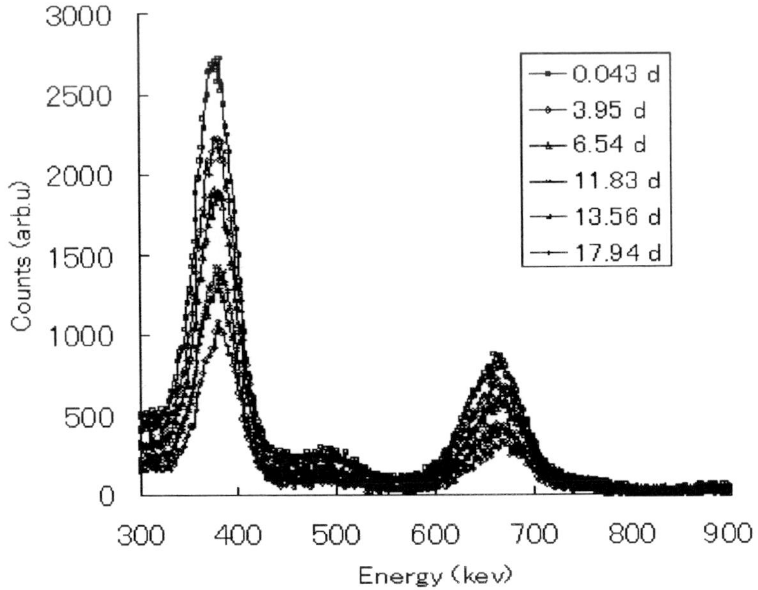

Figure 14. Energy spectrum of decay.

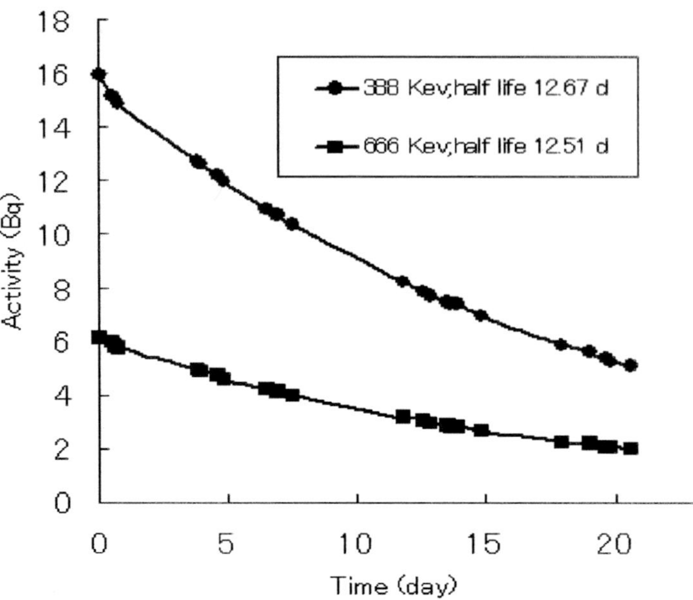

Figure 15. Activity vs. time.

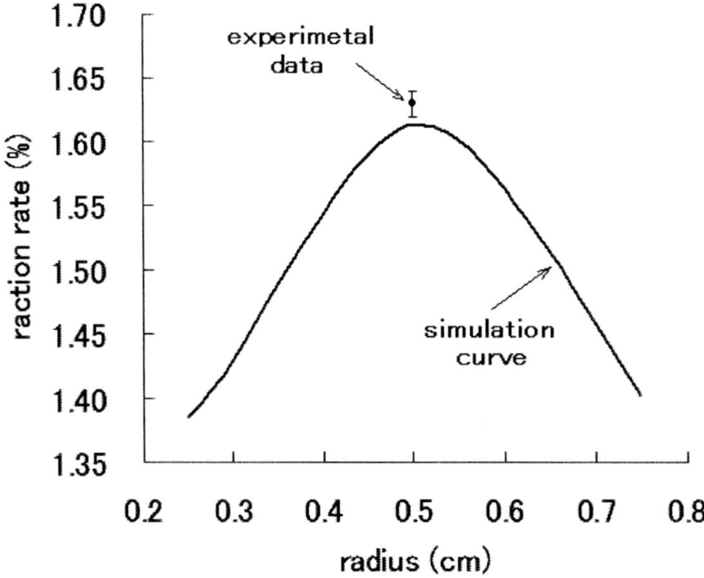

Figure 16. Reaction rate of experimental data.

The simulation curve is also plotted in Figure 16 for comparison. The experimental results illustrate a good agreement with the simulation, and the errors come from the loss evaluation and statistical process. With the promotion of our detecting system, more precise measurement will be reached in the future.

5. SECONDARY TRANSMUTATIONS

In order to improve a cost effectiveness of the transmutation system, a usage of prompt neutron emitted at (γ,n) reaction for another transmutation is proposed. For example, Neptunium-237 is also major long-lived trans-uranium nucleus that is generated in the reactor, and can be arranged as the second target placed around the first target, as shown in Figure 17. Fission reactions induce three or four times by energetic neutrons with chain reaction and make heat which cause the total energy balance of the system.

In this system, this heat energy density is high enough to make energy applications as hydrogen production and so on. This becomes to be out put of the system and may become a recover for initial cost beside the energy balance as shown in Figure 18. This is a rough concept of this scheme. Further investigation is underway.

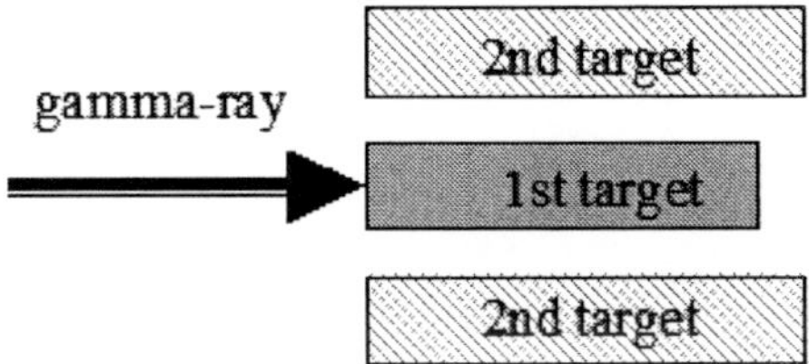

Figure 17. Schematic of complex targets set.

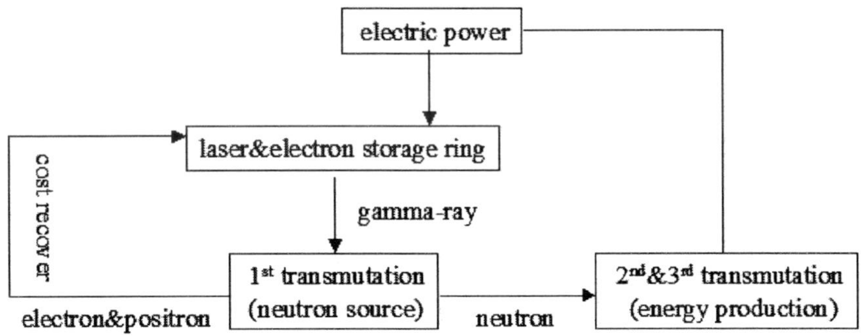

Figure 18. Schematic of energy flow.

6. FURTHER RESEARCHES

The byproduct particles are generated when gamma photons irradiate a target: the electron-positron pair from pair-production reaction, the scattered photon and recoiled electron from Compton effect, and the neutron from the photonuclear reaction. These particles can lead the secondary transmutation. The neutrons can directly induce the transmutation; the electrons can radiate photons through bremsstrahlung, and the photons induce the photonuclear reaction again. To better use the byproduct particles, more information about them should be better understood. Such investigation is going on, and is introduced as following.

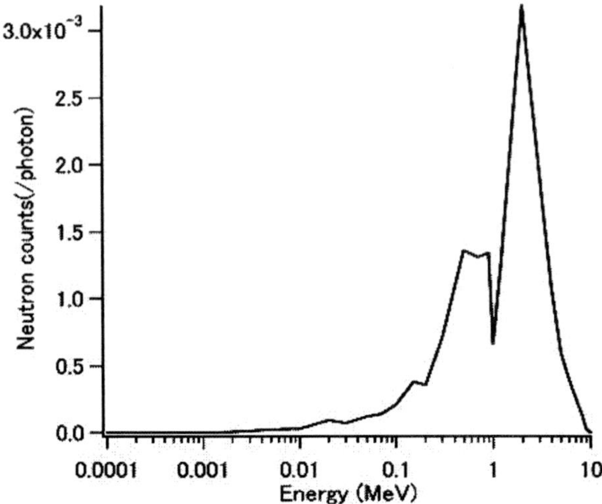

Figure 19. Neutron spectrum from simulation.

6.1. Neutron Generation

We simulated the neutron generation from a cylindrical ^{129}I target, which is 5 cm long and 0.5 cm in radius, by MCNP5 code. Some of the neutrons generated inside the target can come out, and its energy spectrum is as shown in Figure 20. Most of the neutrons concentrate on the energy of 3 MeV, and consequently form a peak in the spectrum. A time-of-flight method is used to measure the neutrons in experiment, and the measurement is underway.

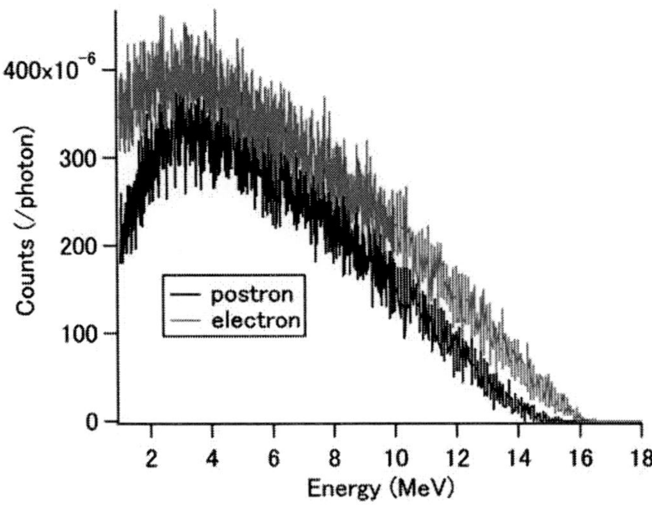

Figure 20. Positron and electron spectrum from simulation.

6.2. Positron Generation

The generated electrons and positrons inside the target are hard to transit and come out of the target. In this simulation, we use the same target mentioned above and the simulation results are given in Figure 20. As is expected, the electrons are more than the positrons, since the electrons are also generated from the Compton effect beside the pair production. The positron measurement will be performed in the near future.

CONCLUSION

A scheme of LCS gamma-ray induced transmutation is introduced. The reaction rate of the transmutation plays an important role in this scheme, and hence is studied through analytical way, simulation and experiments. The LCS gamma-ray facility has been built on NewSUBARU storage ring to perform these experiments. The advantage of the LCS gamma-ray in coupling with the nuclear giant resonance is implied. In order to realize a good energy balance, a complex targets set is proposed to utilize the byproduct particles, and related research is going on.

REFERENCES

[1] D. Li; K. Imasaki; M. Aoki; S. Miyamoto; S. Amano; T. Mochizuki J. *Nucl. Sci. and Tech.* 2002, *39*, 1247-1249.

[2] J.Chen; K. Imasaki; M. Fujita; C. Yamanaka; M. Asakawa; T. Asakuma *Nucl. Instr. Meth.* 1994, *A321*, 346-350.

[3] V. Balakin; V. A. Alexandrov; A. Mikhailinchenko *Phys. Rev. Lett.* 1995, *74*, 2479-2482.

[4] Zhirong Huang, Ronald D. Ruth, *Phys. Rev. Lett.* 1989, *80*, 976-979.

[5] Valery Telnov N*ucl. Instr. and Meth.* 1995, *A355*,3-18

[6] Shuji Miyamoto; Yoshihiro Asano; *Sho Amano Radiation measurements* 2007, *41*, 179-185.

[7] D. Li; K. Imasaki; M. Aoki *Nucl. Instr. and Meth.* 2004, *A528*, 516-519.

[8] D.Li; K. Imasaki; M. Aoki *Int. J. infra. milli.* 2003, *24*, 1301-1305.

In: Nuclear Waste Research: Siting, Technology and Treatment   ISBN 978-1-60456-184-5
Editor: Arnold P. Lattefer, pp. 207-220   © 2008 Nova Science Publishers, Inc.

Chapter 7

EXPLORING THE EARTH'S CRUST AND MANTLE USING SELF-DESCENDING, RADIATION-HEATED, PROBES AND ACOUSTIC EMISSION MONITORING

Michael I. Ojovan and Fergus G. F. Gibb*
Immobilisation Science Laboratory,
Department of Engineering Materials,
University of Sheffield, Mappin Street,
Sheffield S1 3JD, UK.

ABSTRACT

A novel method of exploring the uppermost 100-200 km of the Earth is examined. A small, dense, heat-generating probe melts its way down through the crust and mantle while its position and progress are tracked by acoustic signals generated in the rocks. The data from the descending probe and the signals themselves will yield new insights into the physical properties of the rocks through which they pass. These, when combined with other geophysical methods, should provide unequivocal information on the nature and composition of the Earth's interior. The probe consists of an outer sphere of tungsten ~ 1m in diameter inside which is a ^{60}Co radioactive heat source. We calculate that such a probe will reach the oceanic Moho in less than 6 months and attain minimum depths of well over 100 km in a few decades beneath both oceanic and continental lithosphere.

INTRODUCTION

The experimental and theoretical techniques used to examine our planet have developed significantly over recent years (Vocadlo, Dobson 1999). However our knowledge of the composition and structure of the Earth's interior through direct observation and sampling is limited to the uppermost 12 km. The top 2 to 3 km of the crust have been sampled extensively at outcrop, by excavations, mines and boreholes, but from 3 to 12 km sampling is restricted to a handful of very deep boreholes. Below 12 km we are reliant on inferences from rocks now at the surface but exhumed from greater depths by geological processes or on indirect measurements by geophysical methods. Of the latter, seismology is by far the most important and useful.

The Earth's interior can be explored by observing seismic waves that travel through the Earth. Analysis of these waves showed that the Earth's interior is layered with significant seismic-wave velocity discontinuities at the boundaries between the different layers. The layering of the Earth has been inferred indirectly using the time of travel of refracted and reflected seismic waves created by earthquakes. The core does not allow shear waves to pass through it, while the seismic velocity is different in the other layers. Changes in the seismic velocity between the different layers cause refractions whereas reflections are caused by large increases in seismic velocity. The structure of the Earth is considered in two separate categories: chemically differentiated layers, and layering reflecting the strengths and density of the materials (Jordan 1979). Chemically, the Earth can be divided into the crust, mantle (upper and lower), outer core, and inner core. By material strength, the layering of the earth is categorized as lithosphere, asthenosphere, mesosphere, outer core, and the inner core (Figure 1).

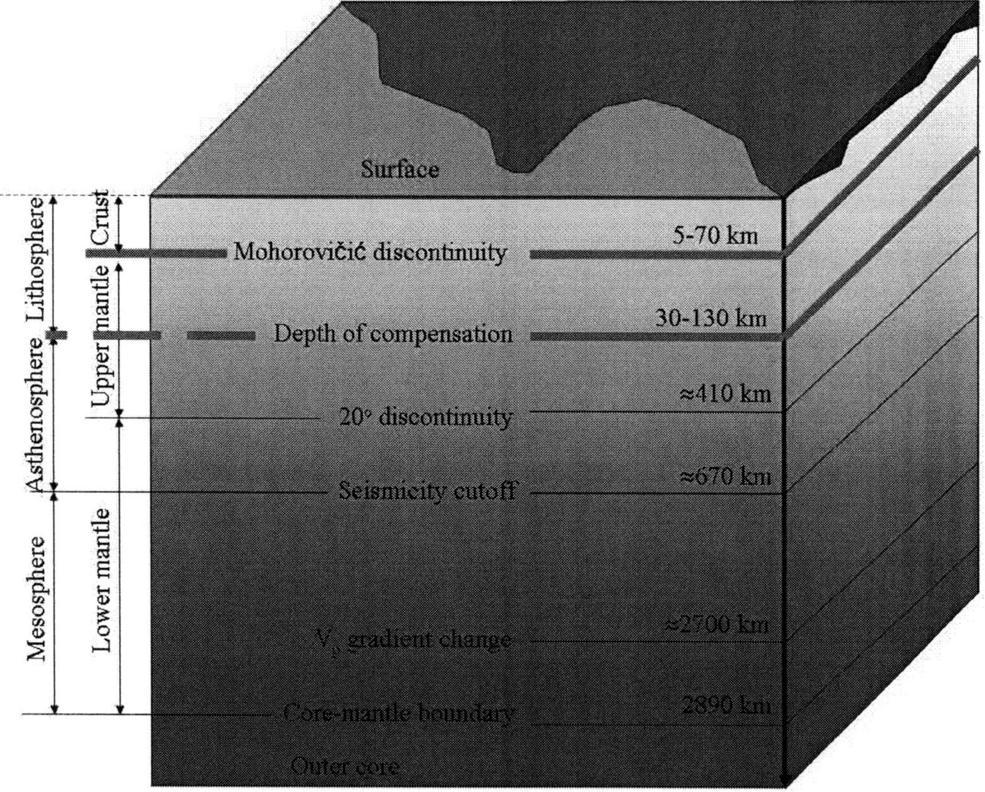

Figure 1. Principal layers of the Earth (after Jordan 1979).

* E-mail: M.Ojovan@sheffield.ac.uk

The Mohorovičić discontinuity, usually referred to as the Moho, is the boundary between the Earth's crust and the mantle. The crust ranges from 5 to 70 km in depth. The thin parts are oceanic crust composed of dense (mafic) iron magnesium silicate rocks and underlie the ocean basins. The thicker crust is continental crust, which is less dense and composed largely of (felsic) sodium potassium aluminium silicate rocks.

Earth's mantle extends to a depth of 2890 km, making it the largest layer of the Earth. The pressure at the bottom of the mantle is ~140 GPa. The mantle is composed of silicate rocks that are rich in iron and magnesium relative to the overlying crust. Although mostly solid, the high temperatures within the mantle cause the silicate material to be sufficiently ductile that it can flow on long timescales and convection of the mantle is expressed at the surface through the motions of tectonic plates. The viscosity of the mantle ranges between 10^{21} and 10^{24} Pa s (Vocadlo and Dobson 1999).

The Earth's core is composed largely of iron (80%), along with nickel and one possibly or more light elements, whereas other dense elements, such as lead and uranium, either are too rare to be significant or tend to bind to lighter elements and thus remain in the crust or mantle. Seismic measurements show that the core is divided into two parts, a solid inner core with a radius of ~1220 km and a liquid outer core extending beyond it to a radius of ~3400 km. The solid inner core is generally believed to be composed primarily of iron and some nickel. The liquid outer core surrounds the inner core and is believed to be composed of iron mixed with nickel and trace amounts of lighter elements.

Although seismic investigations provide relatively unequivocal information about structure, compositional information generally has to be deduced from seismic velocities, which rarely provide a unique solution. Data from an alternative technique that could be used, possibly in conjunction with existing geophysical methods, to ascertain compositions of the crust and mantle would therefore be of immense value. Here we suggest and discuss such a technique.

REACHING DEEP EARTHS LAYERS

Stevenson (2003) proposed a "mission to the Earth's core" in which a small probe embedded in a huge mass of molten iron would descend along a crack propagating under the influence of gravity. The probe would measure properties of the rocks through which it passed, such as temperature, electrical conductivity, etc., and use acoustic communication to transmit its findings to the surface (Figure 2).

The calculated time to reach the core (1 week) and rate of crack propagation (5 m/s) stretch credulity. The enormity of the mass of iron (10^{8}-10^{10} kg) and the explosive energy required to initiate the crack in the earth's crust (a few megatons) render the proposal unrealistic in practice, however interesting in theory. Stevenson did acknowledge that there could be an alternative means of getting the probe down – by melting the rock. However, he dismissed this because "the trip times ... ... are thousands of years".

We contest this last assertion and have investigated the possibility of exploring the deeper reaches of the Earth's crust and upper mantle with a small, self-descending probe that melts the rocks and creates acoustic signals that could be detected at the surface, thus yielding information about the nature of the rocks through which the probe and the signals pass (Ojovan, Gibb, Poluektov and Emets 2005).

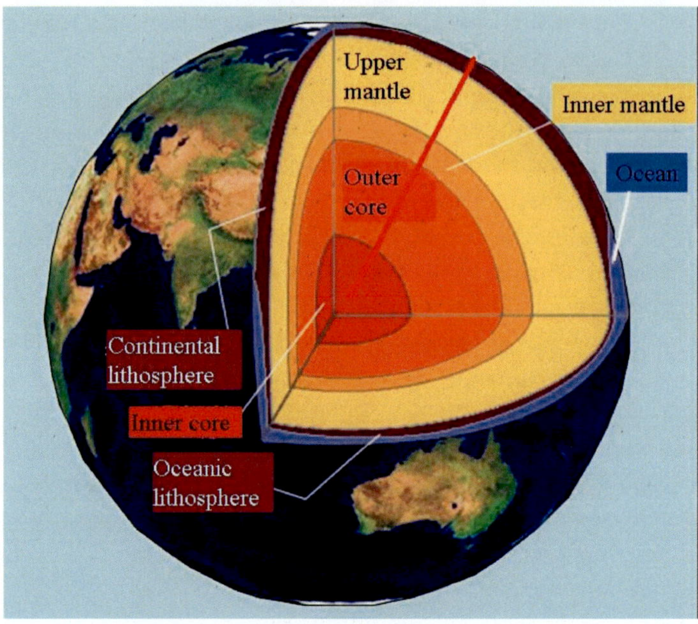

Figure 2. Schematic of the Earth's interior showing Steverson (2003) proposed probe.

Moreover we showed that self-descending radioactive probes can be used for self-burial of dangerous radioactive wastes in deep layers of the Earth preventing any release of radionuclides into the biosphere (Ojovan and Gibb 2005).

SELF-DESCENDING PROBE

The self-descent of a spherical body by melting of the rock through which it passes has been considered in the contexts of nuclear reactor core melt down, the so-called "China syndrome" (Emerman and Turcotte 1983), and the deep self-burial schemes for nuclear wastes first proposed by Logan (Logan 1973; Kascheev *et al.* 1992; Byalko 1994). The mechanism is simple. Heat from a source within the body partially melts the enclosing rock and the relatively low viscosity and density of the silicate melt allow it to be displaced upwards past the heavier body as it sinks (Emerman and Turcotte 1983). Eventually the melt cools and vitrifies or recrystallizes, sealing the route along which the body passed.

From a practical standpoint the probe needs to be small, dense, relatively inexpensive and spherically symmetrical to ensure an even distribution of temperature at its surface. In its simplest form it would consist of two concentric spheres with an outer diameter of less than, say, 1 m (Figure 3).

The outer sphere, or capsule, needs to be made of mechanically strong, dense, refractory material. Its main functions are to protect the inner sphere (the heat source), conduct the heat efficiently to the outer surface of the probe and provide weight.

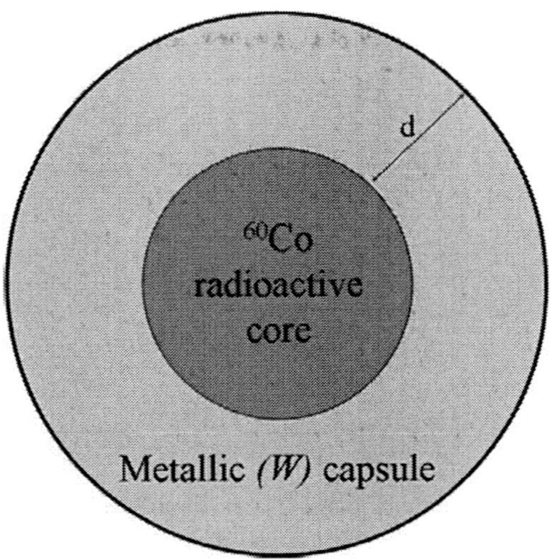

Figure 3. Cross-section of a proposed self-descending radioactive (radiation-heated) probe.

The temperatures required for effective rock melting are well in excess of 1000°C, increasing with pressure, and ideally the capsule needs to withstand temperatures in excess of 2000°C (see below). Ceramic materials, although suitably refractory, tend to be poor heat conductors and would overheat (Kosachevskiy and Sui 1999), thus limiting thermal loading of the probe and hence descent rates and ranges. The capsule should therefore be made of metal and we would propose tungsten (W). W melts at 3410°C, has a specific gravity of 19.3, is relatively inexpensive and is predicted to have a low corrosion rate in silicate liquids at high temperatures and pressures and low oxygen fugacities (i.e., the conditions that would prevail during descent through the crust and mantle). This latter aspect is important. The capsule wall needs to be strong and thick enough to withstand corrosion and abrasion during descent and also to absorb the radiation from the heat source to ensure efficient heating of the probe. Abrasion during sinking through the melted rock is unlikely to be significant but most metals do corrode by reaction with silicate liquids at high temperature. For example, stainless steel in a granitic melt corrodes at a rate of ~ 100 µm/year at temperatures around 800°C and pressure of 150 MPa (Taylor 2003). At the higher temperatures in question stainless steel would be expected to corrode much faster and so would not be a suitable capsule material. The minimum wall thickness needed for adequate absorbtion of β and γ-radiation is ~ 0.1 m.

Several sources of heat for self sinking capsules were recently analysed (Ojovan, Gibb, Poluektov and Emets 2005). Table 1 shows radiogenic heat parameters of the radionuclides that might provide enough heat to maintain the high temperature required by a self-descending probe.

Although the total radionuclide activity in a self-descending probe should be very high, all the radionuclides considered in Table 1 are short-lived and hence do not present an extreme hazard for the environment (Ojovan and Lee 2005). For the inner sphere, or heat source, we would propose ^{60}Co metal with a total initial activity of at least 3.85 x 10^{18}Bq (104 MCi). ^{60}Co is chosen because it has a high specific heat generation, yields a solid daughter decay product (^{60}Ni) and is readily available from the spent sealed radioactive sources (SRS) widely used in industry, medicine and research. Indeed, deep self-burial of SRS has been proposed as a means of safely disposing of them (Ojovan and Gibb 2005).

Table 1. Radiogenic heat parameters of radionuclides

Radionuclide	Half-life, $T_{1/2}$, y	Specific activity*, A, Ci/g	Q factor, mW/Ci	Specific heat release, Q_m, kW/kg
^{60}Co	5.27	1130	15.4	17.4
^{90}Sr	28.5	136	1.16	0.158
^{134}Cs	2.06	1294	10.19	13.2
^{137}Cs	30.17	86.9	6.96	0.61

*1 Ci=3.7 10^{10} Bq (disintegrations per second).

A sphere of ^{60}Co metal with an activity of 3.85 x 10^{18} Bq would yield ~1635 kW of heating power and be about 0.3 m in diameter. Co melts at 1495°C at one atmosphere and, despite the higher pressures, would almost certainly melt at some of the temperatures envisaged within the probe (see below). While this should not greatly affect its capacity for heat generation (other than a small reduction for the latent heat of fusion) it could greatly increase the potential for reaction or alloying with the W, thus possibly reducing the life of the capsule. Any such potential could be readily ascertained by experiment. If it is significant, a suitable, non-reactive refractory lining could be added to the capsule to isolate the Co from the W. Candidate materials for this purpose could be graphite or a thin layer of refractory ceramics.

The probe would have to be launched from a reasonably close-fitting, large-diameter borehole or small shaft. Depending on the local near-surface geology this could be anything between a few tens and a few hundreds of metres deep. Initial heating and melting of the rocks would probably be accompanied by some surface emission of steam as groundwater evaporated but this would soon cease as the probe attained greater depths with higher hydrostatic and lithospheric pressures and the passage became sealed by vitrified or recrystallized rock.

Transport of the completed probe to the launch site would require (radiation) shielding and refrigeration. As an alternative, only the heat source requires special transport arrangements and it could be loaded into the capsule, which is then welded closed, on site immediately prior to launch.

DESCENT RATES

To be of practical use the probe must be capable of attaining reasonable depths (e.g. > 100 km) in realistic times (< 10 to 20 years). The descent rates and ultimate depth range are largely functions of the initial design parameters of the probe.

Assuming the specific gravity of the probe, ρ_p, exceeds that of the melted rock, ρ_m, and the specific heat power of the probe, q (W/m^3), exceeds a threshold value required to melt the rock, q_{th}, the probe will continue to sink. The threshold value is determined by Logan's ratio (Logan 1999):

$$q_{th} = 3\chi(T_m - T_a) / R^2$$

where χ is the thermal conductivity of the rock (W/m °K), T_m is the melting temperature of the rock (°K), T_a is the ambient temperature of the rock beyond the influence of the probe (°K), and R is the radius of the probe (m).

The heat generated by decaying radionuclides, such as ^{60}Co, diminishes with time t (y) according to:

$$q(t) = q_0 \exp(-\lambda t)$$

where q_o is the initial specific heat power of the probe and λ is the decay constant ($\lambda = 0.693/T_{1/2}$ with $T_{1/2}$ the half life of the radionuclide, in years). Consequently, when $q(t)$ equals q_{th} melting of the rock and, hence, descent of the probe will terminate. The descent period τ (y) is therefore given by:

$$\tau = 1.44 T_{1/2} \ln q_0 R^2 / 3\chi(T_m - T_a)$$

and the depth at which the probe ceases to descend, H_τ (m), can be ascertained from:

$$H(\tau) \approx 1.9[1 - q_{th} / q_0] T_{1/2} R q_0 / \rho_m [L + c_p(T_m - T_a)]$$

where L is the heat of fusion (J/kg) and c_p is the heat capacity (J/kg °K) of the rock.

In reality rocks do not melt at a single temperature (T_m) but rather over an interval between the solidus (T_s) and liquidus (T_l) temperatures and, similarly, the specific gravity of the melt fraction (ρ_m) varies slightly over the melting interval. Clearly, the rock would not have to melt completely for the denser probe to sink through it. It is known that the viscosity of magmatic crystal/liquid mushes tends to decrease dramatically when the percentage of crystals falls below that at which rigid crystal networks form on cooling (Philpotts *et al.* 1999) or could be sustained on melting. This has been shown experimentally to be between 30% and 40% crystals (Philpotts and Carroll 1996). Consequently, for the purposes of modelling descent rates and depths, the most appropriate value for T_m would probably be $\sim T_s + 0.4$ (T_l -T_s). A further complication would then arise as to what proportion of the heat of fusion (L) should be used in the calculations. Because rocks are mixtures of different mineral phases that do not melt uniformly over the solidus – liquidus interval, it would not be strictly correct to linearly apportion a fraction of the heat of fusion, although this would probably be a reasonable first approximation.

Reasonably good data are available for the solidus and liquidus temperatures of most common rock types under anhydrous melting conditions but are much sparser for hydrous melting. An even greater constraint applies to heats of fusion – as far as we know, no data exist for hydrous melting of rocks (although they may not be very different from anhydrous melting values). Melting in the presence of water, or even hydrous minerals, greatly lowers both the solidus and liquidus temperatures of silicate rocks. Since at least some of the melting caused by the probe, especially when passing through the middle and lower reaches of continental crust, is likely to be under hydrous conditions, the use of data for anhydrous melting will lead to underestimates of both sinking rates and depths of penetration.

Below we model two cases for a self-descending probe, one through oceanic lithosphere and the other through continental lithosphere. Because of the uncertainties over the effects of partial melting on sinking of the probe and the complications of, at least some, hydrous melting we have chosen to model both cases assuming complete anhydrous melting. Having used T_l and the full heat of fusion, the calculated values for descent rates, terminal depths and total sinking times are absolute minimum values. We accept that these could be out by more than a factor of two but defend the choice on the grounds that conservative estimates are the best demonstration of the feasibility of the proposal.

OCEANIC LITHOSPHERE

The lithosphere beneath the ocean basins is relatively straightforward, consisting of a basaltic crust underlain by peridotitic mantle. Admittedly there are local variations, especially with different varieties of peridotite, but since the thermal properties of these are very similar, the simple model should be a good general case. We have assumed a thickness of 7 km of basaltic crust and that the parameters for the rocks are as given in Table 2.

Table 2. Parameters used in modelling descent through oceanic lithosphere

Depth D (km)	Rock type	Pressure P (kbar)	Temperature (Ambient), (°C)	Density ρ (kg/m^3)	Thermal conductivity λ (W/m, °K)	Heat capacity C_p (J/kg, °K)	Heat of fusion L (J/kg)	Thermal diffusivity K (m^2/s)	Solidus T_S (°C)	Liquidus T_L (°C)
0	Basalt	0.001	25	2900	1.590[b]	782[c]	307730[b]	7.013 10^{-7}	1079	1358
7	Basalt	1.930	128	2900	1.590[b]	895[c]	307730[b]	6.126 10^{-7}	1114	1370
7	Peridotite	1.930	128	3234[a]	3.389[b]	895[c]	383734[b]	1.171 10^{-6}	1134	1721
30	Peridotite	8.250	464	3234[a]	3.138[b]	982[c]	420669[b]	9.881 10^{-8}	1200	1772
100	Peridotite	27.500	1256	3234[a]	2.720[b]	~1067[c]	509324[b]		1453	1901
145	Peridotite	40.000	1544	3234[a]	3.222[b]	???	566893[b]		1628	1958

a = data from (Clark Jr. 1966);
b = data from (Yoder 1976);
c = data corrected for ambient temperature from Fig.4a of (Vorsteen & Schellschmidt 2003);
T_S & T_L from various sources (anhydrous melting).

Some of the parameters are pressure and/or temperature dependent and, where possible, we have applied corrections, For example, thermal conductivity and heat capacity have been corrected for ambient temperature (Vorsteen and Schellschmidt 2003). Unfortunately this has not always been possible due to lack of data but for most such cases, e.g., the effect of pressure on rock density, any resulting errors are likely to be small.

Treating the problem as a series of discrete layers with fixed properties is, of course, only an approximation but again we believe any departure of the results from reality will be insignificant in the context of demonstrating the feasibility of the proposal. For our simple model of oceanic lithosphere the probe will stop descending at a depth of 106.7 km after 39.1 years (Figure 4).

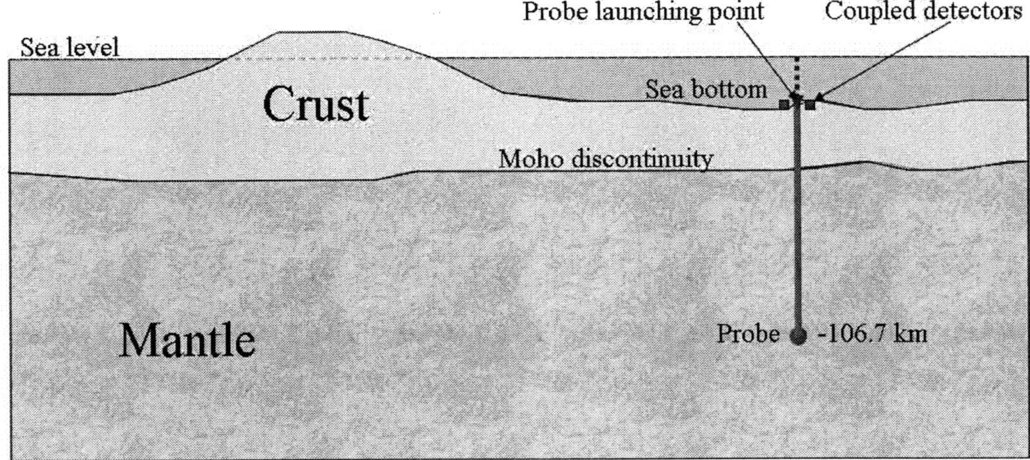

Figure 4. Schematic of acoustic emission probing of the Earth's interior through oceanic lithosphere.

The calculated mean threshold power for sinking, the initial rate of descent at the top of the layer and the total descent time to the base of the layer are given in Table 3 for four depth ranges through the lithosphere.

Table 3. Descent of probe through oceanic lithosphere

Depth range (km)	Rock type	Threshold power q_{th} (W/m^3)	Initial velocity U_0 (m/y)	Descent time (y)
0 – 7	Basalt	24566	16934	0.43
7 – 30	Peridotite	56804	11296	2.80
30 – 100	Peridotite	34322	9955	22.40
100 -106.7	Peridotite	18878	999	39.10

The initial descent through the basaltic layer is remarkably rapid. At just under 2 m/hour this is comparable with conventional rotary drilling of scientific boreholes. The probe would reach the Mohorovicic discontinuity in 5.1 months. Beneath the oceans it would pass through the lithosphere/asthenosphere transition at a depth of less than 100 km. Clearly sinking would be more effective through the already partially molten asthenosphere thus further increasing the extent of our underestimations of descent rate, terminal depth, etc.

CONTINENTAL LITHOSPHERE

By comparison with oceanic lithosphere that beneath the continents is much more complex and subject to much greater lithological variations: there is no such thing as "typical" continental lithosphere. For our purpose we have created a hypothetical section through the European lithosphere that is based largely on data from the Eastern Alpine crust

(Vorsteen and Schellschmidt 2003) and the Ivrea – Verbano crustal zone of N.W. Italy (Rutter *et al.* 2003). The components of this model and the parameters used for the various layers and rock types are given in Table 4.

Table 4. Parameters used in modelling descent through continental lithosphere

Depth	Rock type	Pressure	Temperature	Density	Thermal conductivity	Heat capacity	Heat of fusion	Thermal diffusivity	Solidus	Liquidus
D (km)		P (kbar)	Ambient, (°C)	ρ (kg/m^3)	λ (W/m, °K)	C_p (J/kg, °K)	L (J/kg)	K (m^2/s)	T_S (°C)	T_L (°C)
0	Granite[h]	0.001	25	2600	3.400[d]	782[c]	290000[f]	1.672 10^{-6}	959	1089
10	Granite[h]	2.750	100	2600	2.900[d]	870[c]	290000[f]	1.282 10^{-6}	977	1118
10	Schist[i]	2.750	100	2610	3.405[d]	868[c]	290000[f]	1.503 10^{-6}	977[f]	1118[f]
17	Schist[i]	4.650	183	2610	2.995[d]	926[c]	290000[f]	1.239 10^{-6}	995[f]	1140[f]
17	Metapelite[j]	4.650	183	2700	2.557[d]	926[c]	290000[f]	1.023 10^{-6}	995[f]	1140[f]
22	Metapelite[j]	6.000	242	2700	2.370[d]	955[c]	290000[f]	9.191 10^{-7}	1002[f]	1150[f]
22	Amphibolite[k]	6.000	242	2999[a]	2.698[d]	955[c]	307730[b]	9.420 10^{-7}	750[b]	1117[b]
27	Amphibolite[k]	7.400	300	2999[a]	2.082[d]	970[c]	307730[b]	7.156 10^{-8}	705[b]	1102[b]
27	Mafic granulite	7.400	300	2930[a]	1.992[b]	970[c]	385428[b]	7.010 10^{-8}	1162[e]	1426[e]
30	Mafic granulite	8.250	325	2930[a]	1.992[b]	982[c]	689592[b]	6.925 10^{-7}	1180[e]	1437[e]
30	Peridotite	8.250	325	3234[a]	3.138[b]	982[c]	420669[b]	9.881 10^{-8}	1180[e]	1884[e]
100	Peridotite	27.500	888	3234[a]	2.720[b]	~1067[c]	509324[b]		1402[e]	1963[e]
145	Peridotite	40.000	1168	3234[a]	3.222[b]	???	566893[b]		1561[e]	2016[e]

a = data from (Clark Jr. 1966); b = data from (Yoder 1976); c = data corrected for ambient temperature from Fig.4a of (Vorsteen & Schellschmidt 2003); d = values corrected for ambient temperature using equations 4 & 5 of (Vorsteen & Schellschmidt 2003); e = values from (Wyllie 1971); f = best available estimate.

h = Niznekansky granite (Petrov et. al. 2005); i = Bi-Musc-Kspar schist 14 of (Vorsteen & Schellschmidt 2003); j = Gnt-Bi-Musc orthogneiss 13 of (Vorsteen & Schellschmidt 2003); k = Gnt-Amph-Bi paragneiss 16 of (Vorsteen & Schellschmidt 2003). T_S & T_L from various sources (anhydrous melting except for amphibolite).

Most of the same provisos apply as for the oceanic lithosphere model (above). Also, although many of the parameters are based on good modern data, e.g., Vorsteen and Schellschmidt (2003), some can only be regarded as best available estimates. As before, we have erred on the side of generating minimum values for sinking rates, terminal depth etc.

The results of the modelling (Table 5) indicate that the initial sinking through the granitic upper crust is over 30% faster than through the oceanic basaltic layer and that the probe would reach the mantle in just over a year and a half.

Thereafter it would attain a depth of 101.5 km before stopping after 34.6 years. Beneath the continents, with their thicker crust, less of the later stages of the descent is likely to be through partly molten asthenosphere so reducing the underestimations compared with our oceanic lithosphere model.

Table 5. Descent of probe through continental lithosphere

Depth range (km)	Rock type	Threshold power q_{th} (W/m^3)	Initial velocity U_0 (m/y)	Descent time (y)
0 – 10	Granite	39350	22787	0.45
10 – 17	Schist	37920	20919	0.79
17 – 22	Metapelite	27567	19477	1.06
22 - 27	Amphibolite	24048	17737	1.34
27 - 30	Mafic granulite	26749	14772	1.55
30 -100	Peridotite	46290	9472	28.6
100 – 101.5	Peridotite	34279	312	34.6

ACOUSTIC TRACKING

Stevenson's (2003) proposal involved an active probe that measured properties of the rocks through which it passed and used compressional acoustic radiation to transmit the data to sensors at the Earth's surface. He calculated that a probe power output of 10 W would provide strong enough signals to be detectable at the surface even when his probe reached the core/mantle boundary at a depth around 2900 km. (Admittedly, the detection system would have had to be extremely sensitive and very much "state of the art")

While we would not rule out such an arrangement if technology and cost permitted, we would opt instead for a passive probe that generated acoustic signals by thermo-mechanical and radiation interactions with the rocks through which it passes. Melting and subsequent recrystallization of the rock generate intense acoustic signals over a wide frequency spectrum with a peak pressure of 10^3 Pa (Zhekamukhov and Shokarov 2000). Similar acoustic emissions are also created by intensive irradiation of the rocks (Lyamshev 1996) and this could be enhanced by incorporation of powerful neutron emitters such as [226]Ra and [9]Be into the ^{60}Co heat source. Moreover a source of sound can be incorporated directly into the capsule using frequencies at which attenuation of signals travelling through the rocks is minimal. The source of acoustic signals can be made in the form of a thermo-mechanical generator or the so called Harwell machine, which utilises the principle of the Stirling engine to transform radioactive decay energy into mechanical oscillations (Reader and Hooper 1986). The power of the acoustic signals generated by the sinking 1 m diameter probe with a 3.85×10^{18} Bq source would be $\sim 10^2$ W. The intensity of acoustic emission from a spherical capsule can be assessed (Landau and Lifshits 1986) as

$$I_a = 2\pi\rho_r R^4 a^2 \omega^4 / c_s,$$

where ρ_r is the density of the rock, ω is the frequency and a the amplitude of sound in the vicinity of probe and c_s is the velocity of sound in the rock ($\sim$2.8 -3.3 km s^{-1} (Kikoin 1991). The acoustic field created by the sinking sphere can be approximated as spherically symmetric waves of amplitude $a(r) = aR/r$ at distance r (Landau and Lifshits 1986). At a source frequency ω of $\sim10^3$ s^{-1} and an amplitude a of $\sim10^{-5}$ m for the acoustic signals, the amplitude on the surface of the Earth, e.g. at distances of ~100 km, $a(100$ km$)$ would be $\sim5 \times10^{-11}$ m. Current transducer technology allows detection of acoustic signals with an amplitude as small as 10^{-14} m (Stevenson 2003). Hence the signals from the probe reaching the Earths surface should be readily detectable. Moreover by using frequency discriminators these signals can be distinguished from natural and extraneous artificial noises.

A few suitably positioned detectors continually monitoring the signals reaching the Earth's surface from the probe would provide data on the position and motion of the probe as well as the properties of the rocks through which it was sinking and through which the acoustic signals passed (Figure 5).

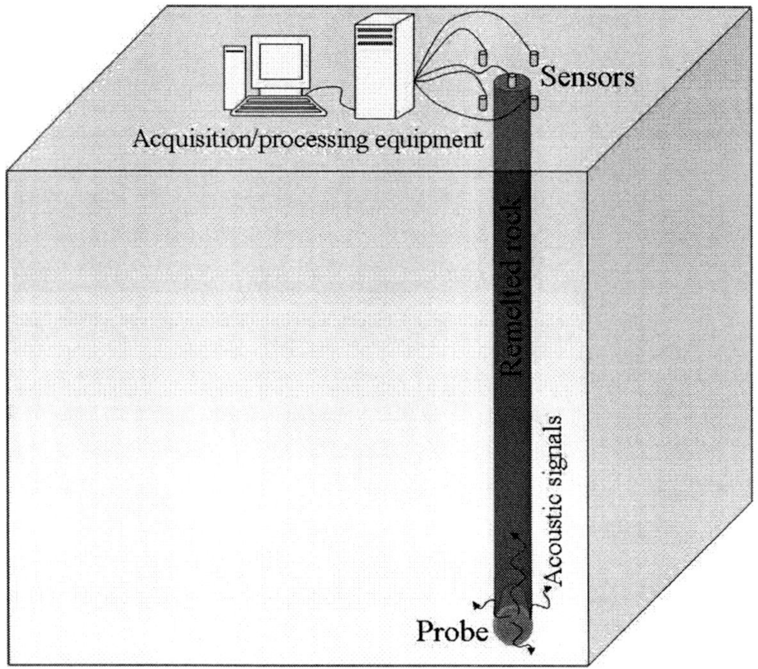

Figure 5. Schematic of acoustic emission monitoring of a self-descending probe.

In order to glean the maximum amount of information from this data, especially about the chemical and mineralogical compositions of the rocks, it might be necessary to undertake some preliminary laboratory calibrations on appropriate rock types over a range of pressures and temperatures.

CONCLUSION

Our calculations based on very conservative models suggest that a small, spherical, heat-emitting probe could reach depths well in excess of 100 km below the surface of both oceanic and continental crust. Initial penetration of the crust would be very rapid and worthwhile depths in the mantle could be reached in ~35 years. The acoustic signals generated during the melting and subsequent recrystallization of the rocks through which the probe descends, possibly enhanced by neutron and other irradiation effects, could be detected at the Earth's surface with suitable equipment. These signals could provide valuable information about the physical properties of the rocks through which the probe and the signals passed and, in conjunction with other techniques, uniquely define the mineralogical and chemical compositions and other properties of the rocks.

The experiment is on a time scale comparable with contemporary robotic missions to explore the solar system and would go some way towards rectifying the well-publicised anomaly that "we know more about outer space than we do about the inside of our own planet".

REFERENCES

Byalko, A.V. 1994. *Nuclear waste disposal: geophysical safety*. CRC Press, London.

Clark, S.P. Jr., 1966. *Handbook of physical constants*. Geological Society of America, Memoir 97, 587 pp.

Emerman, S.H. and D.L.Turcotte, 1983. Stokes's problem with melting. *International. Journal of Heat Mass Transfer*, 26, 1625-1630.

Jordan T. H. Structural Geology of the Earth's Interior, *Proceedings of the National Academy of Science*, 1979, Sept., 76(9): 4192–4200.

Kascheev, V.A., Nikiforov, A.S., Poluektov, P.P. and Polyakov, A.S. 1992. Towards a theory of self-disposal of high-level waste. *Atomnaya Energiya*, 73, 215-221.

Kikoin, I.K. 1991. *Table of physical parameters, Handbook*, Energooatomizdat, Moscow, 1232 pp.

Kosachevskiy, L.Y. and Sui, L.S. 1999. On the "self-burial" of radioactive wastes. *Journal of Technical Physics* 69, 123-127.

Landau, L.D. and Lifshits, E.M. 1986. Hydrodynamics. *Nauka*, Moscow, 736 pp.

Logan, S.E. 1973. Deep self-burial of radioactive wastes by rock melting capsules. *Nuclear Technology*, 21, 111-124.

Logan, S.E. 1999. Deeper geologic disposal: a new look at self-burial. In: *Proceedings of the WM'99 Conference*. Tucson, Arizona.

Lyamshev, L.M. 1996. Radiation Acoustics. *Nauka,* Moscow, 304 pp.

Ojovan, M.I. and Gibb, F.G.F. 2005. Feasibility of very deep self-disposal for sealed radioactive sources In: *Proceedings of the WM'05 Conference*, Tucson, Arizona.

Ojovan M.I., Gibb F.G.F., Poluektov P.P., Emets E.P. 2005. Probing of the interior layers of the Earth with self-sinking capsules. *Atomic Energy*, 99, 556-562.

Ojovan M. I., Lee W.E. 2005. *An Introduction to Nuclear Waste Immobilisation*, Elsevier Science Publishers B.V., Amsterdam, 315p.

Petrov, V.A., Poluektov, V.V., Zharikov, A.V., Nasimov, R.M., Diaur, N.I., Terentiev, V.A., Burmistrov, A.A., Petrunin, G.I., Popov, V.G., Sibgatulin, V.G., Lind, E.N., Grafchikov, A.A. and Shmonov, V.M. 2005. Microstructure, filtration, elastic and thermal properties of granite rock samples: implications for HLW disposal. In: Harvey, P.K., Brewer, T.S., Pezard, P.A. and Petrov, V.A. (eds.). *Physical Properties of Crystalline Rocks*. Geological Society, London, Special Publications, 240, 237-253.

Philpotts, A.R., Brustman, C.M., Shi, J., Carlson, W.D. and Denison, C. 1999. Plagioclase-chain networks in slowly cooled basaltic magma. *American Mineralogist*, 84, 1819-1829.

Philpotts, A.R. and Carroll, M. 1996. Physical properties of partly melted tholeiitic basalt. *Geology*, 24, 1029-1032.

Reader, G.T. and Hooper, C. 1986. *Stirling Engines*. EandF.N. Spon, London. 464 pp.

Rutter, E., Brodie, K., James, T., Blundell, D.J.and Waltham, D.A. 2003. Seismic modelling of lower and mid-crustal structure as exemplified by the Massiccio dei Laghi (Ivrea – Verbano zone and Serie dei Laghi) crustal section, N.W. Italy. In: Goff, J.A. and Holliger, K. (eds.) *Heterogeneity in the crust and upper mantle*. Kluwer, New York, 67-99.

Stevenson, D.J. 2003. Mission to Earth's core – a modest proposal. *Nature*, 423, 239-240.

Taylor, K.J. 2003. Container materials for high-temperature very-deep borehole disposal of radioactive wastes. *BNFL/University Research Alliances Conference*, Sellafield, January 2003.

Vocadlo L., Dobson D. 1999. The Earth's deep interior: advances in theory and experiment. *Phil. Trans. R. Soc. Lond.* A 357, 3335-3357.

Vorsteen, H.D. and Schellschmidt, R. 2003. Influence of temperature on thermal conductivity, thermal capacity and thermal diffusivity for different types of rock. *Physics and Chemistry of the Earth*, 28, 499-509.

Wyllie, P.J. 1971. *The Dynamic Earth*, J. Wiley and Sons, New York , 416 pp.

Yoder, H.S. 1976. *Generation of basaltic magmas*. National Academy of Sciences, Washington, 265 pp.

Zhekamukhov, M.K. and Shokarov, K.B. 2000. Mechanism of initiation of acoustic emission in crystallization and melting of a substance. *Journal of Engineering Physics and Thermophysics*, 73, 1064-1079.

Reviewed by Valentina L. Stolyarova, Institute of Silicate Chemistry of The Russian Academy of Sciences,St. Petersburg, Russia

In: Nuclear Waste Research: Siting, Technology and Treatment   ISBN 978-1-60456-184-5
Editor: Arnold P. Lattefer, pp. 221-238   © 2008 Nova Science Publishers, Inc.

Chapter 8

NUCLEAR WASTE CONTAINMENT BY SOL GEL PROCESS

*Thierry Woignier[*1], Jerome Reynes[2] and Jean Phalippou[3]*

[1]UR Seqbio IRD-PRAM, Le Lamentin, France,
CNRS, Université de Montpellier 2, France
[2]Université de Montpellier 2, France
[3]Université de Montpellier 2, France

ABSTRACT

The usual approach to nuclear waste fixation is to incorporate radioactive elements in borosilicate glasses. However, researches are in progress to incorporate long lived nuclear wastes like actinides in a glass matrix with a higher chemical durability. Good chemical durability is generally achieved by increasing the silica content in the glass composition. Associated to high structural stability and thermal shock resistance, silica glass will optimize the properties which characterize a desirable material for the actinide fixation. But, actinide containment in silica glass involves melting the glass at high temperature ($\approx$ 2000°C) giving rise to various problems (evaporation losses, interaction between crucible and molten materials...).

According to an easy sintering stage, the sol-gel process is a new way to synthesize silica glasses at low temperature ($\approx$1000°C). In the first part of the paper we explain what is the sol-gel process and why it is an interesting way to prepare glasses at a temperature two times lower than by the conventional melting process.

Generally nuclear wastes exist as aqueous salt solutions and in the second part of the paper, we propose to use the totally open pore structure of the gel to allow migration of that liquid throughout the silica gel. We investigate the physical properties (mechanical strength, pore volume, permeability) of different gels porous structure (xerogels, aerogels, and composite gels). With this approach we show that thanks to a higher permeability and good mechanical properties, the porous network of composite gels can be easily used as a host matrix for the actinides simulating salts (Nd and Ce nitrates

---

[*] E-mail : woignier@.univ-montp2.fr

dissolved in water). After soaking and drying, the loaded material is sintered in the temperature range 1100°C-1200°C. Nd_2O_3 and CeO_2 loading in the range 0-20% can be achieved with this process.

The last part of the paper shows that the final structure of the fully sintered materials is that of a glass ceramic (silica glass+lanthanide oxides) and the results prove clearly the improvement of the chemical durability glass ceramic compared to the conventional nuclear glass (borosilicate glass). Owing to its simple structure, the corrosion rate of the glass-ceramics is close to that of the pure silica, almost 2 orders of magnitude lower than that of the borosilicate nuclear glass. Beside the good chemical durability the glass ceramics present also interesting mechanical properties.

1. INTRODUCTION

In an attempt to confine radioactive waste, the French nuclear agency (CEA: Commissariat à l'Energie Atomique) forms nuclear glass. In this process, the radioactive elements are mixed and melted with a glass frit [1]. This glass frit is an alumino borosilicate glass containing sodium and the borosilicate glass is considered as the reference in nuclear waste conditioning [2]. However, to ensure long-term storage, long-life nuclear waste (actinides) must be incorporated in a matrix exhibiting excellent chemical durability. In the case of nuclear waste glasses it is crucial to limit a possible release of radioactive chemical species, if the glass structure is destroyed by a corrosion process. Among the usual glasses, silica glass, which does not contain alkali, and boron are expected to present high durability. Silica glass is also a good candidate because of its good mechanical properties, low thermal expansion, and good thermal shock resistance. However, such a glass preparation requires a high-temperature melting process (2000 °C).

An alternative way is the use of a low temperature sinterable porous matrix. In the literature [3] it is shown that the sol-gel process is an appropriate low temperature way to prepare glasses at around 1000°C, not by melting but by sintering of a porous amorphous silica gel. Consequently, we propose to use the silica gel porous network as a host matrix for the actinides and we will demonstrate the feasibility of the fixation of the actinides in pure silica glass. The actinides will be simulated by lanthanides nitrates (Nd and Ce) in water, which diffuse by capillarity in the porous network. After drying and sintering around 1100°C, the glass-ceramic material is obtained. In the literature, phase separated and leached glasses have been also proposed as host matrix [4] but, such glasses are not pure silica and contain boron in a few weight percent. Boron, being more soluble in water than silica, decreases the chemical durability of the glass.

Because of the microporous texture and the low permeability of the gels, the impregnation step could be long and responsible for significant stresses which will induce fracture. For these reasons and also for safety reasons (the time process must be shortened with respect to nuclear environment), the impregnation step must be optimized which requires a material with high permeability and mechanical properties.

In this chapter we will characterize different families of gels and show that this goal can be achieved from the "composite aerogels". We will follow the mechanical, textural and permeability features of the porous material *versus* the relative density. The relative density can be adjusted by different parameters such as: silica composition, sintering or drying parameters. So, first of all, it is necessary to briefly describe what the sol gel process is and why it is an interesting way to prepare glasses at low temperature.

2. POROUS GLASSES BY THE SOL-GEL PROCESS

The classical procedure for making glasses includes a step at elevated temperature which ensures that the raw materials are dissolved and have reacted. Then the amorphous structure of the liquid is preserved by quenching the melt. In the sol-gel route this high temperature step is avoided. The homogenization is achieved in the solution at room temperature, cross-linking and preservation of the liquid structure is also accomplished at room temperature by the gelling step. Further heat treatments only serve to remove organic species, hydroxyls and porosity. The challenge of the sol - gel process is then to obtain a solid material, mineral and amorphous from room temperature liquid (not a melt) which is generally organic [3]. So, several transformations of the liquid compounds are necessary. The first step is the formation of a gel from the solution and involves the hydrolysis reaction of an organometallic compound ($Si(OR)_4$), dissolved in alcohol in the presence of water.

$$(OR)_3 \, Si - OR + H_2O \rightarrow (OR)_3 \, Si - OH + ROH$$

By condensation reaction two silanol bonds (Si-OH) give rise to a siloxane bond Si-O-Si.

$$(OR)_3 \, Si - OH + HO - Si \, (OR)_3 \rightarrow (OR)_3 \, Si - O - Si \, (OR)_3 + H_2O$$

These two reactions lead to porous and non-crystalline materials containing substantial amounts of water and organic species.

Xerogels

After gelation we obtain a gel which is a two-phase medium containing the solid network and the liquid (alcohol + water). A drying treatment can be carried out at ambient temperature but during this drying stage considerable shrinkage occurs converting the wet and soft gel into a dried and harder porous solid (xerogel). The drying procedure is crucial and must be performed extremely slowly because it induces capillary phenomena which can destroy the gel network and lead to the breaking up of the solid network. Various alternatives drying methods involve a favorable compromise between the capillary forces and the mechanical resistance of the gel network (strengthening the gel by reinforcement, reducing the surface tension, enlarging the pores ...) [3].

In summary, xerogel samples can be synthesized by a careful control of the drying parameters. They are porous glasses with relative density (ρr) ranging from 0.3 to 0.7, mainly containing micropores because of the collapse of the larger pores, during drying. The relative density is defined as the volume ratio occupied by the solid phase and the porosity is equal to 1- ρr.

Aerogels

The goal of supercritical drying (SCD) is to eliminate these capillary forces. The magnitude of those stresses is dependent on the interfacial energy γ of the liquid and it is possible to suppress γ if the pressure and the temperature pass over the critical point of the liquid [5,6]. The supercritical solvent is then isothermally evacuated by condensation outside the autoclave. After supercritical drying, the gel (which is called "aerogel") is a solid material, amorphous but extremely porous, ρr ranging from 0.01 to 0.2 (80 to 99 % of porosity) [6]. The silica aerogel network can be described as an assembly of clusters (~ 50 nm) built by the aggregation of small particles ($\sim$ 1-2 nm) [7,8]. The porosity is totally open and spans over the whole range of pore sizes.

The potential applications of these very porous materials are for catalysis [9,10], insulator [11,12] but also as a glass precursor. By an appropriate schedule of sintering treatments, the silica gels can be easily transformed into silicate glasses [13,14] and glass ceramics [3].

Partially Sintered Aerogels

Another possible interest of aerogels is that they can play the role of a host matrix for the synthesis of multi-phase materials. The large pore volume could be used as a sponge to incorporate chemical species with the objective to get a two phases material. However, the counterpart of this porosity is the poor mechanical properties and the consequence is that very porous glasses crack during filling (once again because of capillary forces). One way to enhance the mechanical properties is the sintering. Depending on the duration of the heat treatment, the microporosity is progressively eliminated and partially sintered samples can be obtained in the range of relative density between 0.15 (lightest aerogel) and 1 (silica glass obtained by a complete sintering) [14].

Composite Aerogels

Another important parameter is the permeability. A high permeability is generally an advantage because it means that the fluid and thus the chemical species of interest migrate easily in the porous network and we can expect a homogeneous repartition of the chemical species.

In ceramic science, it is generally admitted that inclusion of particles or fibers could improve the mechanical properties of composite ceramics and also modify the porous structure. We will show that it is possible to adjust the relative density, the mechanical properties and the permeability of the composite aerogels by the addition of silica powder (silica soot like aerosil), in the monomer solution, just before gelation [15,16].

In summary, mechanical properties, capillary forces and permeability are the most important parameters for the filling of the porous gel network. Macroscopically, xerogels, aerogels, partially sintered aerogels or composite aerogels networks seem rather far from that of the silica glass. However, Raman, infrared and RMN spectroscopy which characterize the molecular structure, put in evidence vibration bands identical to those observed in vitreous silica [3,17]. These different kinds of porous materials can be considered as porous glasses although the very different way used to attain the "glassy "state. These different kinds of gel-derived porous glasses will be tested as host matrix for nuclear waste containment. Surrogate salts of actinides will be deposited on the internal surface of the porous glasses when the salts solution is evaporated. The tailoring of the mean pore size and of the pore size distribution is required to facilitate a homogeneous dispersion of salts within the texture.

3. SYNTHESIS OF GEL DERIVED POROUS GLASSES AS HOST MATRIX

Three different families of porous gels derived glasses (xerogels, partially sintered aerogels, and composite aerogels) have been elaborated for this study. Our first objective is to prepare samples sets which cover a large porosity range and to test their different advantages as host matrix.

In the fist step, the initial wet gels are prepared by hydrolysis and polycondensation reactions of tetraethoxysilane (TEOS). The TEOS is dissolved in various amounts of ethanol thereby adjusting the oxide content of the sol). The wet gels are transformed into xerogels by a slow room temperature drying or aerogels by supercritical evacuation of the solvent. In the literature [15,16], it has been shown that the addition of soot silica such as "aerosil" in the TEOS solution favors the formation of macropores. Soot silica (aerosil OX50 from Degussa) was added to the hydrolyzed solution of TEOS. The aerosil weight percentage (reported to the total silica weight) ranges between 0 and 70 %. These aerogels are labeled as "composite aerogels".

In conclusion we will compare three kinds of porous samples for which the relative density is adjusted by different ways. In the case of xerogels the bulk density is controlled by the amounts of TEOS diluted in ethanol and by the room temperature drying (xerogels set). Another way to increase the relative density is the sintering. Aerogels have been sintered by a heat treatment close to 1000°C. Depending on the duration of the heat treatment, the porosity is progressively eliminated and more or less sintered samples can be obtained (partially sintered aerogels set, PSA). The third way consist of the addition of silica soot particles in the TEOS ethanol solution which will control the relative density (composite aerogels set, CA).

These different gel families cover the range of relative density between 0.15 and 1 which corresponds respectively to a pore volume range between 95 % and 0%. However, the process of impregnation by salt solutions induces high capillary stresses which could damage and destroy the porous network. Consequently it is necessary to characterize the mechanical behavior of the different kinds of gels.

4. MECHANICAL PROPERTIES OF POROUS GLASSES

The mechanical properties such as rupture strength (σ) and elastic modulus (E) are measured in flexion using the three-point bending test. The rupture strength and the elastic modulus are given by the relationships:

$$\sigma = 3Ps/ 2e^2b \qquad\qquad 1$$
$$E = 4Ps^3/3\pi D^4 \qquad \delta \qquad\qquad 2$$

where P is the applied force, e the sample thickness, b the sample width , s the span and δ is the deflection. In figures 1 and 2 are plotted the mechanical properties of the different sets of porous glasses.

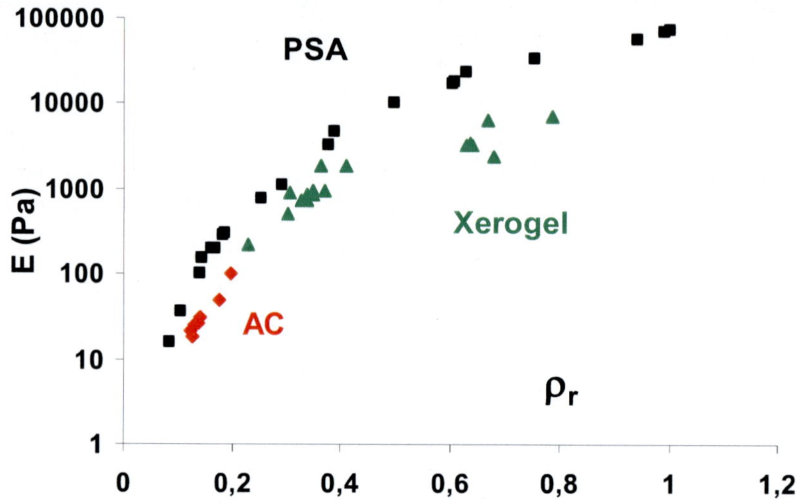

Figure 1. elastic modulus *versus* relative density.

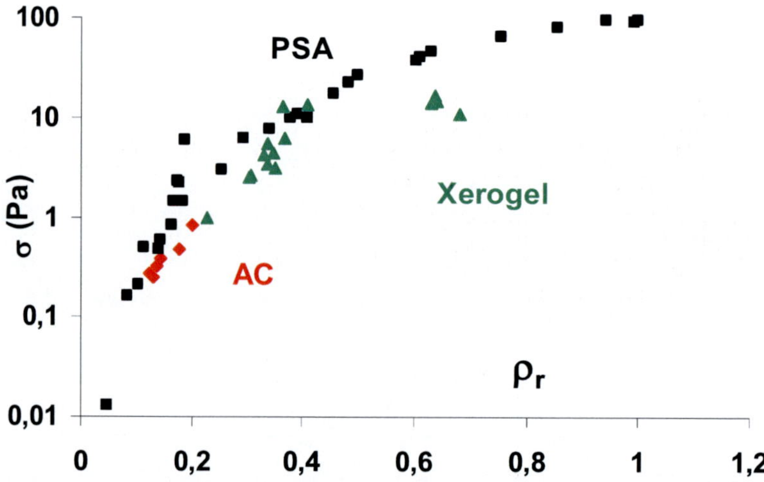

Figure 2. Mechanical strength *versus* relative density.

It appears that PSA, CA and xerogels in the density range 0.15-0.5 show about the same values of mechanical properties. However, if a partially sintered aerogel or a composite aerogel is able to resist the filling by a solution of water, the xerogels generally not [18]. Two reasons explained these different behaviors: The xerogel network is probably locally damaged by the stresses, which occur during the drying. The assumption of the structure damage is deduced from the lower mechanical properties of the xerogels in the density range 0.5-1. During drying, micro-flaws are created but they are not "critical" in the sense of the fracture mechanics of brittle materials. Flaws inside the structure act as stress concentrators and capillary stresses are locally amplified by this effect. In other words the toughness of the xerogels is lower than that of the partially sintered aerogels or composite aerogels. The second explanation is related to the lower pore size of the xerogel structure compared to that of partially sintered aerogel and composites aerogel. During drying, xerogel is submitted to compression force which tends to eliminate the larger pores [19]. Drying shifts the pores size distribution towards the lower pores and capillary stresses during a further filling will increase.

Concerning the PSA set, the mechanical strength increases by 3 orders of magnitude (figure 2) over the whole range of relative density. The most part of the strengthening arises in the density range 0.15-0.5.The heat treatment has two effects: it increases the mechanical properties of the material but also causes the collapse of the smallest pores so as to decrease the effects of the capillary phenomena. During the sintering process, the densification is induced by viscous flow, which tends to reduce the whole sample volume, eliminating the smallest pore [20,21]. Because of the mechanical improvement and the elimination of the micro pores, PSA samples with relative density higher than 0.45, are able to resist the liquid impregnation [14].

The figures 1 and 2 show that the mechanical properties of the composite aerogels are quite close to the data measured on xerogel and PSA sets. The aerosil addition increases the bulk density and therefore the mechanical properties, but also affects the aggregation process, the aerogel structure [22] and the pore size distribution [15]. It homogenizes the pore structure and increases the mean pore diameter in the range of the mesoporoity (50-100 nm). Consequently, the composite aerogel with ρr =0.2 is able to resist the capillary stresses coming from the water and alcohol surface tension, even though their mechanical properties are close to those of the xerogels.

In conclusion, drying, sintering and the addition of silica particles, will improve the gel mechanical properties. However, all of the studied xerogel samples crack during the filling by liquids certainly because of the presence of flaws which weaken the gel network and the small size pores which enhance the capillary forces. We can conclude that this set of samples is not an appropriate candidate as host for nuclear wastes.

5. PERMEABILITY OF POROUS GLASSES

As explained above the second important porous feature of the host matrix is the permeability. In this work, permeability (D) is measured by a method based on the Archimede's principle. The samples ($\sim$ 1cm^3) are dipped in a liquid and initially the resulting Archimede's force corresponds to the liquid weight occupied by the total samples volume. When the impregnation proceeds, as a function of the pore filling, the Archimede's force decreases proportionally to the filled pore volume. Finally because the pore volume is totally open the Archimede's force corresponds to the volume of the sample solid phase. The data are recorded as V_{imp}/V_p = f(t) where V_{imp} is the impregnated volume and V_p is the total porous volume. From the data $V_{imp}/V_p = f(t)$ and physical properties of the liquid and the porous solid it is possible to calculate the permeability of the porous network. Darcy's law for a small-impregnated volume is:

$$dV_{imp} = \frac{D}{\eta} A(t) \frac{\Delta P}{h(t)} dt$$

3

where D is the permeability, A(t) the normal surface of the flow, η the liquid viscosity, h is the liquid impregnation depth and $\Delta P/h(t)$ the pressure gradient in the liquid. Considering $h(0) = 0$ and Φ the porosity.:

$$h(t)^2 = \frac{2D}{\eta} \frac{1}{\phi} \Delta P t$$

4

h(t) can be expressed as a function of V_{imp} / V_p and derived from the Archimede's force experiments (11).

Thus plotting $h^2 = f(t)$, the slope α is equal to:

$$\alpha = \frac{2D}{\eta} \frac{1}{\phi} \Delta P$$

5

The curves are in agreement with the expected linear behavior (Figure 3).

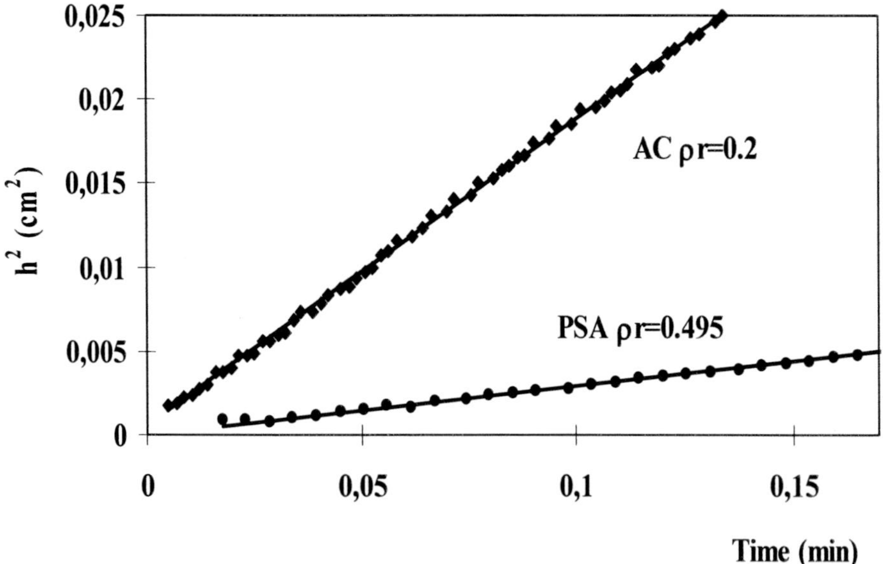

Figure 3. squared impregnation depth *versus* time.

Capillary forces are the driving forces for fluid migration and ΔP is the capillary pressure. So the permeability is

$$D = \alpha\phi\text{rh}\,\frac{\eta}{4\gamma Cos\theta}$$

6

where θ is the wetting angle.

Using η (0.355 10^{-3} Pa.s), γ (0.0725 J.m^{-2}), and cos θ ~1 as data for water, we can estimate D (figure 4) for the PSA set and CA samples.

This figure clearly shows that D strongly decreases with the sintering. As explained above the sintering proceeds by pores collapse, then for the highest densities the mean pore size becomes so small that the permeability tends to 0.

These results are in agreement with the empirical Carman Kozeny relation: $D \propto (1-\rho r)r_h^{\,2}$ [23] where rh is the hydraulic radius (mean pore size) defined as the ratio of pore volume to surface area. According to this relation, because of the decrease of the mean pore size and the increase of the relative density, D decreases. So a compromise should be found in terms of relative density for the use of PSA as a host matrix. The relative density should be high enough to provide a matrix with acceptable mechanical properties but not too, to have a significant permeability, this compromise corresponds to a density in the range 0.4-0.5 [14].

The permeability of the composite aerogels set has been measured by the same method (figure 4) but contrarily to the PSA set the permeability increases with the relative density. This counter intuitive result (with respect to the Carman-Kozeny relation) comes from the fact that while the addition of aerosil particles increases the relative density it also increases the mean pore size [15]. The net result is the strong D increase.

In opposite to the other ways to synthesize host matrices (classical aerogel, PSA, xerogels), the composite aerogels set is the only one which allies the improvement of the mechanical properties, the increase of the mean pore size and consequently of the permeability. This method easy to use, leads to host matrices with a large and accessible porous volume rapidly impregnated by classical liquids such as water or ethanol.

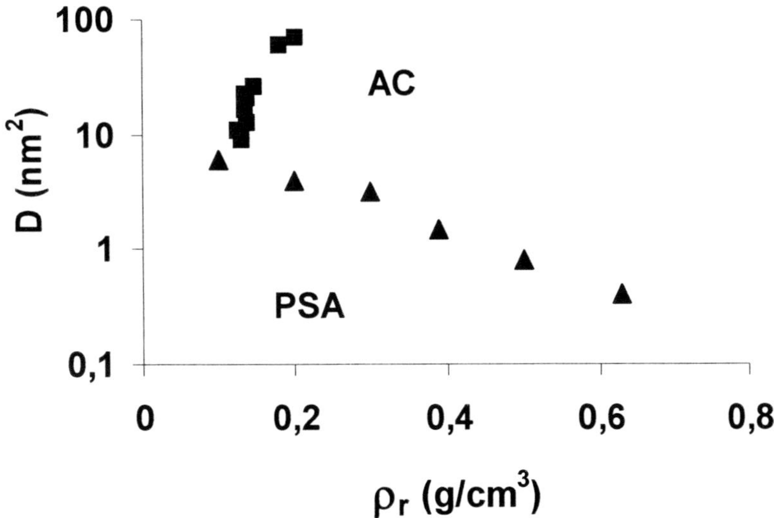

Figure 4. permeability versus the relative density.

6. LOADING OF THE POROUS GLASSES BY ACTINIDES SURROGATES

Now we can test the different porous networks PSA and CA (xerogels has been excluded), as host matrices for long lived nuclides. Generally the nuclear wastes are provided under salt form in aqueous solutions. We will use the totally open pore structure to allow migration of the liquid species (salt in solution) throughout the whole porous volume. Then, the liquid phase is removed; the porous glass containing the salt is fully sintered giving rise to the synthesis of a multi component material. The porous structure is used as a volume host and according to the fine pore structure we expect to prepare a nano composite by a very simple process.

For safety reasons we have worked with lanthanides, Ce and Ne respectively simulating the III and IV valences of actinides [25] (for example Am(III), Cm(III) Np(IV), Pu(IV)). We have also chosen Nd and Ce as actinides surrogates because of their different affinity with silica which will leads to different kinds of glass-ceramics, simulating the possible behaviors of the actinides in the presence of silica. Cerium and neodymium nitrates are dissolved in water which imposes that the relative density of the PSA will be higher than 0.45 [14]. After soaking in an oven (50°C) the samples are dried and calcinated at 600°C in such a way to complete the drying and to decompose the neodynium nitrate. Further heating at 1100°C fully sinters the structure by viscous flow. The dense solid consists of a silica matrix in which lanthanides oxide is trapped. Weight differences before soaking and after sintering allows measuring the surrogate loading in weight percent. Table 1 shows that the neodymium oxide loading measured after soaking obviously increases with the concentration of the neodymium nitrate solution.

Table 1.

Nd(NO$_3$)$_3$ soaking solution (weight %)	Nd$_2$O$_3$ loading after sintering (weight %)
10	2.3
20	3.2
30	5
40	6.6
50	10.8

Loading close to 10 weight % can be achieved with this process but is twice lower than those expected (calculated from the relative density and the solution concentration). We can conclude that during the preliminary sintering step to achieve $\rho_r \approx 0.45$ a part of the porosity becomes closed and the soaking solution cannot invade the whole porosity.

Composite Aerogels

To improve the loading rate it is necessary to propose a porous structure more accessible to the soaking solutions, with a higher permeability. This goal is achieved using the composite aerogels set. After soaking, drying and calcination in the same conditions than the PSA set, the CA samples are fully sintered in the temperature range 1100-1200°C. Because of the lower relative density of CA compared to PSA, waste loading higher than 20% can be achieved with this process (table 2).

Table 2.

$Nd(NO_3)_3$ soaking solution (weight %)	Nd_2O_3 loading after sintering (weight %)
5	3.8
10	8.7
15	14.8
20	20.3

Sintering Behaviour after Soaking

It must be noted that the final sintering procedure is affected by the loading solution. The figure 5 shows the linear expansion curves measured on the samples loaded with solution containing 5 and 15 % of neodymium nitrate.

The curves which express the sintering shrinkage ($\Delta L/L_0$) evidence that the loading increases by 100°C the temperature range where the shrinkage is maximum. This temperature is close to 1100 °C for the CA sample and close to 1200°C for the CA sample soaked with the solution containing 15 % of neodymium n itrate.The loading by the surrogates hinders the sintering, because crystalline phases (CeO_2 and Nd_2O_3) play no role in the viscous flow mechanism responsible for the sintering of amorphous materials. We will see later that the final structure of the fully sintered materials is precisely that of a composite material (oxides +glass). Consequently the sintering step must be fitted to the composition of the loaded porous glass and can vary by 100°C.

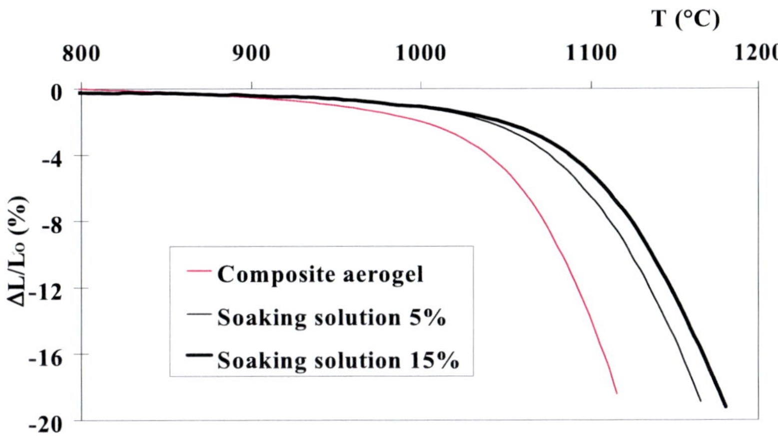

Figure 5. sintering curves for different neodymium loading.

7. CHARACTERIZATION OF THE GLASS CERAMICS

We have demonstrated that these porous matrices can be easily filled with the surrogate solution and fully sintered showing that rapid containment of actinides in silica is possible. After sintering, the dense solids consist of a silica matrix in which the surrogate oxide is trapped. Loading higher than 20% of surrogate can be achieved with this process.

Structure

The final structure of the fully sintered materials is that of a composite material.

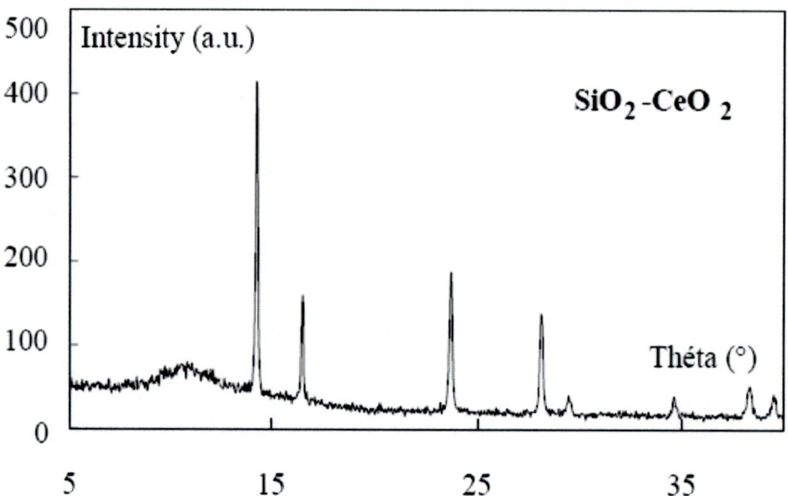

Figure 6. X-rays pattern of the glass ceramic SiO_2-CeO_2 loaded with 15% CeO_2.

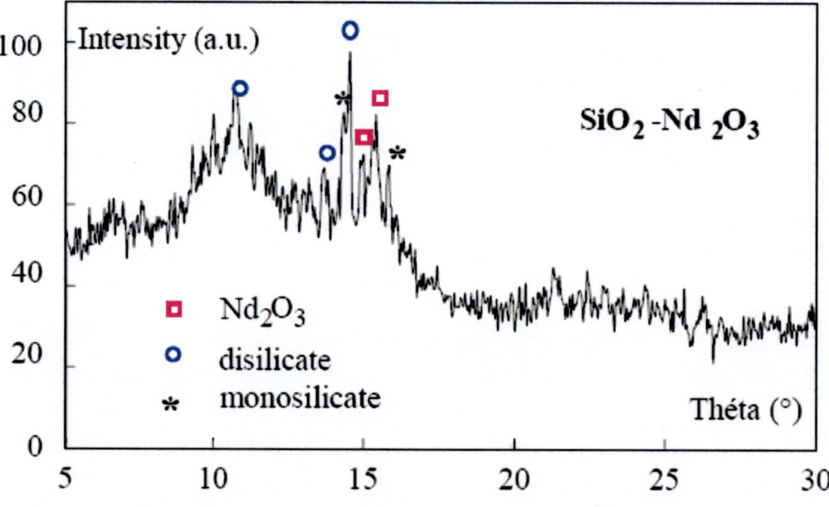

Figure 7. X-rays pattern of the glass ceramic SiO_2- Nd_2O_3 loaded with 14% Nd_2O_3.

The X rays pattern (figures 6 and 7) show that the sintered Ce loaded material is clearly a biphasic compounds (SiO $_2$ -CeO$_2$) otherwise the Nd loaded sample presents three different crystalline phases, neodymium sesquioxide (Nd$_2$O$_3$) but also neodymium mono (Nd$_2$SiO$_5$) and di silicate (Nd$_2$Si$_2$O$_7$).This multiphase structure is confirmed by the bulk density of the Nd loaded materials which are systematically higher than the assumed bulk density calculated from a biphasic description [25].

These structural differences are the result of the Ce and Nd affinity with Si . In the case of Ce, the formation of a binary glass in a melting process is difficult, Ce(IV) generally forms CeO$_2$ crystallites [26]. Moreover the crystalline phases of cerium silicates like Ce$_2$Si$_2$O$_7$ are not stable under 1400°C and transformed into SiO$_2$ and CeO2 [27, 28].

In the case of Nd, glasses with weight percent between 2 and 5 have been obtained [29,30]. The phase diagram shows also that the different Nd silicates (Nd$_2$SiO$_5$ and Nd$_2$Si$_2$O$_7$) are stable at room temperature, besides the oxide (Nd$_2$O$_3$) [31,32].

Mechanical Properties

The mechanical properties of the different glass ceramics obtained from the porous glasses are important to validate the process. Indeed, the mechanical behavior of matrices containing nuclear wastes should be the highest possible because the fracture of the matrix will lead to an increase of the corrosion because of a larger contact surface between the solid and water. The figures 8-9 show the elastic modulus and the rupture strength of the different glass ceramics obtained with Nd and Ce. These figures show that for loading higher than 10-12% the mechanical properties decrease.

This weakening of the materials structure is the result of the thermal expansion coefficients and the elastic modulus of the different compounds present in the glass ceramics.

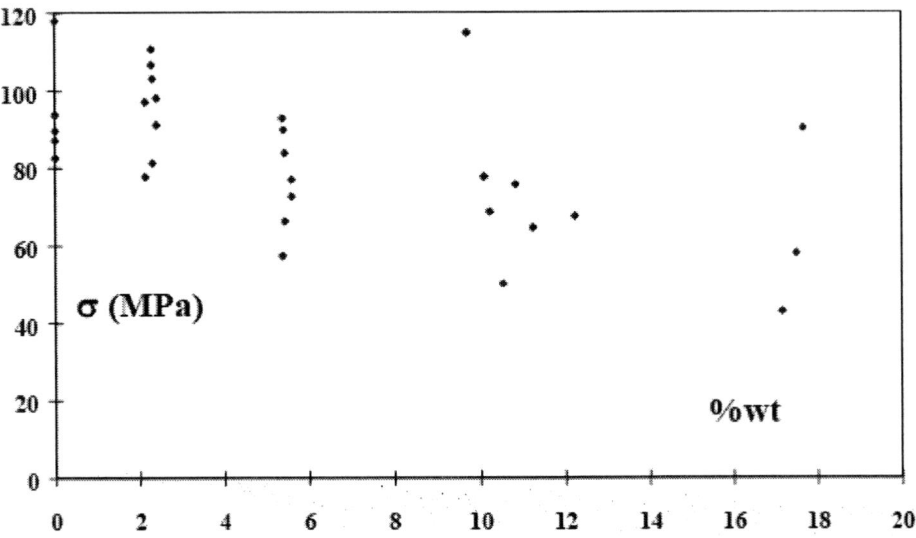

Figure 8. Mechanical strength of Nd glass-ceramic versus the Nd$_2$O$_3$ loading.

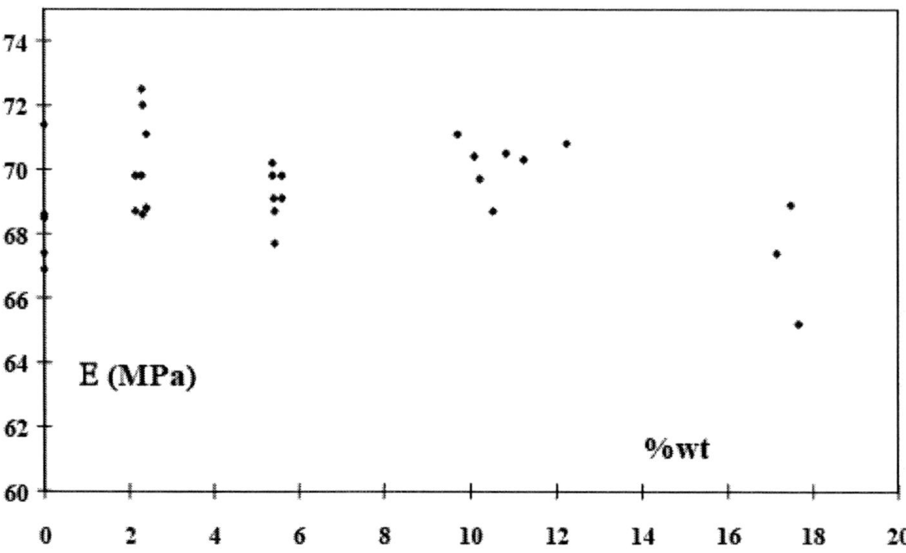

Figure 9. Elastic modulus of Nd glass-ceramic versus the Nd_2O_3 loading.

For example the thermal expansion coefficient of vitreous silica (0.5 $10^{-6}\,°C^{-1}$ [33]) is 20 times lower than the one of the CeO2 (11.5 $10^{-6}\,°C^{-1}$ [34]) and Nd_2O_3 crystallites (12 $10^{-6}\,°C^{-1}$ [35]). The elastic modulus of the silica glass (73 GPa) [33] is much lower than the elastic modulus of CeO_2 (184 GPa [36]) and Nd_2O_3 (418 GPa [36]).After the sintering and during the cooling, local stresses can occur at the boundary of the CeO_2 and Nd_2O_3 crystallites, because of this important differences of thermo elastic properties. The net result is the formation of flaws which will weaken the glass ceramics.

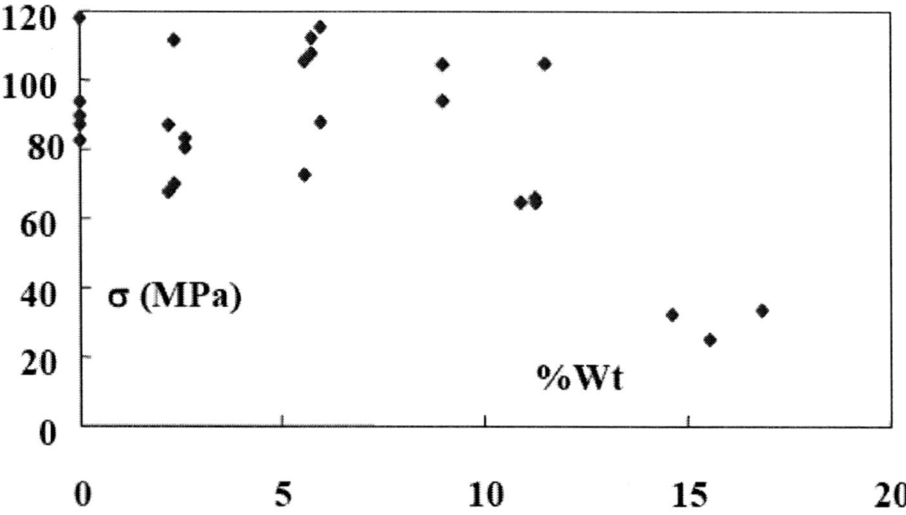

Figure 10. Mechanical strength of Ce glass-ceramic *versus* the CeO_2 loading.

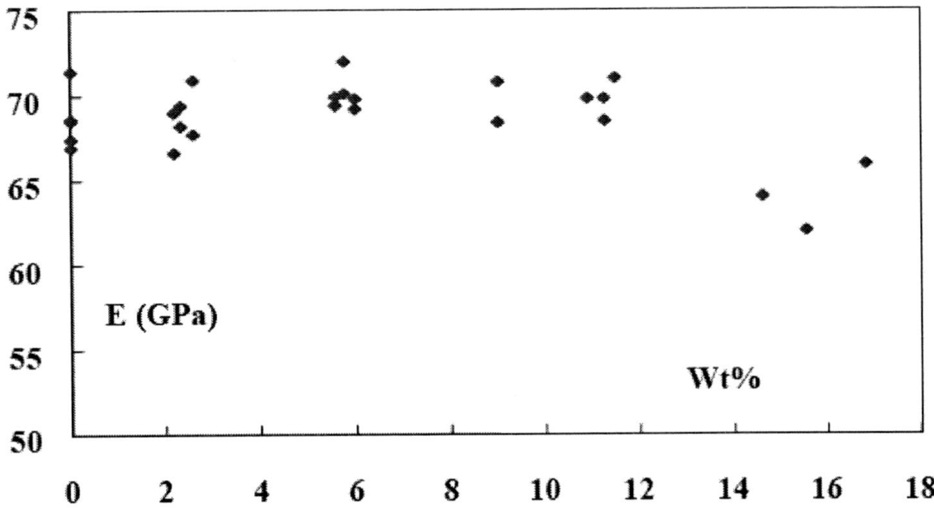

Figure 11. Elastic modulus of Ce glass-ceramic *versus* the CeO$_2$ loading.

Chemical Durability

In the case of long lived nuclear wastes it is crucial to limit a possible release of actinides if the glass structure is attacked by an alteration process [37,40]. The chemical durability characterizes the resistance to the water corrosion. We have seen that the Nd glass ceramics have a structure more complicated than the Ce glass ceramics. Because of these structural differences we can expect different chemical durability behaviors. The chemical durability of the glass ceramics is measured as follows: we use a conventional Soxhlet device consisting of a boiler containing ultra pure water and a condenser system. A monolithic sample was placed in a recipient into which condensed steam flowed continuously. The surface area is close to 10 cm^2. The test was conducted at 100°C, after 28 days of leaching; the normalized mass loss (g/m2) was measured from the weighted difference and from chemical analysis.

It has been explained in the introduction that the vitreous silica has a high chemical durability. It is necessary to compare the mechanisms of alteration of the Nd and Ce glass ceramics with silica and the usual borosilicate glass. For this study we have chosen two glass ceramics loaded respectively with 7.6 weight % of Nd$_2$O$_3$ and 6.6 weigth % of CeO$_2$. Figure 12 shows the normalized SiO$_2$ loss measured on the two kinds of loaded material and on the reference (pure SiO$_2$). The normalized silica loss characterizes the glass network destruction by the glass former dissolution. From this figure we can calculate the aqueous corrosion rate (V$_0$) measured on pure silica glass (as a reference) and on the glass ceramics loaded with the surrogate oxides. We will compare these data to the corrosion rate of the usual nuclear wastes glass (borosilicate). The corrosion rate (V$_0$) for pure silica is equal to 0.015 g/m^2 per day, 100 times lower than the corrosion rate of the usual nuclear wastes glass for which V$_0$ is equal to 2 g/m^2 per day [38]. V$_0$ for the glass ceramics loaded with the Ce and Nd are respectively equal to 0.035 g/m^2 per day and to 0.25 g/m^2 per day. This result evidences the improvement of the chemical durability of the glass ceramics compared to the nuclear waste glass.

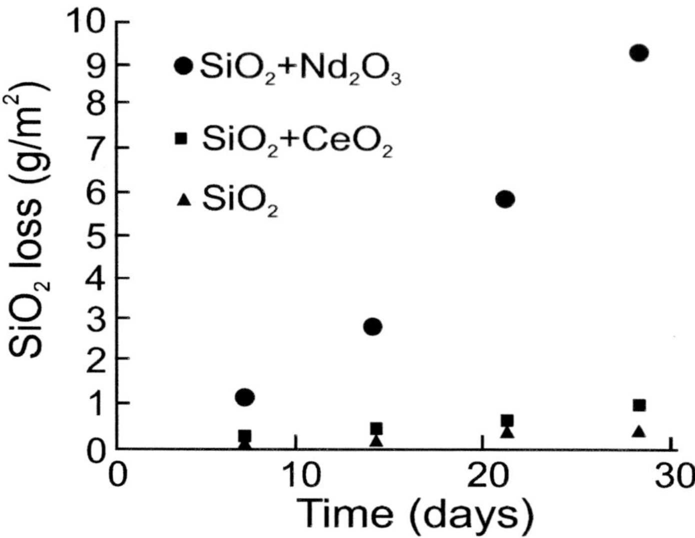

Figure 12. SiO$_2$ loss *versus* the leaching time.

Thanks to its simple structure, the corrosion rate of the Ce-glass ceramics is quite close to that of the pure silica, the small difference can be attributed to a more pronounced silica dissolution at the interface with the cerium oxide domain. The corrosion could be enhanced by the presence of thermo mechanical stresses [39]. The corrosion rate of the Nd-glass ceramic is 10 times higher than SiO$_2$. This result suggests that a part of the glass matrix is made of a neodymium silicate glass less durable than SiO$_2$. Because of their different affinity with silica, the two actinides surrogates used (Nd and Ce) indicate what kind of behavior we can expect, when actinides are respectively able to form silicate phase and/or binary glass and, on the opposite, when no silicate crystalline or glassy phases can exist.

CONCLUSION

The different porous glasses synthesized in this study can be potential candidates to be used as a host matrix for nuclear wastes. To avoid fracture in the materials (because of the capillary stresses undergone by the porous network during the drying and filling), the best compromise is materials with high mechanical properties and permeability. The nuclear wastes are processed to salt form in aqueous solutions. Hence to simulate the actinides solution, the cerium and neodymium nitrates are dissolved in water. Ce and Ne respectively simulate the valences III and IV of the actinides. After soaking the samples are dried and calcinated at 600°C in such a way to complete the drying and to decompose the nitrates. Further heating around 1100°C fully sinters the structure by viscous flow.

Composite silica aerogels are the best candidates to confine actinides. Comparatively to partially sintered aerogels, its pore size distribution is shifted towards the higher pore size. Consequently a high pore volume is available for impregnation and the network permeability is increased. This porous matrix can be easily filled by the surrogate solution and fully sintered, showing that rapid containment of actinide oxides in silica is possible.

For the cerium, the surrogate element is not included in the glass structure but rather embedded in the silica matrix. This structure allows a good chemical durability close to that of the vitreous silica with a aqueous corrosion rate 60 times lower than that of the nuclear borosilicate glass. However, in the case of Nd, the complicated structure made of SiO_2 and Nd_2O_3 but also neodymium silicate glass and crystals favours a higher corrosion compared to pure silica.

Besides the problem of the synthesis of new porous matrices, the influence of the reactivity of the matrix on the physical and chemical properties of the invading species is a research field largely open. We have shown two examples where the reactivity matrix – chemical species are strongly important for the physical properties of the final two phases materials. It must be underlined that this process permits the synthesis of multi phase materials which cannot be done by the melting. The sintering can preserve the heterogeneous structure in the glass matrix, which is not possible, when the melting step dissolves, homogenizes or segregates the different chemical species.

The feasibility of actinide containment in silica glass is demonstrated. The main advantages of the process are the high chemical durability of the matrix and the good mechanical resistance. The rather complicated process technology can be considered as a disadvantage in a radioactive environment and the effect on durability of high level radiation and chemical transmutation must also be investigated to validate the process. But the improvement of the chemical durability by almost two orders of magnitude compared to the nuclear borosilicate glass certainly proves the interest of the process.

REFERENCES

[1]   Sombret C. ; Chotin M. *Clefs CEA,* 1988, *n°10*, 14-19.

[2]   Jacquet-Francillon N. ; Bonniaud R. ; Sombret C. *Radiochimica Acta,* 1978, *25*, 231-236.

[3]   Brinker J.F. ; Scherer G.W. *Sol-Gel Science*, 1990, Academic Press Inc, San Diego CA

[4]   Simmons J.H.;. Macedo P.B; Barkatt A.; Litovits T.A. *Nature* 1979, *278*, 729-731.

[5]   Kistler, S.S. *J. Phys. Chem*, 1932., *34*, 52-56.

[6]   Phalippou J.; Woignier T.; Prassas M. *J. Mater. Sci.,*1990, *25*, 3111 -3117.

[7]   Schaeffer W; Keefer K.D. *Phys. Rev. Lett.* , 1986, *56*, 2199-2201.

[8]   Woignier T. ; Phalippou J. ; Pelous J. ; Courtens E. *J. Non-Cryst. Solids* 1990 ,*121*, 198-206.

[9]   Nicolaon G.A; Teichner S.J., *Bull. Soc. Chim. France,*1968, *5*, 1906-1908.

[10]   Teichner S.J; Vicarini G.A.; Gardes G.E.E., *Adv. Coll. Interface Sci.* , 1976 , *5*, 245-249.

[11]   Gronauer M.; Fricke J., *Acoustica*, 1986, *59*, 169 -17.

[12]   Fricke J.; *J. Non-Cryst. Solids,* 1992 ,*147 –148*, 356- 361.

[13]   Woignier T. ; Phalippou J. ; *Encyclopedia of Materials: Science and Technology,* 2001, Elsevier Science, 3581-3586.

[14]   Woignier T. ; Phalippou J. ; Prassas, J. *Mater. Sci.*1990, *25*,) 3117- 3123.

[15]   Toki M.; Miyashita S.; Takeuchi T.; Kande S.; Kochi A. *J. Non-Cryst. Solids* , 1988, *100*, 479-485.

[16]   Reynes J. ; Woignier T. ; Phalippou J. *J. Non-Cryst. Solids* , 2001, *285*, 353- 358.

[17]   Woignier T. ; Fernandez Lorenzo C. ; Sauvajol J.L. ; Schmidt J.F. Phalippou J. ;
       Sempere R., J *J. Sol-Gel Sci. Techn* ,1995, *5*, 1-7.

[18]   Sakka S. ; Adachi T., *J. Mater. Sci.*, 1990, *25*, 3408- 3415.

[19]   Scherer G.W.; Smith D.M.;Qiu X. ; Anderson J., *J. Non Cryst. Solids*, 1995, *186*, 316-
       321.

[20]   Scherer G.W., *J. Amer. Ceram Soc.* 1977, *60* ,237-244.

[21]   Woignier T. ; Phalippou J. ; Quinson J.F. ; Pauthe M., Repellin-Lacroix M. ; Scherer
       G.W., *J. Sol-Gel Sci. Techn.* , 1994, *2*, 277-281.

[22]   Marlière C. ; Woignier T. ; Dieudonné P. ; Primera J. ; Lamy M. ; Phalippou J., *J. Non-
       Cryst. Solids*, 2001, *285*, 175-181.

[23]   Carman P.C, *Trans. Inst. Chem. Eng. London*, 1937,*15* ,150-160.

[24]   Fillet C.; Marillet J.; Dussossoy J.L.; Pacaud F. ; Jacquet-Francillon N.; Phalippou J.,
       99th Annual Meeting on Ceramics of the American Ceramic Society, 1997.

[25]   Woignier T. ; Reynes J. ; Phalippou J. ; Dussossoy J.L, *J. Sol-Gel Sci. Tech.*,2000, *19*,
       835-840.

[26]   Haire R.G.;, Assefa Z.;, Stump N., *in Scientific Basis for Nuclear Waste Management
       XXI Mat. Res. Symp. Proc*, ed. I.G Mac Kinley, C . Mac Combie , 1998, *506*, 153-157.

[27]   Felsche J.;, Hirsinger W., *J. of the Less-Common Metal*, 1969, *18* ,131-137.

[28]   Van Hal H.A.M., Hintzen H.T.;, J. *of Alloys and Compounds*, 1992, *179* ,77-82.

[29]   Thomas I.M.;Payne S.A.;. Wilke G.D. , *J. Non-Cryst. Solids*,1992, *151*, 183- 190.

[30]   Pope E.J.A.;, J.D.; MacKenzie J.D., *J. Am. Ceram. Soc.*, 1993, *76(5)* , 1325-1330.

[31]   Toropov N.A., *in Transactions of the VIIth International Ceramic Congress, Londres*,
       1960, 435-441.

[32]   Miller R.O.; Rase D.E., J. *Am. Ceram. Soc.*, 1964, *47(12)* 653-657.

[33]   Bansal N.P.; Doremus R.H., *Handbook of Glass Properties*, 1986, Academic press.

[34]   Akopov F.A.;, Poluboraninov B.E., *Refractories (USSR)*, 1965, 4 , 196 -199.

[35]   Warshaw I.E.; Roy R.; *J. Am. Ceram. Soc.*, 1961 , *44(8)* ,421-425.

[36]   Keler E.; Andreeva A.B., *Ogneupory*, 1963, *5* , 224-230.

[37]   Lutze W. ; Ewing R.C., *in Scientific Basis for Nuclear Waste Management XII Mat.
       Res. Symp. Proc*, ed. W Lutze and R.C Ewing , 1989, *127*, 13-24.

[38]   Nogues J.L, *PhD thesis « Les mécanismes de corrosion des verres de confinement des
       produits de fission,*1984 , Montpellier University.

[39]   Bunker R.C., *J.Non-Cryst. Solids*, 1994, *179*, 300- 3006.

[40]   Granbow B.; Muller R., *J. Nuc. Mat.*, 2001, *298*, 112-118.

In: Nuclear Waste Research: Siting, Technology and Treatment   ISBN 978-1-60456-184-5
Editor: Arnold P. Lattefer, pp. 239-248

Chapter 9

SHUTTLING FROM APPLIED TO FUNDAMENTAL RESEARCH: DEVELOPING CALIXBISCROWNS FOR NUCLEAR WASTE TREATMENT AND ELABORATION OF NANO-OBJECTS

Jacques Vicens[*]
Ecole Polymères Chimie Matériaux,
Laboratoire de conception moléculaire, associé au CNRS, 25,
rue Becquerel, F-67087, Strasbourg, Cedex 2, France

ABSTRACT

This commentary describes the development of a family of calixbiscrowns for nuclear waste treatment. The development has been world wide with the involvement of scientists and governments. This development previously devoted to application in nuclear wastes has sparked continuing fundamental works in chemistry. This is one striking example of an applied research that spanned and opened new fundamental research fields. In a more general point of view it is shown how supramolecular chemistry has reached nanochemistry.

Keywords: *caesium removal, calixbiscrowns, nuclear waste treatment, supramolecular chemistry, scientists and governments.*

calixarenes
Originally macrocyclic compounds capable of assuming a basket (or 'calix') shaped conformation. They are formed from p-*hydrocarbyl phenols* and formaldehyde. The term now applies to a variety of derivatives b y substitution of the hydrocarbon cyclo{oligo[(1,3-phenylene)methylene]}
IUPAC Compendium of Chemical Terminology 2nd Edition (1997)

---

[*] E-mail: vicens@chimie.u-strasbg.fr

In a stimulating 1990 article assessing the state of organic chemistry and where its future directions might lie, Seebach [1] stated that ''The primary motivations that once induced chemists to undertake natural products syntheses no longer exist. Instead of target structures themselves, molecular function and activity now occupy centre stage.'' He further noted that the inexorable evolution of chemistry reflected in such a perception had involved the blurring of distinctions between not only long-recognised subdivisions of chemistry but between chemistry and many other areas of Science. One expression of these changes has been the rise of supramolecular chemistry [2]. Supramolecular chemistry, based on the use of labile interactions to control the assembly of kinetically inert molecular species into functional superstructures, has been associated with a new vocabulary of terms such as host–guest chemistry, molecular and ionic recognition, supramolecular catalysis, self organization, self assembly and many others. Particularly important as the building blocks of supramolecular systems has been a variety of molecules with macrocyclic structures, the most important of these being the crown ethers [3], the cryptands [4], the spherands [5], the cyclophanes [6], the calixarenes [7] and the natural cyclodextrins [8]. We consider here an example from the field of calixarene chemistry illustrating the passage of the study of a ''supermolecule'' from synthesis to significant applications.

In the cases where a fundamental discovery finds a true industrial application, one or two decades of development are commonly required. From the field of supramolecular chemistry, a striking illustration of this is provided in the now established use of calixarenes, in particular calix[4]biscrowns, molecules which are in fact a hybrid of two types of macrocycle, in the treatment of nuclear waste. Such waste treatment is of course of enormous significance for a future in which energy sources must become increasingly limited and must necessarily be used with great concern for both efficiency and safety. Nuclear waste management schemes differ from country to country, but the practical approach adopted in most is to dispose of wastes in environmentally acceptable ways that minimise the volume of both low and high-activity materials. To most effectively use the storage space available, chemists and engineers have been working on various methods to segregate wastes of differing activity with maximum efficiency. The idea is to reduce in particular the volume of high-activity wastes, which must be stored with the utmost security and to provide low-activity wastes, which can be consigned to cheaper above-ground storage facilities. To avoid storage, an attractive possibility is the isolation of a given radionuclide in a form sufficiently pure to allow its transmutation to a non-radioactive species. This is a possibility, which can be realised in the case of caesium (^{135}Cs, ^{137}Cs) through the use of calixcrowns. In a recent article entitled: ''Use of macrocycles in nuclear-waste cleanup: a real world application of a calixcrown in caesium separation technology'', Moyer *and coll.* [9] describe one of the United States' Government projects [10], conducted in part at Oak Ridge National Laboratory, to develop reagents for caesium binding and extraction with regards to possible use in waste cleanup. The process developed, termed the Caustic-Side Solvent Extraction (CSSX) process for the separation of caesium, was specifically directed towards the treatment of accumulated high-level wastes stored in underground tanks at U.S. Department of Energy (USDOE) sites. They describe the overall technology itself, both in terms of the underlying chemistry and the engineering that makes the use of CSSX practical. This powerful technology, based on a macrocyclic cation receptor appropriately functionalised so as to be sufficiently lipophilic to function as extractants, came to maturation after 10 years research.

World wide, governments are increasingly aware of the complexity of the energy supply problem for the future and their commitment to understanding all aspects of the use of nuclear energy has provided an enormous impetus to such work. The cesium-selective extractant used in the CSSX process is calix[4]arene-bis(*tert*-octyl-benzocrown-6) (BOBcalixC6) [11,12]. In such a molecule, the calixarene serves as an inherently lipophilic, relatively rigid framework to which polyether rings of a size suitable for Cs^+ binding can be attached. Since polyether chains tend to induce water-solubility, functionalisation with the octylbenzo groups is necessary to regain lipophilicity, though it has other subtle effects in enhancing Cs^+/Na^+ selectivity, a factor, which is also influenced by interactions of the cation with the calixarene phenyl rings. Thus, the final molecular structure reflects the lessons drawn from a wide variety of fundamental studies of the factors influencing both the binding and extraction of metal ions by calixcrowns [13]. The importance of such studies was explicitly recognised by Moyer *et al.* as "lending a great confidence in CSSX from this perspective".

In France, under the «Loi Bataille», brought into force on the 30th of December 1991 by a vote of the French Parliament, the Commissariat à l'Energie Atomique (CEA) started the SPIN program and deposited, on the 6th of November 1992, a French Patent entitled: "Calix[4]arènes-bis-couronnes, leur procédé de préparation et leur utilisation pour l'extraction sélective du césium et des actinides" [14] which became applicable in the USA [15] on the 11th of October 1994. The French Patent (issued from the PhD work of Clément Hill [16]) gave rise to an oral communication at MacroAkron'94, 35th IUPAC International Symposium on Macromolecules the same year, a review article [17] and a full paper 1 year later [18] in which Vicens, Dozol *and coll.* reported the synthesis of seven 1,3-calix[4]biscrowns BCs' and their use as selective carriers of caesium through supported liquid membranes (SLM's).

The application of the Danesi diffusional model allowed the transport isotherms of trace level ^{137}Cs through SLMs (Supported Liquid Membranes) containing calix[4] biscrowns to be determined as a function of the ionic concentration of the aqueous feed solutions.

Compounds BC6 and BC6B (Scheme 3) appeared to be much more efficient than mixtures of crown ethers and acidic exchangers, especially in very acidic media. Decontamination factors greater than 20 were obtained in the treatment of synthetic acidic radioactive wastes. Permeability coefficient measurements were conducted for repetitive transport experiments in order to qualify the SLMs' stability with time. Very good results (stable for more than 50 d) and high decontamination yields were observed with BC6 and BC6B. For example, BC6 and BC6B allowed selective removal of ^{137}Cs from sodium containing solutions. A Cs^+/Na^+ selectivity (needed in the French process in highly acidic conditions) of ~ 20,000 was observed. Less than 100 mg of the ~ 100 g of sodium (initially present in the feed solution) was transported in 24 h by BC6B, whereas more than 95% of trace level ^{137}Cs was concentrated in the stripping solution. On the basis of the syntheses of selected receptors adapted to the technology and the evaluation of thermodynamic parameters for complexation and extraction, a module termed CALIXARENES was developed for incorporation into the French PUREX decontamination process (see Scheme 1 [19]). This established a model of broad utility to all members of the nuclear waste management community. Both of the decontamination schemes described above can be seen as specific applications of some of the fundamental concepts of supramolecular chemistry and can in this sense be traced back to the seminal work of Pedersen in 1967 [3a], ultimately recognised in the award of the 1987 Nobel Prize, on the macrocycles known as "crown ethers".

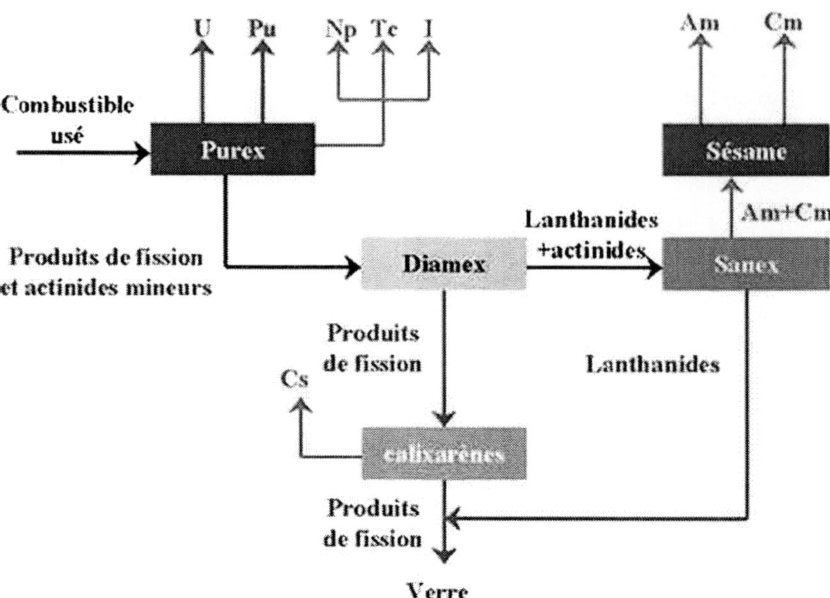

Scheme 1. French decontamination process PUREX [19].

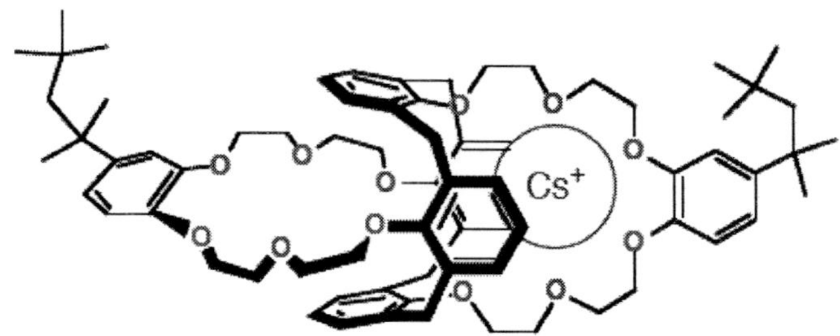

Scheme 2. BOBcalix-C6.Cs$^+$ complex [12].

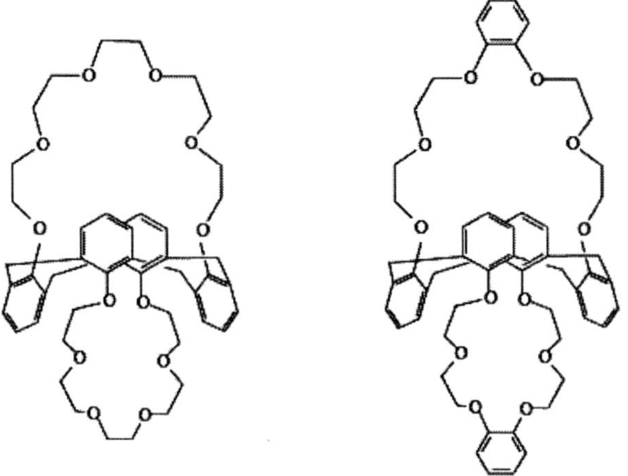

Scheme 3. BC6 (left) and BC6B (right).

Pedersen's work, performed in an industrial environment with particular applications in mind, was developed, refined and extended in academic laboratories, most notably those of his co-recipients of the Nobel Prize, to provide a vast number of new molecules, which in turn have found various applications in addition to nuclear waste treatment. The application in waste treatment, however, provides an outstanding example of the efficacy that is possible in collaborations between industry and research institutions both within and outside universities, the handling of high-level radioactive materials for testing being, of course, something that cannot take place within public universities.

The assertion by Moyer [9] that "this article is intended to serve as an example of how macrocyclic chemistry can be adapted for an industrial purpose use to meet the demanding requirements of real-world-systems" is an affirmation of the value of such collaborations. A congruent view is found in the words of Dautray and Saas in a popular scientific article [20] published in the *Encyclopedia Universalis 2005* entitled: "Stockage des déchets radioactifs". The authors mention the 'realworld exploitation' and the active work devoted to the decontamination process in several countries. «En particulier, pour l'extraction des produits de fission, l'effort porte actuellement sur l'utilisation des macrocycles, par exemple du type éther couronnes ou calixarènes; ces molécules cages peuvent être des extractants très spécifiques avec de forts taux de récupération; certaines d'entre elles pourront donc s'appliquer a` l'extraction du césium, du strontium, voire des lanthanides et des actinides. Le coût élevé actuel de ces extractants conduit à les utiliser en association avec des technologies d'extraction fondées sur l'utilisation de systèmes membranaires (membranes liquides, membranes greffées, etc.). Les recherches sont très actives aux Etats-Unis, au Japon et en Europe».

Due to the scientific achievements and social improvements during the 15 years-period of collaboration existing between national research centers and universities, the French Parliament has voted the 28th of June 2006 a new law on the treatment of nuclear wastes (partly for cesium separation, storage and transmutation) with the aim to pursue and finalize investigations with the objective to propose to the parliament a perennial industrial solution *circa* 2015 [21]. Mention is made in this report [21] of the fruitful comings and goings between governemental institutions, national research centers and an open wide pluridisciplinarity within the academic entities.

The communication of Vicens, Dozol *and coll.* [18] sparked continuing research devoted to the task of finding ever more efficient and discriminating systems for waste treatment. This has ranged from fundamental studies to practical efforts to study and modify the properties of known extractants for improvement. XAFS and computational simulations were used to better characterise the nature of caesium bis(crown)calixarene complexes in solution [22]. Introduction of charge-carrying substituents to calixbiscrowns-6 lead to the production of lariat-like receptors [23]. A combination of dipyridinocalixcrown/diiodoperfluorocarbon acted as an effective 'binary host' system for the selective extraction of CsI from aqueous to fluorous phase [24]. Calix[4]bis-crowns-6 have been submitted to extended radiolysis and were shown to be highly stable as nitric acid slurries in the reprocessing of spent nuclear fuels [25]. In the framework of Actinex program, C.E.A. launched studies for the recovery of long lived nuclides from spent fuel dissolution acidic solutions in order to destroy them by transmutation or to encapsulate them in specific matrixes. Efforts were focused on ^{99}Tc and ^{135}Cs which are among the most harmful elements because of their long half live ($2.1 \cdot 10^5$ and $2.3 \cdot 10^6$ years, respectively) and mobility. Coextraction of the pertechnetate with cesium by calix[4]arenes bis-crown-6 and then to separate it from cesium by using a second calixarene has been achieved [26].

A continuing series of international conferences has always involved extended discussions of the applications of new calixarenes in selective complexation and extraction processes. Very recently, interest has become focussed on "nanomaterials" for use in caesium recovery. For example, novel mesoporous organosilicas containing size-selective micropores from covalently bound calixcrowns have been synthesized. Experiments on the extraction of caesium ions from water in the presence of high concentrations of sodium ions by the insoluble calixcrown-containing mesoporous organosilicas showed good uptake and high caesium selectivity by these novel materials. Simple filtration removes the material containing the extracted caesium ions [27]. A novel extraction chromatographic resin for the separation and preconcentration of cesium from acidic nitrate media comprising an inert polymeric substrate impregnated with BC6B in a chlorinated diluent has been described [28]. Cesium is shown to be both strongly and selectively retained by the resin at low (<1 M) acid concentrations and readily eluted from it using 6 M HNO_3. This is also the case for a resin employing a related macrocyclic extractant, BOBCalix-6, prepared and partly characterized in an effort to overcome certain limitations of the BC6B-based material [28]. Despite this, the resin is shown to be well suited to the isolation of radiocesium from acidic solution for subsequent determination or for the removal of cesium interference in the quantitation of other radionuclides. Further, Bu *et al.* [29] have reported the synthesis from calixbiscrowns of calixdendrimers considered to be potentially caesium selective complexants and absorbents. More recently, the [Cs(BC6B)] 1:1 complex ($K_{eq} = 10^6$ in acetonitrile) became a standard and was used to show that an organo-cyanometallate cage forms a more stable complex with cesium [30]. Another type of organometallic compound BISPHECOSAN has been shown to highly extract cesium cation [31]. However these last examples seem inappropriate to give rise to applications in an industrial process involving drastic media conditions of acidity or basicity. Also research has been extended to the obtention of stable materials such as hollandite ceramics as specific radioactive cesium-host to be used for storage after separation process [32]. As an anecdote not only the chemistry was continued but novel resorcinarene bis-crown ethers BC4 and BC5, prepared for cesium removal, were named according to 'Jacques Vicens' communication' [33] in which BC (or BisCrown) is followed by the number of oxygens in the crown unit [34].

A wide range of analytical sensors have been based on calixarenes and have present application [35]. Concerning the bis(crown)calixarenes, the attachment of 9-cyanoanthracene fluorophores on the benzocrown-6 loops lead to Cs^+-selective PET sensors [36]. The introduction of dioxocoumarins produced a fluorophore system presenting cation tunnelling, photodisruption and photoinduced motions of Cs^+ as fundamental observations in photophysics [37]. A water-soluble fluorescent sensor based on a tetrasulfonated calix[4]arene-bis(crown-6-ether) showed a Cs^+/K^+ selectivity larger than 250 in water [38]. Cs- and F-induced FRET On-Off has been found in a naphthocrown-6 calix[4]arene in the partial cone conformation with two coumarins [39].

Besides these studies concerning cesium removal and reprocessing of spent nuclear fuels, various functionalised calixarenes have been shown to have "supramolecular" properties, in a wide sense, possibly exploitable in what is the nascent "nanoworld" of the future as for instance metal-tunneling, molecular machines or nanotubes in the case of calix[4]biscrowns

[40]. The calix[4]arene moiety can exist under four discrete forms : cone, partial cone, 1,2-alternate, and 1,3-alternate depending on the topology of the four aromatic rings (Scheme 5).

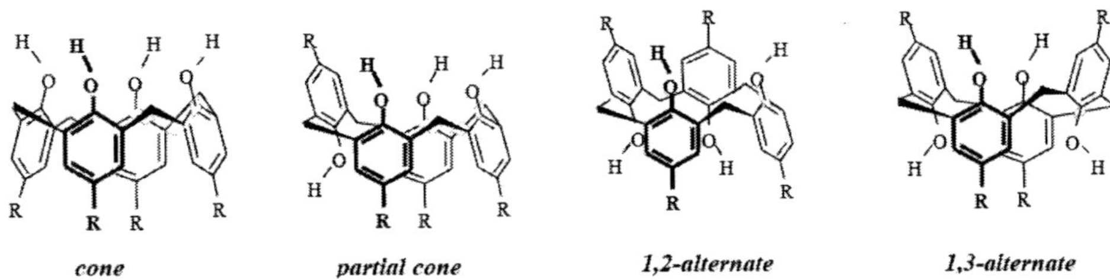

Scheme 5. The four discrete conformations of calix[4]arenes.

The calix[4]biscrowns referring to this paper are in the 1,3-alternate conformation which implies the formation of a square aromatic tunnel allowing the communication between the two crown-ether metal-binding sites. It was shown in several cases the oscillation of cations (alkalis and ammonium) through this π-base tunnel (scheme 6) [40,41].

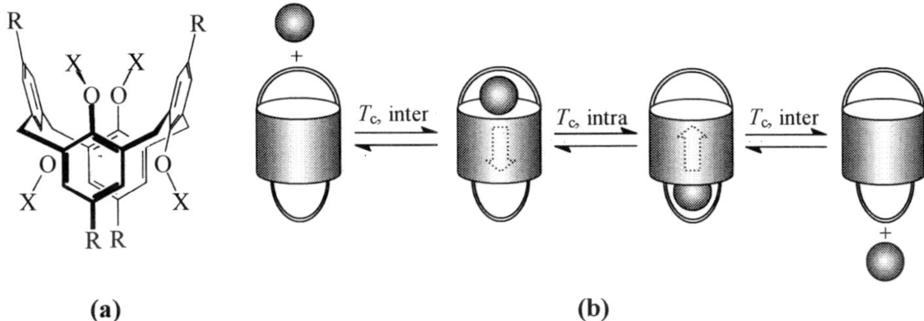

Scheme 6. Representation of a calix[4]arene in the 1,3-alternate conformation (a) and of the metal oscillation through the π-base tunnel (b).

This observation has lead to the obvious design of 1,3-alternate calix tubes as cation- [42] and anion-[43] conducting channels through membranes such as phospholipid bilayers or liposomes. Using the 1,3-alternate calix[4]crown unit as molecular segments it has been built calix[4]arenes nanotubes by modular synthesis with a metal shuttling through several calix crown units [44]. Filling single-walled carbon nanotubes (SWNTs) with foreign guest species is an emerging area of research [45]. In this respect, 1,3-alternate calixarene-based nanomaterials have been prepared with application in sensing, storage and fixation of NO_x gases [46]. Molecular machines are in vogue and scattered examples are found in literature of thermally-, chemically-, electrochemically- and/or photochemically-labile molecular systems or devices, sensors, logic gates etc. [47]. The globular shape of calix[4]arene-bis-crowns has been exploited to prepare molecular mappemondes (globes) and gyroscopes. They are constructed from calix[4]biscrowns-6 held in the arms of a polyether loop via 1+1 condensation [48]. Depending on the length of the polyether arm, a 2+2 dimer was also isolated, leading to a molecular mill [49]. The spinning of the calix unit can be stopped in the presence of large amounts of ammonium picrate.

In conclusion, it is worth reiterating that what might be regarded very much as an applied research problem (although there is a gap, often called "the death valley", between fundamental and applied research), the treatment of nuclear wastes, has been a major stimulus for a wide range of fundamental research activities, some of which have in turn found significant applications and others of which have been important *in reorienting a conceptual understanding of molecular science* and which thus can be seen as having long-term influences.

The involvement of governments and their national research institutes, motivated by concerns for broad social imperatives in which science and technology play fundamental roles, has been crucial in the flowering of this field.

REFERENCES

[1]   D. Seebach: *Angew. Chem. Int. Ed. Engl.* 29, 1320 (1990).

[2]   Supramolecular chemistry and host-guest chemistry are 40-years old. See ref 3(a).(a) B. Dietrich, P. Viout, and J.-M. Lehn: *Macrocyclic Chemistry, Aspects of Organic and Inorganic Supramolecular Chemistry*, VCH, Weinheim, Germany (1992); (b) J.-M. Lehn: *Supramolecular Chemistry*, VCH, Weinheim, Germany (1995).

[3]   (a) C. J. Pedersen: *J. Amer. Chem. Soc.* 89, 2495 (1967); (b) R.M. Izatt and J.S. Bradshaw: *The Pedersen Memorial Issue. Adv. Incl. Sci.*, Kluwer Academic Publishers, Dordrecht, Holland (1992).

[4]   (a) J.-M. Lehn: *Acc. Chem. Res.* 11, 49 (1978); (b) J.-M. Lehn: *Angew. Chem., Int. Ed. Engl.* 27, 89 (1988).

[5]   (a) D. J. Cram: *Angew. Chem., Int. Ed. Engl.* 18, 753 (1979); (b) D.J. Cram: *Angew. Chem., Int. Ed. Engl.* 27, 109 (1988).

[6]   (a) F. Vögtle and P. Neumann: *Top. Curr. Chem.* 48, 67 (1974); (b) F. Diederich: In H.-J. Schneider and H. Dürr (Eds.), *Frontiers in Supramolecular Organic Chemistry and Photochemistry*, VCH Verlagsgesellschaft, Weinheim, pp. 167–192 (1990).

[7]   (a) C. D. Gutsche: *Acc. Chem. Res.* 16, 162 (1983); (b) C. D. Gutsche: *Calixarenes*, The Royal Society of Chemistry, Cambridge (1989); (c) J. Vicens and V. Böhmer: *Calixarenes: A Versatile Class of Macrocyclic Compounds*, Kluwer Academic Publishers, Dordrecht (1991).

[8]   (a) M. L. Bender and M. Komoiyama: *Cyclodextrin Chemistry*, Springer Verlag, Berlin (1978); (b) J. Szejtli: *Cyclodextrin Technology. Topics in Inclusion Science*, Vol. 1, Kluwer Academic Publishers, Dordrecht, (1988).

[9]   B. A. Moyer, J. F. Birdwell, Jr., P. V. Bonnensen, and L. H. Delmau: In K. Gloe (Ed.), *Macrocyclic Chemistry – Current Trends and Future Perspectives*, Springer, Dordrecht (2005) pp 383–405.

[10] B. A. Moyer, P.V. Bonnensen, R. Custelcean, and L.H. Delmau: *Kem. Ind.* 54, 65 (2005).

[11] R. A. Sachleben, A. P. V. Bonnensen, T. Descazeaud, T. J. Haverlock, A. Urvoas, and B. A. Moyer: *Solvent Extr. Ion Exch.* 17, 1445 (1999).

[12] www.osti.gov/bridge/servlets/purl/835027-RDLYUu/native/835027.pdf.

[13] Z. Asfari, S. Wenger, and J. Vicens: *J. Incl. Phenom.* 19, 137 (1994).

[14] Z. Asfari, J. F. Dozol, C. Hill, and J. Vicens: French Patent D. 77583.3 MDI., Paris, 6th of November 1992.

[15] Z. Asfari, J.F. Dozol, C. Hill, and J. Vicens: Patent Applicable to USA B11385.3 MDT, Paris, 11th of October 1994.

[16] C. Hill: *Nouvelle Thèse*, Université Louis Pasteur de Strasbourg, Strasbourg, 25th of May 1994.

[17] C. Hill, J.-F. Dozol, V. Lamare, H. Rouquette, S. Eymard, B. Tournois, J. Vicens, Z. Asfari, C. Bressot, R. Ungaro, and A. Casnati: *J. Incl. Phenom. Mol. Recogn. Chem.* 19, 399 (1994).

[18] Z. Asfari, C. Bressot, J. Vicens, C. Hill, J.-F. Dozol, H. Rouquette, S. Eymard, V. Lamare, and B. Tournois: *Anal. Chem.* 67, 3133 (1995).

[19] J.-F. Dozol and V. Lamare: *Clé-CEA*, numéro 46, 23 (2003).

[20] R. Dautray and A. Saas: *Encyclopedia Universalis* 2005.

[21] http://www.cnrs.fr/publications/imagesdelaphysique/couv-

[22] J.-X. Gao, B.-W. Wang, T. Liu, J.-C. Wang, C.-L. Song, Z.-D. Chen, T.-D. Hu, Y.-N. Xie, J. Zhang, and H. Yang: *J. Synchrotron Rad.* 12, 374 (2005).

[23] (a) H. Zhou, K. Surowiec, D.W. Purkiss, and R.A. Bartsch: *Org. Biomol. Chem.* 3, 1676 (2005); (b) H. Ziou, D. Liu, J. Gega, K. Surowiec, D. W. Purkiss, and R. A. Bartsch: *Org. Biomol. Chem.* 5, 324 (2007), (c) E. Bazelaire, M. G. Gorbunova, P. V. Bonnesen, B. A. Moyer, and L. H. Delmau: *Solvent Extr. Ion Exch.* 22, 637 (2004).

[24] G. Gattuso, A. Pappalardo, M. F. Parisi, I. Pisagatti, F. Crea, R. Liantonio, P. Metrangolo, W. Navarrini, G. Resnati, T. Pilati, and S. Pappalardo: *Tetrahedron* 63, 4951 (2007).

[25] C. K. Jankowski, C. Hocquelet, S. Arseneau, C. Moulin, and L. Mauclaire : *J. Photochem. Photobiol. A : Chemistry* 184, 216 (2006).

[26] M. Grunder, J. F. Dozol, Z. Asfari, J. Vicens: *J. Radioanalytical and Nuclear Chem.* 241, 59 (1999).

[27] C. Liu, J.B. Lambert, and L. Fub: *J. Mater. Chem.* 14, 1303 (2004).

[28] M. L. Dietz, D. D. Ensor, B. Harmon, S. Seekamp: *Sep. Sci. Tech.* 41, 2183 (2006).

[29] J.-H. Bu, Q.-Y. Zheng, C.-F. Chen, and Z.-T. Huang: *Tetrahedron* 61, 897 (2005).

[30] (a) S. C. N. Hsu, M. Ramesh, J. H. Espenson, and T. B. Rauchfuss: *Angew. Chem. Int. Ed.* 42, 2663 (2003);(b) J. L. Boyer, M. L. Kuhlman, and T. B. Rauchfuss: *Acc. Chem. Res.* 40, 233 (2007).

[31] P. Selucký, N. V. Sistková, and J. Rais: *Rad. Nucl. Chem.* 224, 89 (2005).

[32] V. Aubin, F. Studer, D. Caurant, D. Gourier, N. Nguyen, A. Ducouret, N. Baffier, and T. Advocat : Atalante 2004, 032-06, June 21-25, Nîmes, France.

[33] V. Lamare, J. F. Dozol, S. Fuangwasdi, F. Arnaud-Neu, P. Thuéry, M. Nierlich, Z. Asfari, and J. Vicens: *J. Chem. Soc. Perkin 2*, 271 (1999).

[34] K. Salorinne and M. Nissinen : *Org. Lett.* 8, 5473 (2006).

[35] S. H. Kim, H. J. Kim, J. Yoon, and J. S. Kim in *Calixarenes in the Nanoworld*, J. Vicens and J. Harrowfield, Eds, Springer, Dordrecht, The Netherlands (2007).

[36] (a) H. F. Ji, G. M. Brown, and R. Dabestani, *Chem. Commun.* 609 (1999); (b) H. F. Ji, R. Dabestani, G. M. Brown, and R. A. Sachleben, *Chem. Commun.* 833 (2000); (c) H. F. Ji, R. Dabestani, G. M. Brown, and R. L. Hettich, *J. Chem. Soc., Perkin Trans.* 2 585 (2001);(d) J. S. Kim, K. H. Noh, S. H. Lee, S. K. Kim, S. K. Kim, and J. Yoon: *J. Org. Chem.* 68, 597 (2003).

[37]   I. Leray, Z. Asfari, J. Vicens, and B. Valeur: *J. Fluorescence* 14, 451 (2004).

[38]   V. Souchon, I. Leray, and B. Valeur: *Chem. Commun.* 4224 (2006).

[39]   M. H. Lee, D. T. Quang, H. S. Jung, J. Yoon, C.-H. Lee, and J. S. Kim: *J. Org. Chem.* 72, 4242 (2007).

[40]   (a) B. Pulpoka and J. Vicens: *J. Nano Bio-Technol.* 1, 55 (2004); (b) B. Pulpoka and J. Vicens *Coll. Czech. Chem. Com.* 69, 1251 (2004);(c) B. Pulpoka, V. Ruangpornvisuti and J. Vicens: In H. Takemura (ed.), *Cyclophane Chemistry for the 21st Century*, Chap 3 Research Signpost, Kerela, Inde, (2002).

[41]   K. N. Koh, K. Araki, S. Shinkai, Z. Asfari, and J. Vicens: *Tet. Lett.* 36, 6095 (1995).

[42]   (a) J. de Mendoza, F. Cuevas, P. Prados, E. S. Meadows, and G. W. Gokel: *Angew. Chem., Int. Ed.* 37, 1534 (1998).(b) N. Maulucci, F. De Riccardis, C. B. Botta, A. Casapullo, E. Cressina, M. Fregonese, P. Tecilla, and I. Izzo: *Chem. Commun.* 1354 (2005).

[43]   (a) V. Sidorov, F. W. Kotch, G. Abdrakhmanova, R. Mizani, J. C. Fettinger, and J. T. Davis *J. Am. Chem. Soc.* 124, 2267 (2002). (b) V. Sidorov, F. W. Kotch, J. L. Kuebler, Y-F. Lam, and J. T. Davis *J. Am. Chem. Soc.* 125, 2840 (2003).

[44]   S. K. Kim, J. K. Lee, S. H. Lee M. S. Lim S. W. Lee, W. Sim, and J. S. Kim *J. Org. Chem.* 69, 2877 (2004).

[45]   D. A. Britz, A. N. Kholbystov, K. Porfyrakis, A. Arvadan, and G. A. D. Briggs: *Chem. Commun.* 37 (2005).

[46]   (a) E. Wanigasekara, A. V. Leontiev, V. G. Organo, and D. Rudkevich: *Eur. J. Org. Chem.* 2254 (2007);(b) D. M. Rudkevich in *Calixarenes in the Nanoworld*, J. Vicens and J. Harrowfield, Eds, Springer, Dordrecht, The Netherlands (2007).

[47]   (a) C. J. Easton, S. F. Lincoln, L. Barr, and H. Onagi: *Chem. Eur. J.* 10, 3120 (2004);(b) V. Balzani, A. Credi, and M. Venturi: *Chem. Eur. J.* 8, 5525 (2002);(c) Z. Asfari and J. Vicens: *J. Incl. Phenom.* 36, 103 (2000).

[48]   Z. Asfari, C. Naumann, G. Kaufmann, and J. Vicens: *Tetrahedron Lett.* 37, 3325 (1996), highlighted by A. B. Holmes and G. Richard: *Chemistry and Industry* 468 (1996).

[49]   Z. Asfari, C. Naumann, G. Kaufmann, and J. Vicens: *Tetrahedron Lett.* 39, 9007 (1998).

INDEX

D

E

F

G

N

Q

R

S

T

U

V